T0202585

Graduate Texts in Mathematics **197**

Springer
New York
Berlin
Heidelberg
Barcelona
Hong Kong
London
Milan
Paris
Singapore
Tokyo

Graduate Texts in Mathematics

(continued after index)

David Eisenbud
Joe Harris

The Geometry
of Schemes

With 40 Illustrations

 Springer

David Eisenbud
Mathematical Sciences Research
  Institute
1000 Centennial Drive
Berkeley, CA 94720-5070
USA
de@msri.org

Joe Harris
Department of Mathematics
Harvard University
Cambridge, MA 02138
USA
harris@måth.harvard.edu

Mathematics Subject Classification (2000): 14-02, 14A15

Library of Congress Cataloging-in-Publication Data
Eisenbud, David.
       The geometry of schemes / David Eisenbud, Joe Harris.
            p.   cm. – (Graduate texts in mathematics : 197)
       Includes bibliographical references and index.
            ISBN 0-387-98637-5 (softcover : alk. paper) — ISBN 0-387-98638-3
       (hardcover : alk. paper)
            1. Schemes (Algebraic geometry)   I. Harris, Joe.   II. Title.
       QA564.E357   1999
       516.3'5—dc21   99-36219

Printed on acid-free paper.

Production managed by Terry Kornak; manufacturing supervised by Jacqui Ashri.
Photocomposed copy prepared from the authors' LaTeX 2e files.
Printed and bound by Maple-Vail Book Manufacturing Group, York, PA.
Printed in the United States of America.

9 8 7 6 5 4 3 2 (Corrected second printing, 2001)

ISBN 0-387-98637-5 (softcover)   SPIN 10850499
ISBN 0-387-98638-3 (hardcover)   SPIN 10523563

Springer-Verlag   New York Berlin Heidelberg
A member of BertelsmannSpringer Science+Business Media GmbH

... the end of all our exploring
Will be to arrive where we started
And know the place for the first time.

— T. S. Eliot, "Little Gidding" (*Four Quartets*)

Contents

Introduction

What schemes are

The theory of schemes is the foundation for algebraic geometry formulated by Alexandre Grothendieck and his many coworkers. It is the basis for a grand unification of number theory and algebraic geometry, dreamt of by number theorists and geometers for over a century. It has strength-ened classical algebraic geometry by allowing flexible geometric arguments about infinitesimals and limits in a way that the classic theory could not handle. In both these ways it has made possible astonishing solutions of many concrete problems. On the number-theoretic side one may cite the proof of the Weil conjectures, Grothendieck's original goal (Deligne [1974]) and the proof of the Mordell Conjecture (Faltings [1984]). In classical algebraic geometry one has the development of the theory of moduli of curves, including the resolution of the Brill–Noether–Petri problems, by Deligne, Mumford, Griffiths, and their coworkers (see Harris and Morrison [1998] for an account), leading to new insights even in such basic areas as the theory of plane curves; the firm footing given to the classification of algebraic surfaces in all characteristics (see Bombieri and Mumford [1976]); and the development of higher-dimensional classification theory by Mori and his coworkers (see Kollár [1987]).

No one can doubt the success and potency of the scheme-theoretic methods. Unfortunately, the average mathematician, and indeed many a beginner in algebraic geometry, would consider our title, "The Geometry of Schemes", an oxymoron akin to "civil war". The theory of schemes is widely regarded as a horribly abstract algebraic tool that hides the appeal of geometry to promote an overwhelming and often unnecessary generality.

By contrast, experts know that schemes make things simpler. The ideas behind the theory — often not told to the beginner — are directly related to those from the other great geometric theories, such as differential geometry, algebraic topology, and complex analysis. Understood from this perspective, the basic definitions of scheme theory appear as natural and necessary ways of dealing with a range of ordinary geometric phenomena, and the constructions in the theory take on an intuitive geometric content which makes them much easier to learn and work with.

It is the goal of this book to share this "secret" geometry of schemes. Chapters I and II, with the beginning of Chapter III, form a rapid introduction to basic definitions, with plenty of concrete instances worked out to give readers experience and confidence with important families of examples. The reader who goes further in our book will be rewarded with a variety of specific topics that show some of the power of the scheme-theoretic approach in a geometric setting, such as blow-ups, flexes of plane curves, dual curves, resultants, discriminants, universal hypersurfaces and the Hilbert scheme.

What's in this book?

Here is a more detailed look at the contents:

Chapter I lays out the basic definitions of schemes, sheaves, and morphisms of schemes, explaining in each case why the definitions are made the way they are. The chapter culminates with an explanation of fibered products, a fundamental technical tool, and of the language of the "functor of points" associated with a scheme, which in many cases enables one to characterize a scheme by its geometric properties.

Chapter II explains, by example, what various kinds of schemes look like. We focus on affine schemes because virtually all of the differences between the theory of schemes and the theory of abstract varieties are encountered in the affine case — the general theory is really just the direct product of the theory of abstract varieties à la Serre and the theory of affine schemes. We begin with the schemes that come from varieties over an algebraically closed field (II.1). Then we drop various hypotheses in turn and look successively at cases where the ground field is not algebraically closed (II.2), the scheme is not reduced (II.3), and where the scheme is "arithmetic" — not defined over a field at all (II.4).

In Chapter II we also introduce the notion of *families* of schemes. Families of varieties, parametrized by other varieties, are central and characteristic aspects of algebraic geometry. Indeed, one of the great triumphs of scheme theory — and a reason for much of its success — is that it incorporates this aspect of algebraic geometry so effectively. The central concepts of *limits*, and *flatness* make their first appearance in section II.3 and are discussed in detail, with a number of examples. We see in particular how to take flat limits of families of subschemes, and how nonreduced schemes occur naturally as limits in flat families.

In all geometric theories the compact objects play a central role. In many theories (such as differential geometry) the compact objects can be embedded in affine space, but this is not so in algebraic geometry. This is the reason for the importance of projective schemes, which are *proper*—this is the property corresponding to compactness. Projective schemes form the most important family of nonaffine schemes, indeed the most important family of schemes altogether, and we devote Chapter III to them. After a discussion of properness we give the construction of Proj and describe in some detail the examples corresponding to projective space over the integers and to double lines in three-dimensional projective space (in affine space all double lines are equivalent, as we show in Chapter II, but this is not so in projective space). We also discuss the important geometric constructions of tangent spaces and tangent cones, the universal hypersurface and intersection multiplicities.

We devote the remainder of Chapter III to some invariants of projective schemes. We define free resolutions, graded Betti numbers and Hilbert functions, and we study a number of examples to see what these invariants yield in simple cases. We also return to flatness and describe its relation to the Hilbert polynomial.

In Chapters IV and V we exhibit a number of classical constructions whose geometry is enriched and clarified by the theory of schemes. We begin Chapter IV with a discussion of one of the most classical of subjects in algebraic geometry, the flexes of a plane curve. We then turn to blow-ups, a tool that recurs throughout algebraic geometry, from resolutions of singularities to the classification theory of varieties. We see (among other things) that this very geometric construction makes sense and is useful for such apparently non-geometric objects as arithmetic schemes. Next, we study the *Fano schemes* of projective varieties—that is, the schemes parametrizing the lines and other linear spaces contained in projective varieties—focusing in particular on the Fano schemes of lines on quadric and cubic surfaces. Finally, we introduce the reader to the *forms* of an algebraic variety—that is, varieties that become isomorphic to a given variety when the field is extended.

In Chapter V we treat various constructions that are defined locally. For example, Fitting ideals give one way to define the *image* of a morphism of schemes. This kind of image is behind Sylvester's classical construction of resultants and discriminants, and we work out this connection explicitly. As an application we discuss the set of all tangent lines to a plane curve (suitably interpreted for singular curves) called the *dual* curve. Finally, we discuss the double point locus of a morphism.

In Chapter VI we return to the functor of points of a scheme, and give some of its varied applications: to group schemes, to tangent spaces, and to describing moduli schemes. We also give a taste of the way in which geometric definitions such as that of tangent space or of openness can be extended from schemes to certain functors. This extension represents the beginning of the program of enlarging the category of schemes to a more flexible one, which is akin to the idea of adding distributions to the ordinary theory of functions.

Since we believe in learning by doing we have included a large number of exercises, spread through the text. Their level of difficulty and the background they assume vary considerably.

Didn't you guys already write a book on schemes?

This book represents a major revision and extension of our book *Schemes: The Language of Modern Algebraic Geometry*, published by Wadsworth in 1992. About two-thirds of the material in this volume is new. The introductory sections have been improved and extended, but the main difference is the addition of the material in Chapters IV and V, and related material elsewhere in the book. These additions are intended to show schemes at work in a number of topics in classical geometry. Thus for example we define blowups and study the blowup of the plane at various nonreduced points; and we define duals of plane curves, and study how the dual degenerates as the curve does.

What to do with this book

Our goal in writing this manuscript has been simply to communicate to the reader our sense of what schemes are and why they have become the fundamental objects in algebraic geometry. This has governed both our choice of material and the way we have chosen to present it. For the first, we have chosen topics that illustrate the geometry of schemes, rather than developing more refined tools for working with schemes, such as cohomology and differentials. For the second, we have placed more emphasis on instructive examples and applications, rather than trying to develop a comprehensive logical framework for the subject.

Accordingly, this book can be used in several different ways. It could be the basis of a second semester course in algebraic geometry, following a course on classical algebraic geometry. Alternatively, after reading the first two chapters and the first half of Chapter III of this book, the reader may wish to pass to a more technical treatment of the subject; we would recommend Hartshorne [1977] to our students. Thirdly, one could use this book selectively to complement a course on algebraic geometry from a book such as Hartshorne's. Many topics are treated independently, as illustrations, so that they can easily be disengaged from the rest of the text.

We expect that the reader of this book will already have some familiarity with algebraic varieties. Good sources for this include Harris [1995], Hartshorne [1977, Chapter 1], Mumford [1976], Reid [1988], or Shafarevich [1974, Part 1], although all these sources contain more than is strictly necessary.

Beginners do not stay beginners forever, and those who want to apply schemes to their own areas will want to go on to a more technically oriented treatise fairly soon. For this we recommend to our students Hartshorne's book *Algebraic Geometry* [1977]. Chapters 2 and 3 of that book contain many fundamental topics not treated here but essential to the modern uses of the theory. Another classic source, from which we both learned a great deal, is David Mumford's *The Red Book of Varieties and Schemes* [1988]. The pioneering work of Grothendieck [Grothendieck 1960; 1961a; 1961b; 1963; 1964; 1965; 1966; 1967] and Dieudonné remains an important reference.

Who helped fix it

We are grateful to many readers who pointed out errors in earlier versions of this book. They include Leo Alonso, Joe Buhler, Herbert Clemens, Vesselin Gashorov, Andreas Gathmann, Tom Graber, Benedict Gross, Brendan Hassett, Ana Jeremias, Alex Lee, Silvio Levy, Kurt Mederer, Mircea Mustata, Arthur Ogus, Keith Pardue, Irena Peeva, Gregory Smith, Jason Starr, and Ravi Vakil.

Silvio Levy helped us enormously with his patience and skill. He transformed a crude document into the book you see before you, providing a level of editing that could only come from a professional mathematician devoted to publishing.

How we learned it

Our teacher for most of the matters presented here was David Mumford. The expert will easily perceive his influence; and a few of his drawings, such as that of the projective space over the integers, remain almost intact. It was from a project originally with him that this book eventually emerged. We are glad to express our gratitude and appreciation for what he taught us.

David Eisenbud
Joe Harris

I
Basic Definitions

Just as topological or differentiable manifolds are made by gluing together open balls from Euclidean space, schemes are made by gluing together open sets of a simple kind, called *affine schemes*. There is one major difference: in a manifold one point looks locally just like another, and open balls are the only open sets necessary for the construction; they are all the same and very simple. By contrast, schemes admit much more local variation; the smallest open sets in a scheme are so large that a lot of interesting and nontrivial geometry happens within each one. Indeed, in many schemes no two points have isomorphic open neighborhoods (other than the whole scheme). We will thus spend a large portion of our time describing affine schemes.

We will lay out basic definitions in this chapter. We have provided a series of easy exercises embodying and applying the definitions. The examples given here are mostly of the simplest possible kind and are not necessarily typical of interesting geometric examples. The next chapter will be devoted to examples of a more representative sort, intended to indicate the ways in which the notion of a scheme differs from that of a variety and to give a sense of the unifying power of the scheme-theoretic point of view.

I.1 Affine Schemes

An *affine scheme* is an object made from a commutative ring. The relationship is modeled on and generalizes the relationship between an affine variety and its coordinate ring. In fact, one can be led to the definition of scheme in the following way. The basic correspondence of classical algebraic geometry is the bijection

$$\{\text{affine varieties}\} \longleftrightarrow \left\{ \begin{array}{c} \text{finitely generated, nilpotent-free rings} \\ \text{over an algebraically closed field } K \end{array} \right\}$$

Here the left-hand side corresponds to the geometric objects we are naively interested in studying: the zero loci of polynomials. If we start by saying that these are the objects of interest, we arrive at the restricted category of rings on the right. Scheme theory arises if we adopt the opposite point of view: if we do not accept the restrictions "finitely generated," "nilpotent-free" or "K-algebra" and insist that the right-hand side include all commutative rings, what sort of geometric object should we put on the left? The answer is "affine schemes"; and in this section we will show how to extend the preceding correspondence to a diagram

$$\{\text{affine varieties}\} \longleftrightarrow \left\{ \begin{array}{c} \text{finitely generated, nilpotent-free rings} \\ \text{over an algebraically closed field } K \end{array} \right\}$$

$$\downarrow \qquad\qquad\qquad\qquad \downarrow$$

$$\{\text{affine schemes}\} \longleftrightarrow \{\text{commutative rings with identity}\}$$

We shall see that in fact the ring and the corresponding affine scheme are equivalent objects. The scheme is, however, a more natural setting for many geometric arguments; speaking in terms of schemes will also allow us to globalize our constructions in succeeding sections.

Looking ahead, the case of differentiable manifolds provides a paradigm for our approach to the definition of schemes. A differentiable manifold M was originally defined to be something obtained by gluing together open balls — that is, a topological space with an atlas of coordinate charts. However, specifying the manifold structure on M is equivalent to specifying which of the continuous functions on any open subset of M are differentiable. The property of differentiability is defined locally, so the differentiable functions form a subsheaf $\mathscr{C}^\infty(M)$ of the sheaf $\mathscr{C}(M)$ of continuous functions on M (the definition of sheaves is given below). Thus we may give an alternative definition of a differentiable manifold: it is a topological space M together with a subsheaf $\mathscr{C}^\infty(M) \subset \mathscr{C}(M)$ such that the pair $(M, \mathscr{C}^\infty(M))$ is locally isomorphic to an open subset of $\mathbb{R}^n$ with its sheaf of differentiable functions. Sheaves of functions can also be used to define many other kinds of geometric structure — for example, real analytic manifolds, complex analytic manifolds, and Nash manifolds may all be defined in this way. We will adopt an analogous approach in defining schemes: a scheme will be a topological space X with a sheaf $\mathcal{O}$, locally isomorphic to an affine scheme as defined below.

Let R be a commutative ring. The affine scheme defined from R will be called Spec R, the *spectrum* of R. As indicated, it (like any scheme) consists of a set of points, a topology on it called the *Zariski topology*, and a sheaf $\mathcal{O}_{\mathrm{Spec}\,R}$ on this topological space, called the *sheaf of regular functions*, or *structure sheaf* of the scheme. Where there is a possibility of confusion we will use the notation |Spec R| to refer to the underlying set or topological space, without the sheaf; though if it is clear from context what we mean ("an open subset of Spec R," for example), we may omit the vertical bars.

We will give the definition of the affine scheme Spec R in three stages, specifying first the underlying set, then the topological structure, and finally the sheaf.

I.1.1 Schemes as Sets

We define a *point* of Spec R to be a prime — that is, a prime ideal — of R. To avoid confusion, we will sometimes write [$\mathfrak{p}$] for the point of Spec R corresponding to the prime $\mathfrak{p}$ of R. We will adopt the usual convention that R itself is not a prime ideal. Of course, the zero ideal (0) is a prime if R is a domain.

If R is the coordinate ring of an ordinary affine variety V over an algebraically closed field, Spec R will have points corresponding to the points of the affine variety — the maximal ideals of R — and also a point corresponding to each irreducible subvariety of V. The new points, corresponding to subvarieties of positive dimension, are at first rather unsettling but turn out to be quite convenient. They play the role of the "generic points" of classical algebraic geometry.

Exercise I-1. Find Spec R when R is (a) $\mathbb{Z}$; (b) $\mathbb{Z}/(3)$; (c) $\mathbb{Z}/(6)$; (d) $\mathbb{Z}_{(3)}$; (e) $\mathbb{C}[x]$; (f) $\mathbb{C}[x]/(x^2)$.

Each element $f \in R$ defines a "function", which we also write as f, on the space Spec R: if $x = [\mathfrak{p}] \in$ Spec R, we denote by $\kappa(x)$ or $\kappa(\mathfrak{p})$ the quotient field of the integral domain $R/\mathfrak{p}$, called the *residue field* of X at x, and we define $f(x) \in \kappa(x)$ to be the image of f via the canonical maps

$$R \to R/\mathfrak{p} \to \kappa(x).$$

Exercise I-2. What is the value of the "function" 15 at the point $(7) \in$ Spec $\mathbb{Z}$? At the point (5)?

Exercise I-3. (a) Consider the ring of polynomials $\mathbb{C}[x]$, and let $p(x)$ be a polynomial. Show that if $\alpha \in \mathbb{C}$ is a number, then $(x - \alpha)$ is a prime of $\mathbb{C}[x]$, and there is a natural identification of $\kappa((x - \alpha))$ with $\mathbb{C}$ such that the value of $p(x)$ at the point $(x - \alpha) \in$ Spec $\mathbb{C}[x]$ is the number $p(\alpha)$.

(b) More generally, if R is the coordinate ring of an affine variety V over an algebraically closed field K and $\mathfrak{p}$ is the maximal ideal corresponding to a point $x \in V$ in the usual sense, then $\kappa(x) = K$ and $f(x)$ is the value of f at x in the usual sense.

In general, the "function" f has values in fields that vary from point to point. Moreover, f is not necessarily determined by the values of this "function". For example, if K is a field, the ring $R = K[x]/(x^2)$ has only one prime ideal, which is (x); and thus the element $x \in R$, albeit nonzero, induces a "function" whose value is 0 at every point of $\operatorname{Spec} R$.

We define a *regular function* on $\operatorname{Spec} R$ to be simply an element of R. So a regular function gives rise to a "function" on $\operatorname{Spec} R$, but is not itself determined by the values of this "function".

I.1.2 Schemes as Topological Spaces

By using regular functions, we make $\operatorname{Spec} R$ into a topological space; the topology is called the *Zariski topology*. The closed sets are defined as follows. For each subset $S \subset R$, let

$$V(S) = \{x \in \operatorname{Spec} R \mid f(x) = 0 \text{ for all } f \in S\} = \{[\mathfrak{p}] \in \operatorname{Spec} R \mid \mathfrak{p} \supset S\}.$$

The impulse behind this definition is to make each $f \in R$ behave as much like a continuous function as possible. Of course the fields $\kappa(x)$ have no topology, and since they vary with x the usual notion of continuity makes no sense. But at least they all contain an element called zero, so one can speak of the locus of points in $\operatorname{Spec} R$ on which f is zero; and if f is to be like a continuous function, this locus should be closed. Since intersections of closed sets must be closed, we are led immediately to the definition above: $V(S)$ is just the intersection of the loci where the elements of S vanish.

For the family of sets $V(S)$ to be the closed sets of a topology it is necessary that it be closed under arbitrary intersections; from the description above it is clear that for any family of sets S_a we have $\bigcap_a V(S_a) = V(\bigcup_a S_a)$, as required. It is worth noting also that, if I is the ideal generated by S, then $V(I) = V(S)$.

An open set in the Zariski topology is simply the complement of one of the sets $V(S)$. The open sets corresponding to sets S with just one element will play a special role, essentially because they are again spectra of rings; for this reason they get a special name and notation. If $f \in R$, we define the *distinguished* (or *basic*) open subset of $X = \operatorname{Spec} R$ associated with f to be

$$X_f = |\operatorname{Spec} R| \setminus V(f).$$

The points of X_f — that is, the prime ideals of R that do not contain f — are in one-to-one correspondence with the prime ideals of the localization

R_f of R obtained by adjoining an inverse to f, via the correspondence that sends $\mathfrak{p} \subset R$ to $\mathfrak{p}R_f \subset R_f$. We may thus identify X_f with the points of Spec R_f, an indentification we will make implicitly throughout the remainder of this book.

The distinguished open sets form a *base* for the Zariski topology in the sense that any open set is a union of distinguished ones:

$$U = \operatorname{Spec} R \setminus V(S) = \operatorname{Spec} R \setminus \bigcap_{f \in S} V(f) = \bigcup_{f \in S} (\operatorname{Spec} R)_f .$$

Distinguished open sets are also closed under finite intersections; since a prime ideal contains a product if and only if it contains one of the factors, we have

$$\bigcap_{i=1,\ldots,n} (\operatorname{Spec} R)_{f_i} = (\operatorname{Spec} R)_g,$$

where g is the product $f_1 \cdots f_n$. In particular, any distinguished open set that is a subset of the distinguished open set $(\operatorname{Spec} R)_f$ has the form $(\operatorname{Spec} R)_{fg}$ for suitable g.

Spec R is almost never a Hausdorff space — the open sets are simply too large. In fact, the only points of Spec R that are closed are those corresponding to maximal ideals of R. In general, it is clear that the smallest closed set containing a given point $[\mathfrak{p}]$ must be $V(\mathfrak{p})$, so the closure of the point $[\mathfrak{p}]$ consists of all $[\mathfrak{q}]$ such that $\mathfrak{q} \supset \mathfrak{p}$. The point $[\mathfrak{p}]$ is closed if and only if $\mathfrak{p}$ is maximal. Thus in the case where R is the affine ring of an algebraic variety V over an algebraically closed field, the points of V correspond precisely to the closed points of Spec R, and the closed points contained in the closure of the point $[\mathfrak{p}]$ are exactly the points of V in the subvariety determined by $\mathfrak{p}$.

Exercise I-4. (a) The points of Spec $\mathbb{C}[x]$ are the primes $(x-a)$, for every $a \in \mathbb{C}$, and the prime (0). Describe the topology. Which points are closed? Are any of them open?

(b) Let K be a field and let R be the local ring $K[x]_{(x)}$. Describe the topological space Spec R. (The answer is given later in this section.)

To complete the definition of Spec R, we have to describe the *structure sheaf*, or *sheaf of regular functions* on X. Before doing this, we will take a moment out to give some of the basic definitions of sheaf theory and to prove a proposition that will be essential later on (Proposition I-12).

I.1.3 An Interlude on Sheaf Theory

Let X be any topological space. A *presheaf* $\mathscr{F}$ on X assigns to each open set U in X a set, denoted $\mathscr{F}(U)$, and to every pair of nested open sets $U \subset V \subset X$ a *restriction map*

$$\operatorname{res}_{V,U} : \mathscr{F}(V) \to \mathscr{F}(U)$$

satisfying the basic properties that

$$\text{res}_{U,U} = \text{identity}$$

and

$$\text{res}_{V,U} \circ \text{res}_{W,V} = \text{res}_{W,U} \quad \text{for all } U \subset V \subset W \subset X.$$

The elements of $\mathscr{F}(U)$ are called the *sections of* $\mathscr{F}$ *over* U; elements of $\mathscr{F}(X)$ are called *global sections*.

Another way to express this is to define a *presheaf* to be a contravariant functor from the category of open sets in X (with a morphism $U \to V$ for each containment $U \subseteq V$) to the category of sets. Changing the target category to abelian groups, say, we have the definition of a presheaf of abelian groups, and the same goes for rings, algebras, and so on.

One of the most important constructions of this type is that of a *presheaf of modules* $\mathscr{F}$ over a presheaf of rings $\mathscr{O}$ on a space X. Such a thing is a pair consisting of

for each open set U of X, a ring $\mathscr{O}(U)$ and an $\mathscr{O}(U)$-module $\mathscr{F}(U)$

and

for each containment $U \supseteq V$, a ring homomorphism $\alpha : \mathscr{O}(U) \to \mathscr{O}(V)$ and a map of sets $\mathscr{F}(U) \to \mathscr{F}(V)$ that is a map of $\mathscr{O}(U)$-modules if we regard $\mathscr{F}(V)$ as an $\mathscr{O}(U)$-module by means of α.

A presheaf (of sets, abelian groups, rings, modules, and so on) is called a *sheaf* if it satisfies one further condition, called the *sheaf axiom*. This condition is that, for each open covering $U = \bigcup_{a \in A} U_a$ of an open set $U \subset X$ and each collection of elements

$$f_a \in \mathscr{F}(U_a) \quad \text{for each } a \in A$$

having the property that for all $a, b \in A$ the restrictions of f_a and f_b to $U_a \cap U_b$ are equal, there is a unique element $f \in \mathscr{F}(U)$ whose restriction to U_a is f_a for all a.

A trivial but occasionally confusing point deserves a remark. The empty set $\varnothing$ is of course an open subset of Spec R, and can be written as the union of an empty family (that is, the indexing set A in the preceding paragraph is empty). Therefore the sheaf axiom imply that any sheaf has exactly one section over the empty set. In particular, for a sheaf $\mathscr{F}$ of rings, $\mathscr{F}(\varnothing)$ is the zero ring (where $0 = 1$). Note that the zero ring has no prime ideals at all — it is the only ring with unit having this property, if one accepts the axiom of choice — so that its spectrum is $\varnothing$.

Exercise I-5. (a) Let X be the two-element set $\{0, 1\}$, and make X into a topological space by taking each of the four subsets to be open. A sheaf on X is thus a collection of four sets with certain maps between them; describe the relations among these objects. (X is actually Spec R for some rings R; can you find one?)

(b) Do the same in the case where the topology of $X = \{0,1\}$ has as open sets only $\varnothing$, $\{0\}$ and $\{0,1\}$. Again, this space may be realized as Spec R.

If $\mathscr{F}$ is a presheaf on X and U is an open subset of X, we may define a presheaf $\mathscr{F}|_U$ on U, called the *restriction* of $\mathscr{F}$ to U, by setting $\mathscr{F}|_U(V) = \mathscr{F}(V)$ for any open subset V of U, the restriction maps being the same as those of $\mathscr{F}$ as well. It is easy to see that, if $\mathscr{F}$ is actually a sheaf, so is $\mathscr{F}|_U$.

In the sequel we shall work exclusively with presheaves and sheaves of things that are at least abelian groups, so we will usually omit the phrase "of abelian groups". Given two presheaves of abelian groups, one can define their direct sum, tensor product, and so on, open set by open set; thus, for example, if $\mathscr{F}$ and $\mathscr{G}$ are presheaves of abelian groups, we define $\mathscr{F} \oplus \mathscr{G}$ by

$$(\mathscr{F} \oplus \mathscr{G})(U) := \mathscr{F}(U) \oplus \mathscr{G}(U) \quad \text{for any open set } U.$$

This always produces a presheaf, and if $\mathscr{F}$ and $\mathscr{G}$ are sheaves then $\mathscr{F} \oplus \mathscr{G}$ will be one as well. Tensor product is not as well behaved: even if $\mathscr{F}$ and $\mathscr{G}$ are sheaves, the presheaf defined by

$$(\mathscr{F} \otimes \mathscr{G})(U) := \mathscr{F}(U) \otimes \mathscr{G}(U)$$

may not be, and we define the sheaf $\mathscr{F} \otimes \mathscr{G}$ to be the sheafification of this presheaf, as described below.

The simplest sheaves on any topological space X are the sheaves of locally constant functions with values in a set K — that is, sheaves $\mathscr{K}$ where $\mathscr{K}(U)$ is the set of locally constant functions from U to K; if K is a group, we may make $\mathscr{K}$ into a sheaf of groups by pointwise addition. Similarly, if K is a ring and we define multiplication in $\mathscr{K}(U)$ to be pointwise multiplication, then $\mathscr{K}$ becomes a sheaf of rings. When K has a topology, we can define the *sheaf of continuous functions* with values in K as the sheaf $\mathscr{C}$, where $\mathscr{C}(U)$ is the set of continuous functions from U to K, again with pointwise addition. If X is a differentiable manifold, there are also sheaves of differentiable functions, vector fields, differential forms, and so on.

Generally, if $\pi : Y \to X$ is any map of topological spaces, we may define the sheaf $\mathscr{S}$ of sections of π; that is, for every open set U of X we define $\mathscr{S}(U)$ to be the set of continuous maps $\sigma : U \to \pi^{-1}U$ such that $\pi \circ \sigma = 1$, the identity on U (such a map being a *section* of π in the set-theoretical sense: elements of $\mathscr{F}(U)$ for *any* sheaf $\mathscr{F}$ are called sections by extension from this case).

Exercise I-6. (For readers familiar with vector bundles.) Let V be a vector bundle on a topological space X. Check that the sheaf of sections of V is a sheaf of modules over the sheaf of continuous functions on X. (Sheaves of modules in general may in this way be seen as generalized vector bundles.)

Another way to describe a sheaf is by its stalks. For any presheaf $\mathscr{F}$ and any point $x \in X$, we define the *stalk* $\mathscr{F}_x$ of $\mathscr{F}$ at x to be the *direct limit* of the groups $\mathscr{F}(U)$ over all open neighborhoods U of x in X—that is, by definition,

$$\mathscr{F}_x = \varinjlim_{x \in U} \mathscr{F}(U)$$

$$= \left\{ \begin{array}{l} \text{the disjoint union of } \mathscr{F}(U) \text{ over all open sets } U \text{ containing } x, \\ \text{modulo the equivalence relation } \sigma \sim \tau \text{ if } \sigma \in \mathscr{F}(U),\ \tau \in \mathscr{F}(V), \\ \text{and there is an open neighborhood } W \text{ of } x \text{ contained in } U \cap V \\ \text{such that the restrictions of } \sigma \text{ and } \tau \text{ to } W \text{ are equal:} \\ \qquad\qquad \mathrm{res}_{U,W}\sigma = \mathrm{res}_{V,W}\tau. \end{array} \right\}$$

For every $x \in U$ there is a map $\mathscr{F}(U) \to \mathscr{F}_x$, sending a section s to the equivalence class of (U, s); this class is denoted s_x. If $\mathscr{F}$ is a sheaf, a section $s \in \mathscr{F}(U)$ of $\mathscr{F}$ over U is determined by its images in the stalks $\mathscr{F}_x$ for all $x \in U$—equivalently, $s = 0$ if and only if $s_x = 0$ for all $x \in U$. This follows from the sheaf axiom: to say that $s_x = 0$ for all $x \in U$ is to say that for each x there is a neighborhood U_x of x in U such that $\mathrm{res}_{U,U_x}(s) = 0$, and then it follows that $s = 0$ in $\mathscr{F}(U)$.

This notion of stalks has a familiar geometric content: it is an abstraction of the notion of rings of germs. For example, if X is an analytic manifold of dimension n and $\mathscr{O}_X^{\mathrm{an}}$ is the sheaf of analytic functions on X, the stalk of $\mathscr{O}_X^{\mathrm{an}}$ at x is the ring of germs of analytic functions at x—that is, the ring of convergent power series in n variables.

Exercise I-7. Find the stalks of the sheaves you produced for Exercises I-5 and I-6.

Exercise I-8. Topologize the disjoint union $\overline{\mathscr{F}} = \bigcup \mathscr{F}_x$ by taking as a base for the open sets of $\overline{\mathscr{F}}$ all sets of the form

$$\mathscr{V}(U, s) := \{(x, s_x) : x \in U\},$$

where U is an open set and s is a fixed section over U.

(a) Show that the natural map $\pi : \overline{\mathscr{F}} \to X$ is continuous, and that, for U and $s \in \mathscr{F}(U)$, the map $\sigma : x \mapsto s_x$ from U to $\overline{\mathscr{F}}$ is a continuous section of π over U (that is, it is continuous and $\pi \circ \sigma$ is the identity on U).

(b) Conversely, show that any continuous map $\sigma : U \to \overline{\mathscr{F}}$ such that $\pi \circ \sigma$ is the identity on U arises in this way.
    Hint. Take $x \in U$ and a basic open set $\mathscr{V}(V, t)$ containing $\sigma(x)$, where $V \subset U$. What relation does t have to σ?

This construction shows that the sheaf of germs of sections of $\pi : \overline{\mathscr{F}} \to X$ is isomorphic to $\mathscr{F}$, so any sheaf "is" the sheaf of germs of sections of a suitable map. In early works sheaves were defined this way. The topological space $\overline{\mathscr{F}}$ is called the "espace étalé" of the sheaf, because its open sets are "stretched out flat" over open sets of X.

A *morphism* $\varphi : \mathscr{F} \to \mathscr{G}$ of sheaves on a space X is defined simply to be a collection of maps $\varphi(U) : \mathscr{F}(U) \to \mathscr{G}(U)$ such that for every inclusion $U \subset V$ the diagram

$$
\begin{array}{ccc}
\mathscr{F}(V) & \xrightarrow{\;\varphi(V)\;} & \mathscr{G}(V) \\
{\scriptstyle \mathrm{res}_{V,U}}\big\downarrow & & \big\downarrow{\scriptstyle \mathrm{res}_{V,U}} \\
\mathscr{F}(U) & \xrightarrow[\;\varphi(U)\;]{} & \mathscr{G}(U)
\end{array}
$$

commutes. (In categorical language, a morphism of sheaves is just a natural transformation of the corresponding functors from the category of open sets on X to the category of sets.)

A morphism $\varphi : \mathscr{F} \to \mathscr{G}$ induces as well a map of stalks $\varphi_x : \mathscr{F}_x \to \mathscr{G}_x$ for each $x \in X$. By the sheaf axiom, the morphism is determined by the induced maps of stalks: if φ and ψ are morphisms such that $\varphi_x = \psi_x$ for all $x \in X$, then $\varphi = \psi$.

We say that a map $\varphi : \mathscr{F} \to \mathscr{G}$ of sheaves is injective, surjective, or bijective if each of the induced maps $\varphi_x : \mathscr{F}_x \to \mathscr{G}_x$ on stalks has the corresponding property. The following exercises show how these notions are related to their more naive counterparts defined in terms of sections on arbitrary sets.

Exercise I-9. Show that, if $\varphi : \mathscr{F} \to \mathscr{G}$ is a morphism of sheaves, then $\varphi(U)$ is injective (respectively, bijective) for all open sets $U \subset X$ if and only if φ_x is injective (respectively, bijective) for all points $x \in X$.

Exercise I-10. Show that Exercise I-9 is *false* if the condition "injective" is replaced by "surjective" by checking that in each of the following examples the maps induced by φ on stalks are surjective, but for some open set U the map $\varphi(U) : \mathscr{F}(U) \to \mathscr{G}(U)$ is not surjective.

(a) Let X be the topological space $\mathbb{C} \setminus \{0\}$, let $\mathscr{F} = \mathscr{G}$ be the sheaf of nowhere-zero, continuous, complex-valued functions, and let φ be the map sending a function f to f^2.

(b) Let X be the Riemann sphere $\mathbb{CP}^1 = \mathbb{C} \cup \{\infty\}$ and let $\mathscr{G}$ be the sheaf of analytic functions. Let $\mathscr{F}_1$ be the sheaf of analytic functions vanishing at 0; that is, $\mathscr{F}_1(U)$ is the set of analytic functions on U that vanish at 0 if $0 \in U$, and the set of all analytic functions on U if $0 \notin U$. Similarly, let $\mathscr{F}_2$ be the sheaf of analytic functions vanishing at ∞. Let $\mathscr{F} = \mathscr{F}_1 \oplus \mathscr{F}_2$, and let $\varphi : \mathscr{F} \to \mathscr{G}$ be the addition map.

(c) Find an example of this phenomenon in which the set X consists of three points.

These examples are the beginning of the cohomology theory of sheaves; the reader will find more in this direction in the references on sheaves listed on page 18.

If $\mathscr{F}$ is a presheaf on X, we define the *sheafification* of $\mathscr{F}$ to be the unique sheaf $\mathscr{F}'$ and morphism of presheaves $\varphi : \mathscr{F} \to \mathscr{F}'$ such that for all $x \in X$ the map $\varphi_x : \mathscr{F}_x \to \mathscr{F}'_x$ is an isomorphism. More explicitly, the sheaf $\mathscr{F}'$ may be defined by saying that a section of $\mathscr{F}'$ over an open set U is a map σ that takes each point $x \in U$ to an element in $\mathscr{F}_x$ in such a way that σ is *locally induced* by sections of $\mathscr{F}$; by this we mean that there exists an open cover of U by open sets U_i and elements $s_i \in \mathscr{F}(U_i)$ such that $\sigma(x) = (s_i)_x$ for $x \in U_i$. The map $\mathscr{F} \to \mathscr{F}'$ is defined by associating to $s \in \mathscr{F}(U)$ the function $x \mapsto s_x \in \mathscr{F}_x$. The sheaf $\mathscr{F}'$ should be thought of as the sheaf "best approximating" the presheaf $\mathscr{F}$.

Exercise I-11. Here is an alternate construction for $\mathscr{F}'$: topologize the disjoint union $\overline{\mathscr{F}} = \bigcup \mathscr{F}_x$ exactly as in Exercise I-8; then let $\mathscr{F}'$ be the sheaf of sections of the natural map $\pi : \overline{\mathscr{F}} \to X$. Convince yourself that the two constructions are equivalent, and that the result does have the universal property stated at the beginning of the preceding paragraph.

If $\varphi : \mathscr{F} \to \mathscr{G}$ is an injective map of sheaves, we will say that $\mathscr{F}$ is a *subsheaf* of $\mathscr{G}$. We often write $\mathscr{F} \subset \mathscr{G}$, omitting φ from the notation. If $\varphi : \mathscr{F} \to \mathscr{G}$ is any map of sheaves, the presheaf $\operatorname{Ker} \varphi$ defined by $(\operatorname{Ker} \varphi)(U) = \operatorname{Ker}(\varphi(U))$ is a subsheaf of $\mathscr{F}$.

The notion of a quotient is more subtle. Suppose $\mathscr{F}$ and $\mathscr{G}$ are presheaves of abelian groups, where $\mathscr{F}$ injects in $\mathscr{G}$. The quotient of $\mathscr{G}$ by $\mathscr{F}$ as presheaves is the presheaf $\mathscr{H}$ defined by $\mathscr{H}(U) = \mathscr{G}(U)/\mathscr{F}(U)$. But if $\mathscr{F}$ and $\mathscr{G}$ are sheaves, $\mathscr{H}$ will generally *not* be a sheaf, and we must define their quotient as sheaves to be the sheafification of $\mathscr{H}$, that is, $\mathscr{G}/\mathscr{F} := \mathscr{H}'$. The natural map from $\mathscr{H}$ to its sheafification $\mathscr{H}'$, together with the map of presheaves $\mathscr{G} \to \mathscr{H}$, defines the quotient map from $\mathscr{G}$ to $\mathscr{G}/\mathscr{F}$. This map is the *cokernel* of φ.

The significance of the sheaf axiom is that sheaves are defined by local properties. We give two aspects of this principle explicitly.

In our applications to schemes, we will encounter a situation where we are given a *base* $\mathscr{B}$ for the open sets of a topological space X, and we will want to specify a sheaf $\mathscr{F}$ just by saying what the groups $\mathscr{F}(U)$ and homomorphisms $\operatorname{res}_{V,U}$ are for open sets U of our base and inclusions $U \subset V$ of basic sets. The next proposition is exactly the tool that says we can do this.

We say that a collection of groups $\mathscr{F}(U)$ for open sets $U \in \mathscr{B}$ and maps $\operatorname{res}_{V,U} : \mathscr{F}(V) \to \mathscr{F}(U)$ for $V \subset U$ form a *$\mathscr{B}$-sheaf* if they satisfy the sheaf axiom with respect to inclusions of basic open sets in basic open sets and coverings of basic open sets by basic open sets. (The condition in the definition that sections of $U_a, U_b \in \mathscr{B}$ agree on $U_a \cap U_b$ must be replaced by the condition that they agree on any basic open set $V \in \mathscr{B}$ such that $V \subset U_a \cap U_b$.)

Proposition I-12. *Let $\mathscr{B}$ be a base of open sets for X.*

(i) *Every $\mathscr{B}$-sheaf on X extends uniquely to a sheaf on X.*

(ii) *Given sheaves $\mathscr{F}$ and $\mathscr{G}$ on X and a collection of maps*

$$\tilde{\varphi}(U) : \mathscr{F}(U) \to \mathscr{G}(U) \quad \text{for all } U \in \mathscr{B}$$

commuting with restrictions, there is a unique morphism $\varphi : \mathscr{F} \to \mathscr{G}$ of sheaves such that $\varphi(U) = \tilde{\varphi}(U)$ for all $U \in \mathscr{B}$.

Beginning of the proof. For any open set $U \subset X$, define $\mathscr{F}(U)$ as the inverse limit of the sets $\mathscr{F}(V)$, where V runs over basic open sets contained in U:

$$\mathscr{F}(U) = \varprojlim_{V \subset U, V \in \mathscr{B}} \mathscr{F}(V)$$
$$= \left\{ \begin{array}{l} \text{the set of families } (f_V)_{V \subset U, V \in \mathscr{B}} \in \prod_{V \subset U, V \in \mathscr{B}} \mathscr{F}(V) \text{ such} \\ \text{that } \mathrm{res}_{V,W}(f_V) = f_W \text{ whenever } W \subset V \subset U \text{ with } V, W \in \mathscr{B}. \end{array} \right\}$$

The restriction maps are defined immediately from the universal property of the inverse limit. □

Exercise I-13. Complete the proof of the proposition by checking the sheaf axioms and showing that, for $U \in \mathscr{B}$, the new definition of $\mathscr{F}$ agrees with the old one.

The second application, which is really a special case of the first, says that to define a sheaf it is enough to give it on each open set of an open cover, as long as the definitions are compatible.

Corollary I-14. *Let $\mathscr{U}$ be an open covering of a topological space X. If $\mathscr{F}_U$ is a sheaf on U for each $U \in \mathscr{U}$, and if*

$$\varphi_{UV} : \mathscr{F}_U|_{U \cap V} \to \mathscr{F}_V|_{U \cap V}$$

are isomorphisms satisfying the compatibility conditions

$$\varphi_{VW}\varphi_{UV} = \varphi_{UW} \quad \text{on } U \cap V \cap W,$$

for all $U, V, W \in \mathscr{U}$, there is a unique sheaf $\mathscr{F}$ on X whose restriction to each $U \in \mathscr{U}$ is isomorphic to $\mathscr{F}_U$ via isomorphisms $\Psi_U : \mathscr{F}|_U \to \mathscr{F}_U$ compatible with the isomorphisms φ_{UV} — in other words, such that

$$\varphi_{UV} \circ \Psi_U|_{U \cap V} = \Psi_V|_{U \cap V} : \mathscr{F}|_{U \cap V} \to \mathscr{F}_V|_{U \cap V}$$

for all U and V in $\mathscr{U}$.

Proof. The open sets contained in some $U \in \mathscr{U}$ form a base $\mathscr{B}$ for the topology of X. For each such set V we choose arbitrarily a set U that contains it, and define $\mathscr{F}(V) = \mathscr{F}_U(V)$. If for some $W \subset V$ the value $\mathscr{F}(W)$ has been defined with reference to a different $\mathscr{F}_{U'}$, we use the isomorphism $\varphi_{UU'}$ to define the restriction maps. These maps compose correctly because of the compatibility conditions on the isomorphisms $\varphi_{UU'}$. Thus we have a $\mathscr{B}$-sheaf, and therefore a sheaf. □

The pushforward operation on sheaves is so basic (and trivial) that we introduce it here: If $\alpha : X \to Y$ is a continuous map on topological spaces and $\mathscr{F}$ is a presheaf on X, we define the *pushforward* $\alpha_* \mathscr{F}$ of $\mathscr{F}$ by α to be the presheaf on Y given by

$$\alpha_* \mathscr{F}(V) := \mathscr{F}(\alpha^{-1}(V)) \quad \text{for any open } V \subset Y.$$

Of course, the pushforward of a sheaf of abelian groups (rings, modules over a sheaf of rings, and so on) is again of the same type.

Exercise I-15. Show that the pushforward of a sheaf is again a sheaf.

References for the Theory of Sheaves. Serre's landmark paper [1955], which established sheaves as an important tool in algebraic geometry, is still a wonderful source of information. Godement [1964] and Swan [1964] are more systematic introductions. Hartshorne [1977, Chapter II] contains an excellent account adapted to the technical requirements of scheme theory; it is a simplified version of that found in Grothendieck [1961a; 1961b; 1963; 1964; 1965; 1966; 1967]. Some good references for the analytic case are Forster [1981] (especially for an introduction to cohomology) and Gunning [1990].

I.1.4   Schemes as Schemes (Structure Sheaves)

We return at last to the definition of the scheme $X = \operatorname{Spec} R$. We will complete the construction by specifying the structure sheaf $\mathscr{O}_X = \mathscr{O}_{\operatorname{Spec} R}$. As indicated above, we want the relationship between $\operatorname{Spec} R$ and R to generalize that between an affine variety and its coordinate ring; in particular, we want the ring of global sections of the structure sheaf $\mathscr{O}_X$ to be R.

We thus wish to extend the ring R of functions on X to a whole sheaf of rings. This means that for each open set U of X, we wish to give a ring $\mathscr{O}_X(U)$; and for every pair of open sets $U \subset V$ we wish to give a restriction homomorphism

$$\operatorname{res}_{V,U} : \mathscr{O}_X(V) \to \mathscr{O}_X(U)$$

satisfying the various axioms above. It is quite easy to say what the rings $\mathscr{O}_X(U)$ and the maps $\operatorname{res}_{V,U}$ should be for distinguished open sets U and V: we set

$$\mathscr{O}_X(X_f) = R_f.$$

If $X_f \supset X_g$, some power of g is a multiple of f (recall that the radical of (f) is the intersection of the primes containing f). Thus the restriction map $\operatorname{res}_{X_f, X_g}$ can be defined as the localization map $R_f \to R_{fg} = R_g$. By Proposition I-12, this will suffice to define the structure sheaf $\mathscr{O}$, as long as we verify that it satisfies the sheaf axiom with respect to coverings of distinguished opens by distinguished opens. Before doing this, in Proposition I-18 below, we exhibit a simple but fundamental lemma that describes the coverings of affine schemes by distinguished open sets.

Lemma I-16. *Let $X = \operatorname{Spec} R$, and let $\{f_a\}$ be a collection of elements of R. The open sets X_{f_a} cover X if and only if the elements f_a generate the unit ideal. In particular, X is quasicompact as a topological space.*

Recall that *quasicompact* means that every open cover has a finite sub-cover; the *quasi* is there because the space is not necessarily Hausdorff. In fact, schemes are almost never Hausdorff! Unfortunately, this fact vitiates most of the usual advantages of compactness. For example, in contrast to the situation for compact manifolds, say, the continuous image of one affine scheme in another need not be closed. For this reason, we will discuss in Section III.1 a better "compactness" notion, called *properness*, which will play just as important a role as compactness does in the usual geometric theories.

Proof. The X_{f_a} cover X if and only if no prime of R contains all the f_a, which happens if and only if the f_a generate the unit ideal; this proves the first statement. To prove the second, note first that every open cover has a refinement of the form $X = \bigcup X_{f_a}$, where each $f_a \in R$. Since the X_{f_a} cover X, the f_a generate the unit ideal, so the element 1 can be written as a linear combination — necessarily finite — of the f_a. Taking just the f_a involved in this expansion of 1, we see that the cover $X = \bigcup X_{f_a}$, and with it the original cover, has a finite subcover. □

Exercise I-17. If R is Noetherian, every subset of $\operatorname{Spec} R$ is quasicompact.

Proposition I-18. *Let $X = \operatorname{Spec} R$, and suppose that X_f is covered by open sets $X_{f_a} \subset X_f$.*

(a) *If $g, h \in R_f$ become equal in each R_{f_a}, they are equal.*

(b) *If for each a there is $g_a \in R_{f_a}$ such that for each pair a and b the images of g_a and g_b in $R_{f_a f_b}$ are equal, then there is an element $g \in R_f$ whose image in R_{f_a} is g_a for all a.*

Equivalently, if $\mathscr{B}$ is the collection of distinguished open sets $\operatorname{Spec} R_f$ of $\operatorname{Spec} R$, and if we set $\mathscr{O}_X(\operatorname{Spec} R_f) := R_f$, then $\mathscr{O}_X$ is a $\mathscr{B}$-sheaf. By Proposition I-12, $\mathscr{O}_X$ extends uniquely to a sheaf on X.

Definition I-19. The sheaf $\mathscr{O}_X$ defined in the proposition is called the *structure sheaf* of X or the *sheaf of regular functions on X.*

Proof of Proposition I-18. We begin with the case $f = 1$, so $R_f = R$ and $X_f = X$.

For the first part, observe that if g and h become equal in each X_{f_a} then $g - h$ is annihilated by a power of each f_a. Since by Lemma I-16 we may assume that the cover is finite, this implies that $g - h$ is annihilated by a power of the ideal generated by all the f_a^N for some N. But this ideal contains a power of the ideal generated by all the f_a, which is the unit ideal. Thus $g = h$ in R.

For part (b), we will use an argument analogous to the classical partition of unity to piece together the elements g_a into a single element $g \in R$. For large N the product $f_a^N g_a \in R_{f_a}$ is the image of an element $h_a \in R$. By Lemma I-16 we may assume the covering $\{X_{f_a}\}$ is finite, and it follows that one N will do for all a. Next, since g_a and g_b become equal in $X_{f_a f_b}$, we must have

$$f_b^N h_a = (f_a f_b)^N g_a = (f_a f_b)^N g_b = f_a^N h_b$$

for large N. Again, since we have assumed the covering $\{X_{f_a}\}$ is finite, one N will do for all a and b. By Lemma I-16 the elements $f_a \in R$ generate the unit ideal, and hence so do the elements f_a^N, and we may write

$$1 = \sum_a e_a f_a^N$$

for some collection $e_a \in R$; this is our partition of unity. We claim that

$$g = \sum_a e_a h_a$$

is the element of R we seek. Indeed, for each b, we have in R_{f_b}

$$f_b^N g = \sum_a f_b^N e_a h_a = \sum_a f_a^N e_a h_b = h_b = f_b^N g_b,$$

so g becomes equal to g_b on X_{f_b}, as required.

Returning to the case of arbitrary f, set $X' = X_f$, $R' = R_f$, $f_a' = f f_a$; then $X' = \operatorname{Spec} R'$ and $X'_{f_a'} = X_{f_a}$, so we can apply the case already proved to the primed data.   □

The proposition is still valid, and has essentially the same proof, if we replace R_f and R_{f_a} by M_f and M_{f_a} for any R-module M.

Exercise I-20. Describe the points and the sheaf of functions of each of the following schemes.

(a) $X_1 = \operatorname{Spec} \mathbb{C}[x]/(x^2)$.   (b) $X_2 = \operatorname{Spec} \mathbb{C}[x]/(x^2 - x)$.

(c) $X_3 = \operatorname{Spec} \mathbb{C}[x]/(x^3 - x^2)$.   (d) $X_4 = \operatorname{Spec} \mathbb{R}[x]/(x^2 + 1)$.

In contrast with the situation in many geometric theories (though similar to the situation in the category of complex manifolds), there may be really rather few regular functions on a scheme. For example, when we define arbitrary schemes, we shall see that the schemes that are the analogues of compact manifolds may have no nonconstant regular functions on them at all. For this reason, partially defined functions on a scheme X—that is, elements $\mathscr{O}_X(U)$ for some open dense subset U—play an unusually large role. They are called *rational* functions on X because in the case $X = \operatorname{Spec} R$ with R a domain, and $U = X_f$, the elements of $\mathscr{O}_X(X_f) = R_f$ are ratios of elements in R. In the cases of most interest, we shall see that every nonempty open set is dense in X, so the behavior of rational functions reflects the properties of X as a whole.

Exercise I-21. Let $\mathcal{U}$ be the set of open and dense sets in X. Compute the *ring of rational functions*

$$\varinjlim_{U \in \mathcal{U}} \mathcal{O}_X(U) :=$$

$$\left\{\begin{array}{l} \text{the disjoint union of } \mathcal{O}_X(U) \text{ for all } U \in \mathcal{U}, \text{ modulo the equiva-} \\ \text{lence relation } \sigma \sim \tau \text{ if } \sigma \in \mathcal{O}_X(U),\ \tau \in \mathcal{O}_X(V), \text{ and the restric-} \\ \text{tions of } \sigma \text{ and } \tau \text{ are equal on some } W \in \mathcal{U} \text{ contained in } U \cap V \end{array}\right\},$$

first in the case where R is a domain and then for an arbitrary Noetherian ring.

Example I-22. Another very simple example will perhaps help to fix these ideas. Let K be a field, and let $R = K[x]_{(x)}$, the localization of the polynomial ring in one variable X at the maximal ideal (x). The scheme $X = \operatorname{Spec} R$ has only two points, the two prime ideals (0) and (x) of R. As a topological space, it has precisely three open sets,

$$\varnothing \subset U := \{(0)\} \subset \{(0), (x)\} = X.$$

U and $\varnothing$ are distinguished open sets, since $\{(0)\} = X_x$. The sheaf $\mathcal{O}_X$ is thus easy to describe. It has values $\mathcal{O}_X(X) = R = K[x]_{(x)}$ and $\mathcal{O}_X(U) = K(x)$, the field of rational functions. The restriction map from the first to the second is the natural inclusion.

Exercise I-23. Give a similarly complete description for the structure sheaf of the scheme $\operatorname{Spec} K[x]$. (The answer is given in Chapter II.)

I.2   Schemes in General

After this lengthy description of affine schemes, it is easy to define schemes in general. A *scheme* X is simply a topological space, called the *support* of X and denoted $|X|$ or $\operatorname{supp} X$, together with a sheaf $\mathcal{O}_X$ of rings on X, such that the pair $(|X|, \mathcal{O}_X)$ is *locally affine*. Locally affine means that $|X|$ is covered by open sets U_i such that there exist rings R_i, and homeomorphisms $U_i \cong |\operatorname{Spec} R_i|$ with $\mathcal{O}_X|_{U_i} \cong \mathcal{O}_{\operatorname{Spec} R_i}$.

To better understand this definition, we must identify the key properties of the structure sheaf of an affine scheme. Let X be any topological space and let $\mathcal{O}$ be a sheaf of rings on it. We call the pair $(X, \mathcal{O})$ a *ringed space*, and ask when it is isomorphic to an affine scheme $(|\operatorname{Spec} R|, \mathcal{O}_{\operatorname{Spec} R})$. Note that if $(X, \mathcal{O})$ were an affine scheme then it would have to be the scheme $\operatorname{Spec} R$.

Now let $(X, \mathcal{O})$ be any ringed space, and let $R = \mathcal{O}(X)$. For any $f \in R$ we can define a set $U_f \subset X$ as the set of points $x \in X$ such that f maps to a unit of the stalk $\mathcal{O}_x$. If $(X, \mathcal{O})$ is an affine scheme we must have:

(i) $\mathcal{O}(U_f) = R[f^{-1}]$.

However, this condition is not enough; it does not even force the existence of a map between X and $|\operatorname{Spec} R|$. To give such a map, we need to assume a further condition on $\mathcal{O}$ that is posessed by affine schemes:

(ii) The stalks $\mathcal{O}_x$ of $\mathcal{O}$ are local rings.

A ringed space $(X, \mathcal{O})$ satisfying (ii) is often called a *local ringed space*.

If $(X, \mathcal{O})$ satisfies (ii), there is a natural map $X \to |\operatorname{Spec} \mathcal{O}(X)|$ that takes $x \in X$ to the prime ideal of $\mathcal{O}(X)$ that is the preimage of the maximal ideal of $\mathcal{O}_x$. The third condition for $(X, \mathcal{O})$ to be an affine scheme is this:

(iii) The map $X \to |\operatorname{Spec} \mathcal{O}(X)|$ is a homeomorphism.

Given these considerations, we say that a pair $(X, \mathcal{O})$ is *affine* if it satisfies (i)–(iii). The definition of scheme given above now becomes: A pair $(X, \mathcal{O})$ is a scheme if it is locally affine.

Again, where there is no danger of confusion, we will use the same letter X to denote the scheme and the underlying space $|X|$, as in the construction "let $p \in X$ be a point."

Exercise I-24. (a) Take $Z = \operatorname{Spec} \mathbb{C}[x]$, let X be the result of identifying the two closed points (x) and $(x-1)$ of $|Z|$, and let $\varphi : Z \to X$ be the natural projection. Let $\mathcal{O}$ be $\varphi_* \mathcal{O}_Z$, a sheaf of rings on X. Show that $(X, \mathcal{O})$ satisfies condition (i) above for all elements $f \in \mathcal{O}(X) = \mathbb{C}[x]$, but does not satisfy condition (ii). Note that there is no natural map $X \to |\operatorname{Spec} \mathbb{C}[x]|$.

(b) Take $Z = \operatorname{Spec} \mathbb{C}[x, y]$, the scheme corresponding to the affine plane, and let X be the open subset obtained by leaving out the origin in the plane, that is, $X = |Z| - \{(x, y)\}$. Let $\mathcal{O}$ be the sheaf $\mathcal{O}_Z |_X$ (that is, $\mathcal{O}(V) = \mathcal{O}_Z(V)$ for any open subset $V \subset X \subset |Z|$.) Show that $\mathcal{O}(X) = \mathbb{C}[x, y]$, that $X, \mathcal{O}$ satisfies condition (i) and (ii), and that the natural map $X \to |\operatorname{Spec} \mathcal{O}(X)|$ is the inclusion $X \subset |Z|$.

Some notation and terminology are in order at this point.

A *regular function* on an open set $U \subset X$ is a section of $\mathcal{O}_X$ over U. A *global regular function* is a regular function on X.

The stalks $\mathcal{O}_{X,x}$ of the structure sheaf $\mathcal{O}_X$ at the points $x \in X$ are called the *local rings* of $\mathcal{O}_X$. The residue field of $\mathcal{O}_{X,x}$ is denoted by $\kappa(x)$. Just as in the situation of Section I.1.1, a section of $\mathcal{O}_X$ can be thought of as a "function" taking values in these fields $\kappa(x)$: if $f \in \mathcal{O}_X(U)$ and $x \in U$, the image of f under the composite

$$\mathcal{O}_X(U) \to \mathcal{O}_{X,x} \to \kappa(x)$$

is the value of f at x.

Exercise I-25 (the smallest nonaffine scheme). Let X be the topological space with three points p, q_1, and q_2. Topologize X by making $X_1 := \{p, q_1\}$

and $X_2 := \{p, q_2\}$ open sets (so that, in addition, $\varnothing, \{p\}$, and X itself are open). Define a presheaf $\mathcal{O}$ of rings on X by setting

$$\mathcal{O}(X) = \mathcal{O}(X_1) = \mathcal{O}(X_2) = K[x]_{(x)}, \qquad \mathcal{O}(\{p\}) = K(x),$$

with restriction maps $\mathcal{O}(X) \to \mathcal{O}(X_i)$ the identity and $\mathcal{O}(X_i) \to \mathcal{O}(\{p\})$ the obvious inclusion. Check that this presheaf is a sheaf and that $(X, \mathcal{O})$ is a scheme. Show that it is not an affine scheme. (Geometrically, the scheme $(X, \mathcal{O})$ is the "germ of the doubled point" in the scheme called X_1 in Exercise I-44.)

I.2.1 Subschemes

Let U be an open subset of a scheme X. The pair $(U, \mathcal{O}_X|_U)$ is again a scheme, though this is not completely obvious. To check it, note that at least a *distinguished* open set of an affine scheme is again an affine scheme: if $X = \operatorname{Spec} R$ and $U = X_f$, then $(U, \mathcal{O}_X|_U) = \operatorname{Spec} R_f$. Since the distinguished open sets of X that are contained in U cover U, this shows that $(U, \mathcal{O}_X|_U)$ is covered by affine schemes, as required. An open subset of a scheme is correspondingly referred to as an *open subscheme* of X, with this structure understood.

The definition of a closed subscheme is more complicated; it is not enough to specify a closed subspace of X, because the sheaf structure is not defined thereby.

Consider first an affine scheme $X = \operatorname{Spec} R$. For any ideal I in the ring R, we may make the closed subset $V(I) \subset X$ into an affine scheme by identifying it with $Y = \operatorname{Spec} R/I$. This makes sense because the primes of R/I are exactly the primes of R that contain I taken modulo I, and thus the topological space $|\operatorname{Spec} R/I|$ is canonically homeomorphic to the closed set $V(I) \subset X$. We define a *closed subscheme* of X to be a scheme Y that is the spectrum of a quotient ring of R (so that the closed subschemes of X by definition correspond one to one with the ideals in the ring R).

We can define in these terms all the usual operations on and relations between closed subschemes of a given scheme $X = \operatorname{Spec} R$. Thus, we say that the closed subscheme $Y = \operatorname{Spec} R/I$ of X *contains* the closed subscheme $Z = \operatorname{Spec} R/J$ if Z is in turn a closed subscheme of Y—that is, if $J \supset I$. This implies that $V(J) \subset V(I)$, but the converse is not true.

Exercise I-26. The schemes X_1, X_2, and X_3 of Exercise I-20 may all be viewed as closed subschemes of $\operatorname{Spec} \mathbb{C}[x]$. Show that

$$X_1 \subset X_3 \quad \text{and} \quad X_2 \subset X_3,$$

but no other inclusions $X_i \subset X_j$ hold, even though the underlying sets of X_2 and X_3 coincide and the underlying set of X_1 is contained in the underlying set of X_2.

The *union* of the closed subschemes $\operatorname{Spec} R/I$ and $\operatorname{Spec} R/J$ is defined as $\operatorname{Spec} R/(I \cap J)$, and their *intersection* as $\operatorname{Spec} R/(I+J)$. It is important to note that the notions of containment, intersection, and union do *not* satisfy all the usual properties of their set-theoretical counterparts: for example, we will see on page 69 an example of closed subschemes X, Y, Z of a scheme such that $X \cup Y = X \cup Z$ and $X \cap Y = X \cap Z$ but $Y \neq Z$.

We would now like to generalize the notion of closed subscheme to an arbitrary scheme X. To do this, the first step must be to replace the ideal $I \subset R$ associated to a closed subscheme Y of an affine scheme $X = \operatorname{Spec} R$ by a sheaf, which we do as follows. We define $\mathscr{I} = \mathscr{I}_{Y/X}$, the *ideal sheaf of Y in X*, to be the sheaf of ideals of $\mathcal{O}_X$ given on a distinguished open set $V = X_f$ of X by $\mathscr{I}(X_f) = I R_f$. Now we can identify the structure sheaf $\mathcal{O}_Y$ of $Y = \operatorname{Spec} R/I$ — more precisely, the pushforward $j_* \mathcal{O}_Y$, where j is the inclusion map $|Y| \hookrightarrow |X|$ — with the quotient sheaf $\mathcal{O}_X/\mathscr{I}$. (You should spell out this identification.) The sheaf of ideals $\mathscr{I}$ may be recovered as the kernel of the restriction map $\mathcal{O}_X \to j_* \mathcal{O}_Y$.

One subtle point requires mention: not all sheaves of ideals in $\mathcal{O}_X$ arise from ideals of R. For example, in the case of $R = K[x]_{(x)}$ considered in Example I-22, we may define a sheaf of ideals by

$$\mathscr{I}(X) = 0, \qquad \mathscr{I}(U) = \mathcal{O}_X(U) \quad \text{for } U = \{(0)\}.$$

For a sheaf of ideals $\mathscr{I}$ coming from an ideal of R we would have

$$\mathscr{I}(U) = \mathscr{I}(X)_x = \mathscr{I}(X) K(x),$$

so $\mathscr{I}$ does not come from any ideal of R. In the definition of a closed subscheme above, we are only interested in sheaves of ideals that do come from ideals of R. The theory obviously needs a name for such sheaves: they are called *quasicoherent* sheaves of ideals. (This seems a poor name for such a basic and simple object, but it is firmly rooted in the literature. It comes from the fact that a sheaf on the spectrum of a Noetherian ring that corresponds to a finitely generated module has a property called coherence; it was thus natural to say that the sheaf coming from a finitely generated module is *coherent*, and that coming from an arbitrary module is quasicoherent.)

More generally, a *quasicoherent sheaf of ideals* $\mathscr{I} \subset \mathcal{O}_X$ on an arbitrary scheme X is a sheaf of ideals $\mathscr{I}$ such that, for every open affine subset U of X, the restriction $\mathscr{I}|_U$ is a quasicoherent sheaf of ideals on U.

Now we are ready to define a closed subscheme of an arbitrary scheme as something that looks locally like a closed subscheme of an affine scheme:

Definition I-27. If X is an arbitrary scheme, a *closed subscheme* Y of X is a closed topological subspace $|Y| \subset |X|$ together with a sheaf of rings $\mathcal{O}_Y$ that is a quotient sheaf of the structure sheaf $\mathcal{O}_X$ by a quasicoherent sheaf of ideals $\mathscr{I}$, such that the intersection of Y with any affine open subset $U \subset X$ is the closed subscheme associated to the ideal $\mathscr{I}(U)$.

If $V \subset X$ is any open set, we say that a regular function $f \in \mathcal{O}_X(V)$ *vanishes on Y* if $f \in \mathcal{J}(V)$.

In fact, $|Y|$ is uniquely determined by $\mathcal{J}$, so closed subschemes of X are in one-to-one correspondence with the quasicoherent sheaves of ideals $\mathcal{J} \subset \mathcal{O}_X$.

The notion of quasicoherence arises in a more general context as well. We similarly define a *quasicoherent sheaf* $\mathcal{F}$ on X to be a sheaf of $\mathcal{O}_X$-modules (that is, $\mathcal{F}(U)$ is an $\mathcal{O}_X(U)$-module for each U) such that for any affine set U and distinguished open subset $U_f \subset U$, the $\mathcal{O}_X(U_f) = \mathcal{O}_X(U)_f$-module $\mathcal{F}(U_f)$ is obtained from $\mathcal{F}(U)$ by inverting f — more precisely, the restriction map $\mathcal{F}(U) \to \mathcal{F}(U_f)$ becomes an isomorphism after inverting f. $\mathcal{F}$ is called *coherent* if all the modules $\mathcal{F}(U)$ are finitely generated. (A more restrictive use of the word coherence is also current, but coincides with this one in the case where X is covered by finitely many spectra of Noetherian rings, the situation of primary interest.) One might say informally that quasicoherent sheaves are those sheaves of modules whose restrictions to open affine sets are modules (finitely generated in the case of coherent sheaves) on the corresponding rings. This is the right analogue in the context of schemes of the notion of module over a ring; for most purposes, one should think of them simply as modules.

Exercise I-28. To check that a sheaf of ideals (or any sheaf of modules) is quasicoherent (or for that matter coherent), it is enough to check the defining property on each set U of a fixed open affine cover of X.

One of the most important closed subschemes of an affine scheme X is X_{red}, the *reduced scheme associated to X*. This may be defined by setting $X_{\mathrm{red}} = \operatorname{Spec} R_{\mathrm{red}}$, where R_{red} is R modulo its *nilradical* — that is, modulo the ideal of nilpotent elements of R. Recall that the nilradical of a ring R equals the intersection of all the primes of R (in fact, the intersection of all minimal primes). Therefore $|X|$ and $|X_{\mathrm{red}}|$ are identical as topological spaces.

Exercise I-29. X_{red} may also be defined as the topological space $|X|$ with structure sheaf $\mathcal{O}_{X_{\mathrm{red}}}$ associating to every open subset $U \subset X$ the ring $\mathcal{O}_X(U)$ modulo its nilradical.

To globalize this notion, we may define for any scheme X a sheaf of ideals $\mathcal{N} \subset \mathcal{O}_X$, called the *nilradical*; this is the sheaf whose value on any open set U is the nilradical of $\mathcal{O}_X(U)$. Because the construction of the nilradical commutes with localization, $\mathcal{N}$ is a quasicoherent sheaf of ideals. The associated closed subscheme of X is called the *reduced scheme associated to X* and denoted X_{red}. We say that X is *reduced* if $X = X_{\mathrm{red}}$.

Irreducibility is another possible property of schemes; in spite of the name, it is independent of whether the scheme is reduced. A scheme X is *irreducible* if $|X|$ is not the union of two properly contained closed sets.

Here are some easy but important remarks about reduced and irreducible schemes.

Exercise I-30. A scheme is irreducible if and only if every open subset is dense.

Exercise I-31. An affine scheme $X = \operatorname{Spec} R$ is reduced and irreducible if and only if R is a domain. X is irreducible if and only if R has a unique minimal prime, or, equivalently, if the nilradical of R is a prime.

Exercise I-32. A scheme X is reduced if and only if every affine open subscheme of X is reduced, if and only if every local ring $\mathcal{O}_{X,p}$ is reduced for closed points $p \in X$. (A ring is called reduced if its only nilpotent element is 0.)

Exercise I-33. How do you define the disjoint union of two schemes? Show that the disjoint union of two affine schemes $\operatorname{Spec} R$ and $\operatorname{Spec} S$ may be identified with the scheme $\operatorname{Spec} R \times S$.

Exercise I-34. An arbitrary scheme X is irreducible if and only if every open affine subset is irreducible. If it is connected (in the sense that the topological space $|X|$ is connected), then it is irreducible if and only if every local ring of $\mathcal{O}_X$ has a unique minimal prime.

We have now introduced the notion of open subscheme and closed subscheme of a scheme X. A further generalization, a *locally closed subscheme* of X, is immediate: it is simply a closed subscheme of an open subscheme of X. This is as general a notion as we will have occasion to consider in this book; so that when we speak just of a subscheme of X, without modifiers, we will mean a locally closed subscheme.

Exercise I-35. Let X be an arbitrary scheme and let Y, Z be closed subschemes of X. Explain what it means for Y to be contained in Z. Same question if Y, Z are only locally closed subschemes.

Given a locally closed subscheme $Z \subset X$ of a scheme X, we define the *closure* $\bar{Z}$ of Z to be the smallest closed subscheme of X containing Z; that is, the intersection of all closed subschemes of X containing Z. Equivalently, if Z is a closed subscheme of an open subscheme $U \subset X$, the closure $\bar{Z}$ is the closed subscheme of X defined by the sheaf of ideals consisting of regular functions whose restrictions to U vanish on Z.

I.2.2   The Local Ring at a Point

The Noetherian property is fundamental in the theory of rings, and its extension is equally fundamental in the theory of schemes: we say that a scheme X is *Noetherian* if it admits a finite cover by open affine subschemes, each the spectrum of a Noetherian ring. As usual, one can check that this is independent of the cover chosen.

There is a good notion of the germ of a scheme X at a point $x \in X$ which is the intersection, in a natural sense, of all the open subschemes containing the point. This is embodied in the *local ring* of X at x, defined earlier as

$$\mathscr{O}_{X,x} := \varinjlim_{x \in U} \mathscr{O}_X(U).$$

The maximal ideal $\mathfrak{m}_{X,x}$ of this local ring is the set of all sections that vanish at x. The local ring is a simple object: to compute it (and to show in particular that it is a local ring, with the given maximal ideal), we may begin by replacing X by an affine open neighborhood of x, thus assuming that $X = \operatorname{Spec} R$ and $x = [\mathfrak{p}]$. We may next restrict the open subsets U in the direct limit to the distinguished open sets $\operatorname{Spec} R_f$ such that $f(x) \neq 0$ — that is, $f \notin \mathfrak{p}$. Thus

$$\mathscr{O}_{X,x} := \varinjlim_{f \notin \mathfrak{p}} R_f = R_{\mathfrak{p}}$$

and

$$\mathfrak{m}_{X,x} := \varinjlim_{f \notin \mathfrak{p}} \mathfrak{p} R_f = \mathfrak{p} R_{\mathfrak{p}},$$

the localization of R at $\mathfrak{p}$. We can think of the germ of X at x as being $\operatorname{Spec} \mathscr{O}_{X,x}$; we will study some schemes of this type in the next chapter.

This notion of the local ring of a scheme at a point is crucial to the whole theory of schemes. We give a few illustrations, showing how to define various geometric notions in terms of the local ring. Let X be a scheme.

(1) The *dimension of X at a point* $x \in X$, written $\dim(X, x)$, is the (Krull) dimension of the local ring $\mathscr{O}_{X,x}$ — that is, the supremum of lengths of chains of prime ideals in $\mathscr{O}_{X,x}$. (The length of a chain is the number of strict inclusions.) The *dimension* of X, or $\dim X$, itself is the supremum of these local dimensions.

Exercise I-36. The underlying space of a zero-dimensional Noetherian scheme is finite.

(2) The *Zariski cotangent space* to X at x is $\mathfrak{m}_{X,x}/\mathfrak{m}_{X,x}^2$, regarded as a vector space over the residue field $\kappa(x) = \mathscr{O}_{X,x}/\mathfrak{m}_{X,x}$. The dual of this vector space is called the *Zariski tangent space* at x.

To understand this definition, consider first a complex algebraic variety X that is nonsingular. In this setting the notion of the tangent space to X at a point p is unambiguous: it may be taken as the vector space of derivations from the ring of germs of analytic functions at the point into $\mathbb{C}$. If $\mathfrak{m}_{X,p}$ is the ideal of regular functions vanishing at p, then such a derivation induces a $\mathbb{C}$-linear map $\mathfrak{m}_{X,p}/\mathfrak{m}_{X,p}^2 \to \mathbb{C}$, and the tangent space may be identified in this way with $\operatorname{Hom}_{\mathbb{C}}(\mathfrak{m}_{X,p}/\mathfrak{m}_{X,p}^2, \mathbb{C}) = (\mathfrak{m}_{X,p}/\mathfrak{m}_{X,p}^2)^*$. See Eisenbud [1995, Ch. 16]. It was Zariski's insight that this latter vector space is the correct analogue of the tangent space for any point, smooth or singular, on any variety; Grothendieck subsequently carried the idea over to the context of schemes, as in the definition given above. We shall return to this construction, from a new point of view, in Chapter VI.

Exercise I-37. If K is a field, the Zariski tangent space to the scheme $\operatorname{Spec} K[x_1, \ldots, x_n]$ at $[(x_1, \ldots, x_n)]$ is n-dimensional.

(3) X is said to be *nonsingular* (or *regular*) at $x \in X$ if the Zariski tangent space to X at x has dimension equal to $\dim(X, x)$; else the dimension of the Zariski tangent space must be larger, and we say that X is *singular* at x. Thus in the case of primary interest, when X is Noetherian, X is nonsingular at x if and only if the local ring $\mathcal{O}_{X,x}$ is a regular local ring. This fundamental notion represents, historically, one of the important steps toward the algebraization of geometry. It was taken by Zariski in his classic paper [1947] (remarkably, this was some years after Krull had introduced the notion of a regular local ring to generalize the properties of polynomial rings, one of the rare cases in which the algebraists beat the geometers to a fundamental geometric notion).

Exercise I-38. A zero-dimensional Noetherian scheme is nonsingular if and only if it is the union of reduced points.

I.2.3 Morphisms

We will next define morphisms of schemes. In the classical theory a regular map of affine varieties gives rise, by composition, to a map of coordinate rings going in the opposite direction. This correspondence makes the two kinds of objects — regular maps of affine varieties and algebra homomorphisms of their coordinate algebras — equivalent. The definition given below generalizes this: we will see that maps between affine schemes are simply given by maps of the corresponding rings (in the opposite direction).

Given the simple description of morphisms of affine schemes in terms of maps of rings, it is tempting just to define a morphism of schemes to be something that is "locally a morphism of affine schemes." One can make sense of this, and it gives the correct answer, but it leads to awkward problems of checking that the definition is independent of the choice of an affine cover. For this reason, we give a definition below that works without the choice of an affine cover. Although it may at first appear complicated, it is quite convenient in practice. It also has the advantage of working uniformly for all "local ringed spaces" — structures defined by a topological space with a sheaf of rings whose stalks are local rings.

To understand the motivation behind this definition, consider once more the case of differentiable manifolds. A continuous map $\psi : M \to N$ between differentiable manifolds is differentiable if and only if, for every differentiable function f on an open subset $U \subset N$, the pullback $\psi^{\#} f := f \circ \psi$ is a differentiable function on $\psi^{-1} U \subset M$. We can express this readily enough in the language of sheaves. Any continuous map $\psi : M \to N$ induces a map of sheaves on N

$$\psi^{\#} : \mathscr{C}(N) \longrightarrow \psi_* \mathscr{C}(M)$$

sending a continuous function $f \in \mathscr{C}(N)(U)$ on an open subset $U \subset N$ to the pullback $f \circ \psi \in \mathscr{C}(M)(\psi^{-1}U) = (\psi_*\mathscr{C}(M))(U)$. In these terms, a differentiable map $\psi : M \to N$ may be defined as a continuous map $\psi : M \to N$ such that the induced map $\psi^\#$ carries the subsheaf $\mathscr{C}^\infty(N) \subset \mathscr{C}(N)$ into the subsheaf $\psi_*\mathscr{C}^\infty(M) \subset \psi_*\mathscr{C}(M)$. That is, we require that there be a commutative diagram

$$
\begin{array}{ccc}
\mathscr{C}(N) & \xrightarrow{\ \psi^\#\ } & \psi_*\mathscr{C}(M) \\
\uparrow & & \uparrow \\
\mathscr{C}^\infty & \xrightarrow{\ \psi^\#\ } & \psi_*\mathscr{C}^\infty(M)
\end{array}
$$

We'd like to adapt this idea to the case of schemes. The difference is that the structure sheaf $\mathscr{O}_X$ of a scheme X is *not* a subsheaf of a predefined sheaf of functions on X. Thus, in order to give a map of schemes, we have to specify *both* a continuous map $\psi^\# : X \to Y$ on underlying topological spaces *and* a pullback map

$$
\psi^\# : \mathscr{O}_X \to \psi_*\mathscr{O}_Y .
$$

Of course, some compatibility conditions have to be satisfied by $\psi^\#$ and ψ. The problem in specifying them is that a section of the structure sheaf $\mathscr{O}_Y$ does not take values in a fixed field but in a field $\kappa(q)$ that varies with the point $q \in Y$; in particular, it doesn't make sense to require that the value of $f \in \mathscr{O}_Y(U)$ at $q \in U \subset Y$ agree with the value of $\psi^\# f \in \psi_*\mathscr{O}_X(U) = \mathscr{O}_X(\psi^{-1}U)$ at a point $p \in \psi^{-1}U \subset X$ mapping to q (which is in effect how $\psi^\#$ was defined in the case of differentiable functions), since these "values" lie in different fields. About all that does make sense is to require that f vanish at q if and only if $\psi^\# f$ vanishes at p — and this is exactly what we do require. We thus make the following definition.

Definition I-39. A *morphism*, or *map*, between schemes X and Y is a pair $(\psi, \psi^\#)$, where $\psi : X \to Y$ is a continuous map on the underlying topological spaces and

$$
\psi^\# : \mathscr{O}_Y \to \psi_*\mathscr{O}_X
$$

is a map of sheaves on Y satisfying the condition that for any point $p \in X$ and any neighborhood U of $q = \psi(p)$ in Y a section $f \in \mathscr{O}_Y(U)$ vanishes at q if and only if the section $\psi^\# f$ of $\psi_*\mathscr{O}_X(U) = \mathscr{O}_X(\psi^{-1}U)$ vanishes at p.

This last condition has a nice reformulation in terms of the local rings $\mathscr{O}_{X,p}$ and $\mathscr{O}_{Y,q}$. Any map of sheaves $\psi^\# : \mathscr{O}_Y \to \psi_*\mathscr{O}_X$ induces on passing to the limit a map

$$
\mathscr{O}_{Y,q} = \varinjlim_{q \in U \subset Y} \mathscr{O}_Y(U) \to \varinjlim_{q \in U \subset Y} \mathscr{O}_X(\psi^{-1}U),
$$

and this last ring naturally maps to the limit

$$
\varinjlim_{p \in V \subset X} \mathscr{O}_X(V)
$$

over all open subsets V containing p, which is $\mathscr{O}_{X,p}$. Thus $\psi^{\#}$ induces a map of the local rings $\mathscr{O}_{Y,q} \to \mathscr{O}_{X,p}$. Saying that a section $f \in \mathscr{O}_Y(U)$ vanishes at q if and only if $\psi^{\#} f \in \psi_* \mathscr{O}_X(U) = \mathscr{O}_X(\psi^{-1}U)$ vanishes at p is saying that this map $\mathscr{O}_{Y,q} \to \mathscr{O}_{X,p}$ sends the maximal ideal $\mathfrak{m}_{Y,q}$ into $\mathfrak{m}_{X,p}$ — in other words, that it is a *local homomorphism* of local rings.

As we mentioned above, a morphism of affine schemes

$$\psi : X = \operatorname{Spec} S \longrightarrow \operatorname{Spec} R = Y$$

is the same as a homomorphism of rings $\varphi : R \to S$. Here is the precise result, along with an important improvement that describes maps from an arbitrary scheme to an affine scheme.

Theorem I-40. *For any scheme X and any ring R, the morphisms*

$$(\psi, \psi^{\#}) : X \longrightarrow \operatorname{Spec} R$$

are in one-to-one correspondence with the homomorphisms of rings

$$\varphi : R \to \mathscr{O}_X(X)$$

by the association

$$\varphi = \psi^{\#}(\operatorname{Spec} R) : R = \mathscr{O}_{\operatorname{Spec} R}(\operatorname{Spec} R) \to \psi_*(\mathscr{O}_X)(\operatorname{Spec} R) = \mathscr{O}_X(X).$$

Proof. We describe the inverse association. Set $Y = \operatorname{Spec} R$, and let $\varphi : R \to \mathscr{O}_X(X)$ be a map of commutative rings. If $p \in |X|$ is a point, the preimage of the maximal ideal under the composite $R \to \mathscr{O}_X(X) \to \mathscr{O}_{X,p}$ is a prime ideal, so that φ induces a map of sets

$$\psi : |X| \to |Y|,$$

which is easily seen to be continuous in the Zariski topology. Next, for each basic open set $U = \operatorname{Spec} R_f \subset Y$, define the map $\psi^{\#} : R_f = \mathscr{O}_Y(U) \to (\psi_* \mathscr{O}_X)(U)$ to be the composite

$$R_f \to \mathscr{O}_X(X)_{\varphi(f)} \to \mathscr{O}_X(\psi^{-1}U)$$

obtained by localizing ψ. By Proposition I-12(ii) this is enough to define a map of sheaves. Localizing further, we see that if $\psi(p) = q$, then $\psi^{\#}$ defines a local map of local rings $R_q \to \mathscr{O}_{X,p}$, and thus $(\psi, \psi^{\#})$ is a morphism of schemes. Clearly, the induced map satisfies

$$\psi^{\#}(Y) = \varphi,$$

so the construction is indeed the inverse of the given one. $\qquad\square$

Of course this result says in particular that all the information in the category of affine schemes is already in the category of commutative rings.

Corollary I-41. *The category of affine schemes is equivalent to the category of commutative rings with identity, with arrows reversed, the so-called opposite category.*

Exercise I-42. (a) Using this, show that there exists one and only one map from any scheme to $\operatorname{Spec}\mathbb{Z}$. In the language of categories, this says that $\operatorname{Spec}\mathbb{Z}$ is the *terminal object* of the category of schemes.

(b) Show that the one-point set is the terminal object of the category of sets.

For example, each point $[\mathfrak{p}]$ of $X = \operatorname{Spec} R$ corresponds to a scheme $\operatorname{Spec}\kappa(\mathfrak{p})$ that has a natural map to X defined by the composite map of rings

$$R \to R_\mathfrak{p} \to R_\mathfrak{p}/\mathfrak{p}_\mathfrak{p} = \kappa(\mathfrak{p})$$

Of course, the inclusion makes $[\mathfrak{p}]$ a closed subscheme if and only if $\mathfrak{p}$ is a maximal ideal of R (in general, $[\mathfrak{p}]$ is an infinite intersection of open subschemes of a closed subscheme).

If $\psi : Y \to X$ is a morphism of affine schemes, $X = \operatorname{Spec} R$ and $Y = \operatorname{Spec} T$, and X' is a closed subscheme of X, defined by an ideal I in R, then we define the preimage (sometimes, for emphasis, the "scheme-theoretic preimage") $\psi^{-1}X'$ of ψ over X' to be the closed subscheme of Y defined by the ideal $\varphi(I)T$ in T. If X' is a closed point p of X, we call $\psi^{-1}p$ the *fiber* over X'. (We will soon see how to define fibers over arbitrary points.) The underlying topological space of the preimage is just the set-theoretic preimage, while the scheme structure of the preimage gives a subtle and useful notion of the "correct multiplicity" with which to count the points in the preimage. The simplest classical example is given later in Exercise II-2; here we give two others.

Exercise I-43. (a) Let $\varphi : X \to Y$ be the map of affine schemes illustrated by

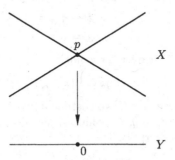

That is, $X = \operatorname{Spec} K[x, u]/(xu)$ is the union of two lines meeting in a point $p = (x, u)$, while $Y = \operatorname{Spec} K[t]$ is a line, and the map is an isomorphism on each of the lines of X; for example, it might be given by the map of rings

$$K[t] \to K[x, u]/(xu),$$
$$t \mapsto x + u.$$

Show that the fiber over the point $q_a = (t-a)$, with $a \neq 0$, is the scheme $\operatorname{Spec}(K \times K)$ consisting of two distinct points, while the fiber over q_0 — that is, the fiber containing the double point p — is isomorphic to $\operatorname{Spec} K[x]/(x^2)$. The fact that the algebra $K[x]/(x^2)$ is two-dimensional (as a vector space over K) reflects the structure of the map locally at p.

(b) Let $\varphi : X \to Y$ be the map of affine schemes illustrated by

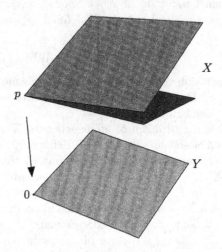

That is, $X = \operatorname{Spec} K[x,y,u,v]/((x,y) \cap (u,v))$ is the union of two planes in four-space meeting in a single point $p = (x,y,u,v)$, while $Y = \operatorname{Spec} K[s,t]$ is a plane, and the map is an isomorphism on each of the planes of X; for example, it might be given by the map of rings

$$K[s,t] \to K[x,y,u,v],$$
$$t \mapsto x + u,$$
$$s \mapsto y + v.$$

Show that the fiber over the point

$$q_{a,b} = (s-a, t-b)$$

is the scheme $\operatorname{Spec}(K \times K)$ consisting of two distinct points if a or $b \neq 0$, while the fiber over $q_{0,0}$ — that is, the fiber containing the "double point" p — is isomorphic to

$$\operatorname{Spec} K[x,y]/(x^2, xy, y^2).$$

The fact that the algebra $K[x,y]/(x^2, xy, y^2)$ is a three-dimensional vector space over K instead of a two-dimensional vector space (as one might expect by analogy with the previous example) reflects a deep fact about the variety X (that it is not "locally Cohen–Macaulay"). This example will be taken up again, from the point of view of flatness, in section II.3.4.

I.2.4   The Gluing Construction

Using the notion of morphism, we can construct more complicated schemes (for example, nonaffine schemes) by identifying simpler schemes along open subsets. This is a basic operation, called the *gluing construction*.

Suppose we are given a collection of schemes $\{X_\alpha\}_I$, and an open set $X_{\alpha\beta}$ in X_α for each $\beta \neq \alpha$ in I. Suppose also that we are given a family of isomorphisms of schemes

$$\psi_{\alpha\beta} : X_{\alpha\beta} \to X_{\beta\alpha} \quad \text{for each } \alpha \neq \beta \text{ in } I,$$

satisfying the conditions $\psi_{\beta\alpha} = \psi_{\alpha\beta}^{-1}$ for all α and β,

$$\psi_{\alpha\beta}(X_{\alpha\beta} \cap X_{\alpha\gamma}) = X_{\beta\alpha} \cap X_{\beta\gamma} \quad \text{for all } \alpha, \beta, \gamma,$$

and the *compatibility condition*

$$\psi_{\beta\gamma} \circ \psi_{\alpha\beta}|_{(X_{\alpha\beta} \cap X_{\alpha\gamma})} = \psi_{\alpha\gamma}|_{(X_{\alpha\beta} \cap X_{\alpha\gamma})}.$$

Under these circumstances we may define a scheme X by gluing the X_α along the $\psi_{\alpha\beta}$ in an obvious way — that is to say, there exists a (unique) scheme X with a covering by open subschemes isomorphic to the X_α such that the identity maps on the intersections $X_\alpha \cap X_\beta \subset X$ correspond to the isomorphisms $\psi_{\alpha\beta}$.

This construction can be used, for example, to define projective schemes out of affine ones. Another use is in the theory of toric varieties; see, for example, Kempf et al. [1973].

In these and indeed in almost all applications, we don't really need to give the maps $\psi_{\alpha\beta}$ explicitly: we are actually given a topological space $|X|$ and a family of open subsets $|X_\alpha|$, each endowed with the structure of an affine scheme — that is, with a structure sheaf $\mathscr{O}_{X_\alpha}$ — in such a way that $\mathscr{O}_{X_\alpha}(X_\alpha \cap X_\beta)$ is *naturally* identified with $\mathscr{O}_{X_\beta}(X_\alpha \cap X_\beta)$. For example, they might both be given as subsets of a fixed set. Under these circumstances it is immediate that the conditions of Corollary I-14 are satisfied, so that there is a uniquely defined sheaf $\mathscr{O}_X$ on X extending all the $\mathscr{O}_{X_\alpha}$. The pair $(|X|, \mathscr{O}_X)$ is then a scheme.

Probably the simplest example of this is the definition of *affine space* $\mathbb{A}_S^n$ over an abitrary scheme S. To begin with, for any affine scheme $X = \operatorname{Spec} R$ we define *affine n-space over X* to be simply $\operatorname{Spec} R[x_1, \ldots, x_n]$; this is denoted by either $\mathbb{A}_X^n$ or $\mathbb{A}_R^n$. (The geometry of affine spaces and their subschemes will be taken up in Chapter II.) Next, we note that any morphism $X \to Y$ of affine schemes induces a natural map $\mathbb{A}_X^n \to \mathbb{A}_Y^n$. As a consequence, we may apply the gluing construction as follows: If S is an arbitrary scheme covered by affine schemes $U_\alpha = \operatorname{Spec} R_\alpha$, we define *affine space $\mathbb{A}_S^n$ over S* to be the union of the affine spaces $\mathbb{A}_{U_\alpha}^n$, with the gluing maps induced by the identity maps on $U_\alpha \cap U_\beta$.

We will see two other ways of defining affine space $\mathbb{A}_S^n$ over an arbitrary base S in Exercises I-47 and I-54 below.

The following exercise illustrates some of the dangers of the gluing construction: we can, by inappropriate (but legal) gluing, create schemes that do not arise in any geometric setting.

Exercise I-44. Put $Y = \operatorname{Spec} K[s]$ and $Z = \operatorname{Spec} K[t]$. Let $U \subset Y$ be the open set Y_s and let $V \subset Z$ be the open set Z_t. Let $\psi : V \to U$ be the isomorphism corresponding to the map

$$\mathscr{O}_Y(U) = K[s, s^{-1}] \to K[t, t^{-1}] = \mathscr{O}_Z(V)$$

sending s to t, and let γ be the map sending s to t^{-1}. Let X_1 be the scheme obtained by gluing together Y and Z along ψ, and let X_2 be the scheme obtained by gluing along γ instead.

Show that X_1 is not isomorphic to X_2. In fact, X_2 is the scheme corresponding to the projective line $\mathbb{P}^1_K$ (which we will describe in the next section), while X_1 is the affine space with a doubled origin:

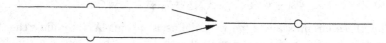

In Chapter III we will introduce a condition, called *separatedness*, that will preclude schemes such as this X_1.

Projective Space. An important example of a scheme constructed by gluing is *projective n-space over a ring R*, denoted $\mathbb{P}^n_R$. It is made by gluing $n + 1$ copies of *affine space*

$$\mathbb{A}^n_R = \operatorname{Spec} R[x_1, \ldots, x_n]$$

over R. An extensive treatment of projective schemes will begin in Chapter III. Here we will use the idea only as an illustration of gluing.

The construction is exactly parallel to the classical construction of projective space as a variety over a field. Although not logically necessary, it is convenient to work as follows. Start with the polynomial ring in $n + 1$ variables $R[X_0, \ldots, X_n]$ and form the localization

$$A := R[X_0, X_0^{-1}, \ldots, X_n, X_n^{-1}].$$

Recall that the ring A has a natural *grading*, that is, a direct-sum decomposition (as an abelian group) into subgroups $A^{(n)}$, for $n \in \mathbb{Z}$, such that

$$A^{(n)} A^{(m)} \subset A^{(m+n)};$$

here $A^{(n)}$ is spanned by monomial rational fractions of degree n. In particular, the degree 0 part $A^{(0)}$ is a subring of A. Now take the rings of our defining affine covering to be R-subalgebras of $A^{(0)}$, the i-th subring being the subalgebra A_i consisting of all polynomials $P/X_i^{\deg(P)}$, where P is a homogeneous element of $R[x_0, \ldots, x_n]$. Clearly, A_i is generated over R by the n algebraically independent elements

$$X_0/X_i, \ldots, \widehat{X_i/X_i}, \ldots, X_n/X_i,$$

where the hat denotes as usual an element omitted from the list. A_i is thus isomorphic to the polynomial ring in n variables over R. Further, for $i \neq j$ we have

$$A_i[(X_j/X_i)^{-1}] = A_j[(X_i/X_j)^{-1}]$$

as subsets of A; both may be described as the subalgebra of all degree 0 elements having denominator of the form $X_i^a X_j^b$. If we use the identity maps as gluing maps, the compatibility conditions are obvious.

If $X = \operatorname{Spec} R$ is an affine scheme, we will often write $\mathbb{P}_X^n$ instead of $\mathbb{P}_R^n$, and refer to the space as *projective space over* X. Any morphism $X \to Y$ of affine schemes induces a natural map $\mathbb{P}_X^n \to \mathbb{P}_Y^n$. As a consequence, we may apply the gluing construction again to define *projective space* $\mathbb{P}_S^n$ *over an arbitrary scheme* S as well. This is straightforward: if S is covered by affine schemes $U_\alpha = \operatorname{Spec} R_\alpha$, we define projective space $\mathbb{P}_S^n$ to be the union of the projective spaces $\mathbb{P}_{U_\alpha}^n$, with the gluing maps induced by the identity maps on $U_\alpha \cap U_\beta$.

I.3 Relative Schemes

I.3.1 Fibered Products

There is an extremely important generalization of the idea of preimage of a set under a function in the notion of the fibered product of schemes. To prepare for the definition, we first recall the situation in the category of sets.

The fibered product of two sets X and Y over a third set S — that is, of a diagram of maps of sets

$$
\begin{array}{ccc}
 & & X \\
 & & \downarrow \varphi \\
Y & \xrightarrow{\ \psi\ } & S
\end{array}
$$

is by definition the set

$$X \times_S Y = \{(x,y) \in X \times Y : \varphi x = \psi y\}.$$

The fibered product is sometimes called the *pullback* of X (or of $X \to S$) to Y. This construction generalizes several more elementary ones in a very useful way:

If S is a point, it gives the usual direct product.

If X, Y are both subsets of S and φ, ψ are the inclusions, it gives the intersection.

If $Y \subset S$ and ψ is the inclusion, it gives the preimage of Y in X.

If $X = Y$, it gives the set on which the maps φ, ψ are equal, the *equalizer* of the maps.

Exercise I-45. Check these assertions!

Note that $X \times_S Y$ comes with natural projection maps to X and Y making the diagram

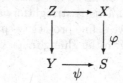

commute. Indeed, the set $X \times_S Y$ may be defined by the following universal property: among all sets Z with given maps to X and Y making the diagram

$$
\begin{array}{ccc}
Z & \longrightarrow & X \\
\downarrow & & \downarrow \varphi \\
Y & \xrightarrow{\psi} & S
\end{array}
$$

commute, $X \times_S Y$ with its projection maps is the unique "most efficient" choice in the sense that, given the diagram with Z above, there is a unique map $Z \to X \times_S Y$ making the diagram

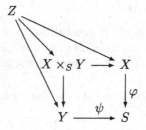

commute.

In the category of schemes we simply *define* the fibered product to be a scheme with this universal property — the universal property guarantees in particular that such a thing, with its projections to X and Y, will be unique. We can then define products, intersections, preimages, and equalizers in terms of the fibered product! However, this begs the question of whether any such object as the fibered product exists in the category of schemes. It does, and we will now describe the construction.

First, we treat the affine case. Recall that the category of affine schemes is opposite to the category of commutative rings, by Corollary I-41. Therefore, if we have schemes

$$X = \operatorname{Spec} A, \qquad Y = \operatorname{Spec} B, \qquad S = \operatorname{Spec} R,$$

where X and Y map to S (so that A and B are R-algebras), we must define the fibered product $X \times_S Y$ to be

$$X \times_S Y = \operatorname{Spec}(A \otimes_R B).$$

This is because the natural diagram

$$
\begin{array}{ccc}
A \otimes_R B & \longleftarrow & A \\
\uparrow & & \uparrow \varphi \\
B & \longleftarrow & R
\end{array}
$$

has, in an obvious sense, the opposite universal property to the one desired for the fibered product. In fancy language, the tensor product is a *fibered coproduct*, or *fibered sum*, in the category of commutative rings.

To check that this definition is reasonable, one may note that in the situation where Y is a closed subscheme of S defined by an ideal I, so that $B = R/I$, we have $A \otimes_R B = A/IA$. Thus $X \times_S Y = \operatorname{Spec} A/IA$ is the same as the preimage of Y in X, as previously defined.

Exercise I-46. A few simple special cases are a great help when computing fibered products. Prove the following facts directly from the universal property of the tensor product of algebras:

(a) For any R-algebra S we have $R \otimes_R S = S$.

(b) If S, T are R-algebras and $I \subset S$ is an ideal, then

$$(S/I) \otimes_R T = (S \otimes_R T)/(I \otimes 1)(S \otimes_R T).$$

(c) If $x_1, \ldots, x_n, y_1, \ldots, y_m$ are indeterminates then

$$R[x_1, \ldots, x_n] \otimes_R R[y_1, \ldots, y_m] = R[x_1, \ldots, x_n, y_1, \ldots, y_m].$$

Use these principles to solve the remainder of this exercise.

(d) Let m, n be integers. Compute the fibered product

$$\operatorname{Spec} \mathbb{Z}/(m) \times_{\operatorname{Spec} \mathbb{Z}} \operatorname{Spec} \mathbb{Z}/(n).$$

(e) Compute the fibered product $\operatorname{Spec} \mathbb{C} \times_{\operatorname{Spec} \mathbb{R}} \operatorname{Spec} \mathbb{C}$.

(f) Show that for any polynomial rings $R[x]$ and $R[y]$ over a ring R, we have

$$\operatorname{Spec} R[x] \times_{\operatorname{Spec} R} \operatorname{Spec} R[y] = \operatorname{Spec} R[x, y].$$

Note that in example (d) the underlying set of the fibered product is the fibered product of the underlying sets, but this is not true in (e) and (f).

(g) Consider the ring homomorphisms

$$R[x] \to R; \quad x \mapsto 0$$

and

$$R[x] \to R[y]; \quad x \mapsto y^2.$$

Show that with respect to these maps we have

$$\operatorname{Spec} R[y] \times_{\operatorname{Spec} R[x]} \operatorname{Spec} R = \operatorname{Spec} R[y]/(y^2).$$

In the general case, we cover S by affine schemes $\operatorname{Spec} R_\rho$, and cover their preimages in X and Y by affine schemes $\operatorname{Spec} A_{\rho\alpha}$ and $\operatorname{Spec} B_{\rho\beta}$, respectively, so that in a suitable sense the diagram

$$
\begin{array}{ccc}
 & & X \\
 & & \downarrow \varphi \\
Y & \xrightarrow{\ \psi\ } & S
\end{array}
$$

is covered by diagrams of the form

$$
\begin{array}{ccc}
 & & \operatorname{Spec} A_{\rho\alpha} \\
 & & \downarrow \varphi_{\rho\alpha} \\
\operatorname{Spec} B_{\rho\beta} & \xrightarrow{\ \psi_{\rho\beta}\ } & \operatorname{Spec} R_\rho
\end{array}
$$

Of course, we already know that the fiber product of this last diagram is $\operatorname{Spec}(A_{\rho\alpha} \otimes_{R_\rho} B_{\rho\beta})$. Using the idea of gluing explained at the end of the preceding section, it is easy but tedious to check that these schemes agree on overlaps and patch together to form the scheme $X \times_S Y$ as required; we omit the computation. A different approach will be sketched in section VI.2.1.

One immediate use of the notion of product is an alternative description of affine space $\mathbb{A}_S^n$ over a scheme S:

Exercise I-47. Let S be any scheme. Let $\mathbb{A}_{\mathbb{Z}}^n = \operatorname{Spec} \mathbb{Z}[x_1, \ldots, x_n]$ be affine space over $\operatorname{Spec} \mathbb{Z}$, as defined above (this scheme will be discussed in detail in the next chapter). Show that affine space $\mathbb{A}_S^n$ over S may be described as a product: $\mathbb{A}_S^n = \mathbb{A}_{\mathbb{Z}}^n \times_{\operatorname{Spec} \mathbb{Z}} S$.

We can also use the fibered product to define the *fiber* of a morphism $\psi : Y \to X$ over an arbitrary point of arbitrary schemes: if p is a point of X corresponding to a prime ideal $\mathfrak{p}$ of R, then the fiber of ψ over p is the fibered product of Y and the one-point scheme $\operatorname{Spec} \kappa(p)$. In the case where X and Y are affine — say, $Y = \operatorname{Spec} T$ and $X = \operatorname{Spec} R$ — we get

$$
\psi^{-1}(p) = \operatorname{Spec} \kappa(p) \times_X Y = \operatorname{Spec}(R_{\mathfrak{p}}/\mathfrak{p}_{\mathfrak{p}} \otimes_R T) = \operatorname{Spec}(R_{\mathfrak{p}}/\mathfrak{p}_{\mathfrak{p}} \otimes_R T/\mathfrak{p}T)
$$

as a point set; this is the set of primes of T whose preimages in R are equal to $\mathfrak{p}$. More generally, we define the *preimage*, or *inverse image* of a closed subscheme X' of X under ψ to be the fibered product $X' \times_X Y$.

Just as in the affine case treated above, the preimage $\psi^{-1}X'$ of X' is a closed subscheme of Y. Using the $\mathscr{O}_X$-algebra structure on $\mathscr{O}_Y$, the ideal sheaf of the preimage may be written as $\mathscr{I}_{\psi^{-1}X'} = \mathscr{I}_{X'} \cdot \mathscr{O}_Y$.

Another typical use of the fibered product is in studying the behavior of varieties under extension of a base field (one usually speaks in this context of a "base change" rather than a fibered product). In this setting, of which we will see some examples in the following chapter, the notion is responsible for the great flexibility and convenience of the theory of schemes in handling arithmetic questions.

As in examples (b) and (c) of Exercise I-46, the set of points of the fibered product of schemes $X \times_S Y$ is usually *not* equal to the fibered product (in the category of sets) of the sets of points of X and Y. This is no terrible pathology but simply reflects the fact that the theory of functions $f(x, y)$ of two variables is much richer than the theory of functions of the form $g(x)h(y)$. In any case, the definitions of Chapter VI provide a viewpoint from which this oddity disappears.

I.3.2   The Category of S-Schemes

Just as in the case of sets, we can use the fibered product to define an absolute product by taking S to be a terminal object in the category of schemes — that is, a scheme such that every scheme has a unique map to S. By Exercise I-42 the terminal object in the category of schemes is Spec $\mathbb{Z}$. However, the absolute product has some rather surprising properties. We have already seen in Exercise I-46(d) cases (when m and n are relatively prime) where the product in this sense of nonempty sets may be empty! There are other peculiarities as well: for example, the dimension of an irreducible scheme can be defined as the Krull dimension of the coordinate ring of any of its affine open sets. One might expect the product

$$X \times Y = X \times_{\mathrm{Spec}\,\mathbb{Z}} Y$$

of two schemes to have dimension equal to the sum of the dimensions of X and Y. But in fact we have the result in the next exercise.

Exercise I-48. Show that if $X = \mathrm{Spec}\,\mathbb{Z}[x]$ and $Y = \mathrm{Spec}\,\mathbb{Z}[y]$, then

$$\dim X \times Y = \dim X + \dim Y - \dim \mathrm{Spec}\,\mathbb{Z} = \dim X + \dim Y - 1.$$

This oddity and many like it can be eliminated by a simple but convenient generalization of our definitions: we often wish to work with *schemes X over a given field (or ring) K*, or *K-schemes*. Of course, we will then use only morphisms that respect this structure. Informally, this just means that we consider X together with a K-algebra structure on $\mathcal{O}_X(X)$ and morphisms respecting these structures. In this category, Spec K is the terminal object and the absolute product is the fibered product over Spec K. If K is a field, the product in the category of K-schemes behaves more in accord with elementary geometric intuition. For example:

Exercise I-49. Let K be a field. If X and Y are nonempty K-schemes, then the product $X \times Y = X \times_{\mathrm{Spec}\,K} Y$ in the category of K-schemes is nonempty.

Further, in this case the dimension of Spec K is 0, and one can check that for schemes built up from spectra of finitely generated K-algebras the dimension of products is additive, as it should be.

In order to accommodate families of schemes, we may extend this notion a little further. A K-algebra structure on $\mathcal{O}_X(X)$ is nothing but a homomorphism of rings from K to $\mathcal{O}_X(X)$, and by Theorem I-40 this is exactly the same as a map $X \to \operatorname{Spec} K$. Replacing $\operatorname{Spec} K$ by an arbitrary scheme S, we define a *scheme over* S, or S-*scheme*, to be a scheme X together with a morphism $X \to S$. We may think of a scheme over S informally as a family of schemes "parametrized by points of S" — for each point of S we have the fiber over that point. A *morphism of schemes over* S (or S-*morphism*) is a commutative diagram

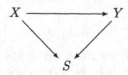

If X and Y are schemes over S, then we write $\operatorname{Mor}_S(X, Y)$ for the set of S-morphisms. Note that the fibered product $X \times_S Y$ of schemes over S is precisely the ordinary direct product in the category of schemes over S.

As usual, if $S = \operatorname{Spec} R$ is affine, we will use the terms "R-scheme" and "the category of R-schemes" interchangeably with "S-scheme" and "the category of S-schemes".

Introducing the category of schemes over S may seem to add a layer of complication, but in reality it more often removes one. For example, if we want to do classical algebraic geometry over the complex numbers in scheme language, it is necessary to work in the category of schemes over $\mathbb{C}$. To see that this is so, note that in any reasonable sense the point $\operatorname{Spec} \mathbb{C}$ should have no nontrivial automorphisms, and the scheme $\operatorname{Spec} \mathbb{C}[x]/(x^2 + 1)$ consisting of a pair of points should have automorphism group $\mathbb{Z}/(2)$. This is in fact the case in the category of schemes over $\mathbb{C}$. In the category of all schemes, however, the automorphism group of the point $\operatorname{Spec} \mathbb{C}$ is huge: it is the Galois group of $\mathbb{C}$ over $\mathbb{Q}$, and the automorphism group of $\operatorname{Spec} \mathbb{C}[x]/(x^2 + 1)$ is worse. Thus, working in the category of schemes over $\mathbb{C}$ removes the (presumably unwanted) extra structure of the Galois group $\operatorname{Gal}(\mathbb{C}/\mathbb{Q})$.

Exercise I-50. Find the automorphism groups of the schemes X_1 and X_3 of Exercise I-20 in the category of schemes over $\mathbb{C}$.

I.3.3   Global Spec

If $S = \operatorname{Spec} R$ is an affine scheme, an affine S-scheme is simply the spectrum of an R-algebra. We will now extend this construction to describe analogous objects in the category of S-schemes for arbitrary S.

To begin with, for any scheme S we define a *quasicoherent sheaf of $\mathcal{O}_S$-algebras*. This is, as you might expect, a sheaf $\mathscr{F}$ of $\mathcal{O}_S$-algebras, such that for any affine open $U = \operatorname{Spec} R \subset S$ and distinguished open subset

$U' = \operatorname{Spec} R_f \subset U$, we have

$$\mathscr{F}(U') = \mathscr{F}(U) \otimes_R \mathscr{O}_S(U') = \mathscr{F}(U) \otimes_R R_f$$

as $R = \mathscr{O}_S(U)$-algebras. We then associate to any quasicoherent sheaf $\mathscr{F}$ of $\mathscr{O}_S$-algebras on a scheme S a scheme $X = \operatorname{Spec}\mathscr{F}$, together with a structure morphism $X \to S$, such that in case $S = \operatorname{Spec} R$ is affine we get simply $X = \operatorname{Spec}\mathscr{F}(S)$ together with the structure morphism $X \to S$ induced by the $R = \mathscr{O}_S(S)$-algebra structure on $\mathscr{F}(S)$.

There are a couple ways to do this. One is simply to use the gluing construction again: we cover S by affine open subsets $U_\alpha = \operatorname{Spec} R_\alpha$, and define X to be the union of the schemes $\operatorname{Spec}\mathscr{F}(U_\alpha)$, with attaching maps induced by the restrictions maps $\mathscr{F}(U_\alpha) \to \mathscr{F}(U_\alpha \cap U_\beta)$. This works, but it's a mess to verify that the resulting space $\operatorname{Spec}\mathscr{F}$ is independent of the choice of cover, and has the further drawback that it can be awkward to describe the set of points of $\operatorname{Spec}\mathscr{F}$. We will give here instead an alternative construction.

We start with a definition: given a quasicoherent sheaf $\mathscr{F}$ of $\mathscr{O}_S$-algebras, we define a *prime ideal sheaf* in $\mathscr{F}$ to be a quasicoherent sheaf of ideals $\mathscr{I} \subsetneq \mathscr{F}$, such that for each affine open subset $U \subset S$, the ideal $\mathscr{I}(U) \subset \mathscr{F}(U)$ is either prime or the unit ideal. (Observe that for any affine scheme X, the points of X are simply the prime ideal sheaves of $\mathscr{O}_X$.) Now, we will define $X = \operatorname{Spec}\mathscr{F}$ in three stages, as we did the spectrum of a ring. First, as a set, X is the set of prime ideal sheaves in $\mathscr{F}$. Second, as a topological space: for every open $U \subset S$ (not necessarily affine) and section $\sigma \in \mathscr{F}(U)$, let $V_{U,\sigma} \subset X$ be the set of prime ideal sheaves $\mathscr{P} \subset \mathscr{F}$ such that $\sigma \notin \mathscr{P}(U)$; take these as a basis for the topology. Finally, we define the structure sheaf $\mathscr{O}_X$ on basis open sets by setting

$$\mathscr{O}_X(V_{U,\sigma}) = \mathscr{F}(U)[\sigma^{-1}].$$

As for the morphism $f : X \to S$: as a set, we associate to a prime ideal sheaf $\mathscr{P} \subset \mathscr{F}$ its inverse image in $\mathscr{O}_S \to \mathscr{F}$; and the pullback map on functions

$$f^\# : \mathscr{O}_S(U) \to \mathscr{O}_X(f^{-1}(U)) = \mathscr{F}(U)$$

is just the structure map $\mathscr{O}_S \to \mathscr{F}$ on U.

Exercise I-51. Show that the points of an affine scheme X are in one-to-one correspondence with the set of prime ideal sheaves in $\mathscr{O}_X$.

Exercise I-52. Show that if $f : Y \to X$ is a morphism and $\mathscr{P}$ is a prime ideal sheaf of $\mathscr{O}_Y$, then $f_*(\mathscr{P})$ is a prime ideal sheaf in $f_*\mathscr{O}_Y$.

Exercise I-53. Show that if $f : Y \to X$ is a morphism, the map on sets corresponding to f sends $\mathscr{P} \subset \mathscr{O}_Y$ to $(f^\#)^{-1}(f_*(\mathscr{P})) \subset \mathscr{O}_X$.

The simplest example of global Spec gives us yet another construction of affine space over an arbitrary scheme S:

Exercise I-54. Let S be any scheme. Show that affine space $\mathbb{A}^n_S$ over S may be constructed as a global Spec:

$$\mathbb{A}^n_S = \operatorname{Spec}\left(\operatorname{Sym}(\mathscr{O}^{\oplus n}_S)\right).$$

I.4   The Functor of Points

One of the intriguing things about schemes is precisely that they have so much structure that is not conveyed by their underlying sets, so that the familiar operations on sets such as taking direct products require vigilant scrutiny lest they turn out not to make sense. It is therefore remarkable that many of the set-theoretic ideas can be restored through a simple device, the functor of points. This point of view, while initially adding a layer of complication to the subject, is often extremely illuminating; as a result it and its attendant terminology have become pervasive. We will give a brief introduction to the necessary definitions here and use them occasionally in the following chapters before returning to them in detail in Chapter VI.

We start with the observation that the points of a scheme do not in general look anything like one another: we have nonclosed points as well as closed ones; and if we are working over a non-algebraically closed field, then even closed points may be distinguished by having different residue fields. Similarly, if we are working over $\mathbb{Z}$, different points may have residue fields of different characteristic; and if we extend the notion of point to "closed subscheme whose underlying topological space is a point," we have an even greater variety. And, of course, a morphism between schemes will not at all be determined by the associated map on underlying point sets.

There is, however, a way of looking at a scheme — via its *functor of points* — that reduces it in effect to a set. More precisely, we may think of a scheme as an organized collection of sets, a functor on the category of schemes, on which the familiar operations on sets behave as usual. In this section we will examine this functorial description. A big payoff is that we will see the category of schemes embedded in a larger category of functors, in which many constructions are much easier. The advantage of this is something like the advantage in analysis of working with distributions, not just ordinary functions; it shifts the problem of making constructions in the category of schemes to the problem of understanding which functors come from schemes. Further, many geometric constructions that arise in the category of schemes can be extended to larger categories of functors in a useful way.

To introduce the notion of the functor of points, we start out in a general categorical setting. To begin with, in many categories whose objects are sets with additional structure, the underlying set $|X|$ of an object X may be described as the set of morphisms from a universal object to X; for example:

(a) In the category of differentiable manifolds, if Z is the manifold consisting of one point, then for any manifold X we have $|X| = \operatorname{Hom}(Z, X)$.

(b) In the category of groups, for any group X we have $|X| = \operatorname{Hom}(\mathbb{Z}, X)$.

(c) In the category of rings with unit and unit-preserving homomorphisms, if we set $Z = \mathbb{Z}[x]$, then for any ring X with unit we have $|X| = \operatorname{Hom}(Z, X)$.

In general, for any object Z of a category $\mathscr{X}$ the association

$$X \mapsto \operatorname{Hom}_{\mathscr{X}}(Z, X)$$

defines a functor φ from the category $\mathscr{X}$ to the category of sets. As indicated in the first paragraph above, however, it is not really satisfactory to call the set $\varepsilon(X) = \operatorname{Hom}_{\mathscr{X}}(Z, X)$ the set of points of the object X unless this functor is *faithful* — that is, unless for any pair of objects X_1 and X_2 of $\mathscr{X}$ a morphism

$$f : X_1 \to X_2$$

is determined by the map of sets

$$f' : \operatorname{Hom}_{\mathscr{X}}(Z, X_1) \to \operatorname{Hom}_{\mathscr{X}}(Z, X_2).$$

It may not always be possible to satisfy this condition. For example, let (Hot) be the category of CW-complexes, where $\operatorname{Hom}_{(\mathrm{Hot})}(X, Z)$ is the set of homotopy classes of continuous maps from X to Z. If Z is the one-point complex, then

$$\operatorname{Hom}_{(\mathrm{Hot})}(Z, X) = \pi_0(X)$$

the set of connected components of X, and this does not give a faithful functor. Nor is it possible to chose a better object Z. Likewise, in the category of schemes, there is no one object Z that will serve in this capacity.

Grothendieck's ingenious idea was to remedy this situation by considering not just one set $\operatorname{Mor}(Z, X)$ but all at once! That is, we associate to each scheme X the "structured set" consisting of all the sets $\operatorname{Mor}(Z, X)$, together with, for each morphism $f : Z \to Z'$, the mapping from $\operatorname{Mor}(Z', X)$ obtained by composing with f.

To put this more formally, the *functor of points* of a scheme X is the "representable" functor determined by X; that is, the functor

$$h_X : (\text{schemes})^\circ \to (\text{sets}),$$

where $(\text{schemes})^\circ$ and (sets) represent the category of schemes with the arrows reversed and the category of sets, respectively; h_X takes each scheme Y to the set

$$h_X(Y) = \operatorname{Mor}(Y, X)$$

and each morphism $f : Y \to Z$ to the map of sets

$$h_X(Z) \to h_X(Y)$$

defined by sending an element $g \in h_X(Z) = \mathrm{Mor}(Z, X)$ to the composition $g \circ f \in \mathrm{Mor}(Y, X)$. The reason for the name "representable functor" is that we say this functor is *represented* by the scheme X. The set $h_X(Y)$ is called the set of Y-*valued points of* X (if $Y = \mathrm{Spec}\, T$ is affine, we will often write $h_X(T)$ instead of $h_X(\mathrm{Spec}\, T)$ and call it the set of T-valued points of X).

To introduce one more layer of abstraction, note that this construction defines a functor

$$h : (\text{schemes}) \to \mathrm{Fun}((\text{schemes})^{\circ}, (\text{sets}))$$

(where morphisms in the category of functors are natural transformations), sending

$$X \mapsto h_X$$

and associating to a morphism $f : X \to X'$ the natural transformation $h_X \to h_{X'}$ that for any scheme Y sends $g \in h_X(Y) = \mathrm{Mor}(Y, X)$ to the composition $f \circ g \in h_{X'}(Y) = \mathrm{Mor}(Y, X')$.

Of course, when we want to work with schemes over a given base S, we should take morphisms over S as well. The situation is completely analogous to that above: we describe in this way a functor

$$X \mapsto h_X$$

from the category of S-schemes to the category

$$\mathrm{Fun}((S\text{-schemes})^{\circ}, (\text{sets})).$$

The apparently abstract idea of the functor of points has its root in the study of solutions of equations. Let $X = \mathrm{Spec}\, R$ be an affine scheme, where $R = \mathbb{Z}[x_1, x_2, \ldots]/(f_1, f_2, \ldots)$. If T is any other ring (one should think of $T = \mathbb{Z}$, $\mathbb{Z}/(p)$, $\mathbb{Z}_{(p)}$, $\hat{\mathbb{Z}}_{(p)}$, $\mathbb{Q}_p$, $\mathbb{R}$, $\mathbb{C}$, and so on), then a morphism from $\mathrm{Spec}\, T$ to $\mathrm{Spec}\, R$ is the same as a ring homomorphism from R to T, and this is determined by the images a_i of the x_i. Of course, a set of elements $a_i \in T$ determines a morphism in this way if and only if they are solutions to the equations $f_i = 0$. We have shown that

$$h_X(T) = \left\{ \begin{array}{c} \text{sequences of elements } a_1, \ldots \in T \text{ that} \\ \text{are solutions of the equations } f_i = 0 \end{array} \right\}.$$

Similarly, if X is an arbitrary scheme, so that X is the union of affine schemes X_a meeting along open subsets, then a map from an affine scheme Y to X may be described by giving a covering of Y by distinguished affine open subsets Y_{f_a} and maps from Y_{f_a} to X_a for each a, agreeing on open sets (some of the Y_{f_a} may, of course, be empty). Thus an element of $h_X(Y)$ may be described even in this general context as a set of solutions to systems of equations, corresponding to some of the X_a, with compatibility conditions satisfied by the solutions on the sets where certain polynomials are non-zero.

Even with this interpretation, the notion of the functor of points may seem an arid one: while we can phrase problems in this new language, it's far from clear that we can solve them in it. The key to being able to work in this setting is the fact that many apparently geometric notions have natural extensions from the category of schemes to larger categories of functors. Thus, for example, we can talk about an open subfunctor of a functor, a closed subfunctor, a smooth functor, the tangent space to a functor, and so on. These notions will be developed in Chapter VI, where we will also give a better idea of how they are used.

In this chapter we have used the word "point" in two different ways: we have both the points of a scheme X, and, for any scheme Y, the set of Y-valued points of X. It is important not to let this double usage cause confusion. The two notions are of course very different: for example, if $Y = \operatorname{Spec} L$ for some finite extension L of $\mathbb{Q}$, then we have a map

$$\{Y\text{-valued points of } X\} \longrightarrow |X|$$

but this map is in general neither injective or surjective: the image will be the subset of points $p \in X$ whose residue field $\kappa(p)$ is a subfield of L, and the fiber of the map over such a point p will be the set of ring homomorphisms from $\kappa(p)$ to L. Another distinction is that while the set $|X|$ of points of X is absolute, the set of Y-valued points is relative in the sense that it may depend on the specification of a base scheme S and the structure morphism $X \to S$. Finally, in case $S = \operatorname{Spec} K$, the set of K-valued points of X— that is, the subset of points $p \in X$ such that $\kappa(p) = K$—is often called the set of K-*rational points* of X.

Each of the two notions of "point" has some (but not all) of the properties we might expect from the behavior of points in the category of sets. For example, the set of Y-valued points of a product $X_1 \times X_2$ is the product of the sets of Y-valued points of X_1 and X_2. However, it is not the case that the set of Y-valued points of a union $X = U \cup V$ is the union of the sets of Y-valued points of U and V (for example, the identity map $X \to X$ is an X-valued point of X not in general contained in U or V). By contrast, exactly the opposite situation holds for the set $|X|$ of points of a scheme X in the ordinary sense.

We have now outlined the basic definitions in the theory of schemes. In the next chapter we will give many examples, from which the reader may form some idea of the "look and feel" of schemes.

II
Examples

II.1 Reduced Schemes over Algebraically Closed Fields

We will start our series of examples with the one that the concept of scheme is intended to generalize: the classical notion of an affine variety over an algebraically closed field K. In our present context, this means considering schemes of the form $\operatorname{Spec} R$, where R is the coordinate ring of a variety X — that is, a finitely generated, reduced algebra over K. (Recall that "reduced" means nilpotent-free.) $\operatorname{Spec} R$ is sometimes called the scheme *associated to the variety* X: such schemes are sometimes referred to just as varieties. In later sections we will consider the ways in which schemes may differ from this basic model.

The K-scheme associated to an affine variety over an algebraically closed field K is an equivalent object to the variety; either one determines and is determined by its coordinate algebra, which is the same for both. But already in this case, classical notions such as the intersection of varieties and the fibers of maps are given a more precise meaning in the theory of schemes. We will see examples of this phenomenon in this and succeeding sections.

II.1.1 Affine Spaces

We start with the scheme $\mathbb{A}_K^n := \operatorname{Spec} K[x_1, \ldots, x_n]$, with K an algebraically closed field. This scheme is called *affine n-space* over K.

We will make use of a standard but nontrivial result from algebra, a form of Hilbert's Nullstellensatz; see Eisenbud [1995], for example.

Theorem II-1 (Nullstellensatz). *Let K be any field. If $\mathfrak{m}$ is a maximal ideal of a polynomial ring $K[x_1, \ldots, x_n]$ (or, geometrically, p is a closed point of any subvariety of an affine space over a field K), then*

$$K[x_1, \ldots, x_n]/\mathfrak{m} = \kappa(p)$$

is a finite-dimensional vector space over K.

In our case, with K algebraically closed, this implies that $\kappa(p) = K$. Thus, writing λ_i for the image of x_i in $\kappa(p)$, we see that

$$\mathfrak{m} = (x_1 - \lambda_1, \ldots, x_n - \lambda_n).$$

In this way the closed points of $\mathbb{A}_K^n$ correspond to n-tuples of elements of K, as one should expect. We will sometimes refer to "the point $(\lambda_1, \ldots, \lambda_n)$" instead of "the point $[(x_1 - \lambda_1, \ldots, x_n - \lambda_n)]$."

To begin with dimension 1, the affine line

$$\mathbb{A}_K^1 = \operatorname{Spec} K[x]$$

looks almost exactly like its classical counterpart, the algebraic variety also called the affine line. It contains one closed point for each value $\lambda \in K$. The Zariski topology on the set of closed points is the same as the classical Zariski topology on the variety: the open sets are the complements of finite sets. The scheme $\mathbb{A}^1$ differs from the variety only in that the scheme contains one more point, called the *generic point* of $\mathbb{A}^1$, corresponding to the ideal (0).

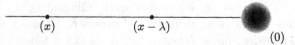

(x) $\qquad\qquad$ $(x - \lambda)$ $\qquad\qquad\qquad$ (0)

The closure of the point (0) is all of $\mathbb{A}_K^1$, so that the closed subsets of $\mathbb{A}_K^1$ are exactly the finite subsets of $\mathbb{A}_K^1 - \{(0)\}$.

The affine plane $\mathbb{A}_K^2 = \operatorname{Spec} K[x, y]$ is also similar to its counterpart variety, but now the additional points of the scheme are more numerous and behave in more interesting ways. We have as before closed points, coming from the maximal ideals $(x - \lambda, y - \mu)$, which correspond to the points (λ, μ) in the ordinary plane. There are now, however, two types of nonclosed points. To begin with, for each irreducible polynomial $f(x, y) \in K[x, y]$ we have a point corresponding to the prime ideal $(f) \subset K[x, y]$, whose closure consists of the point itself and all the closed points (λ, μ) with $f(\lambda, \mu) = 0$. The point (f) is called the *generic point* of this set; more generally, any point in a scheme is called the generic point of its closure. As compared to the variety $\mathbb{A}_K^2$, we have added one more point for every irreducible plane curve. This new point lies in the closure of (the set of closed points on) that curve, and its closure contains this set of closed points. Finally, we have as before a point corresponding to the zero ideal, the generic point of $\mathbb{A}_K^2$, whose closure is all of $\mathbb{A}_K^2$.

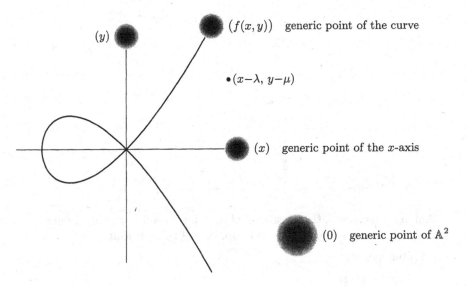

(y)

$(f(x,y))$ generic point of the curve

$\bullet (x-\lambda, \, y-\mu)$

(x) generic point of the x-axis

(0) generic point of $\mathbb{A}^2$

Since $K[x,y] = K[x] \otimes_K K[y]$, we have by definition

$$\mathbb{A}_K^2 = \mathbb{A}_K^1 \times_{\operatorname{Spec} K} \mathbb{A}_K^1.$$

Even here, though it's clear that the fibered product is the correct notion of product, the set of points of the fibered product is not the fibered product of the sets of points of the factors.

The situation with the affine spaces $\mathbb{A}_K^n = \operatorname{Spec} K[x_1, \ldots, x_n]$ is a straightforward extension of the last case: geometrically, we can see the scheme $\mathbb{A}_K^n$ as the classical affine n-space, with one point p_Σ added for every positive-dimensional irreducible subvariety Σ of n-space. As above, p_Σ will lie in the closure of the locus of closed points in Σ and contain in its closure all these points, as well as the generic points of the subvarieties of Σ.

More generally, suppose $X \subset \mathbb{A}_K^n$ is any affine variety, with ideal $I \subset K[x_1, \ldots, x_n]$ and coordinate ring $R = K[x_1, \ldots, x_n]/I$. We can associate to X the affine scheme $\operatorname{Spec} R$; the quotient map $K[x_1, \ldots, x_n] \to R$ expresses this as a subscheme of $\mathbb{A}_K^n$. This scheme is, as in the case of $\mathbb{A}_K^n$ itself, just like the variety X except that we have added one new generic point p_Σ for every positive-dimensional irreducible subvariety $\Sigma \subset X$.

Fibers, and more generally preimages, are among the most common ways that schemes other than varieties may arise even in the context of classical geometry.

Exercise II-2. Consider the map of the affine line $\operatorname{Spec} K[x]$ to itself induced by the ring homomorphism $K[x] \to K[x]$ mapping x to x^2. Show that the scheme-theoretic fiber over the point 0 is the subscheme of the line defined by the ideal (x^2).

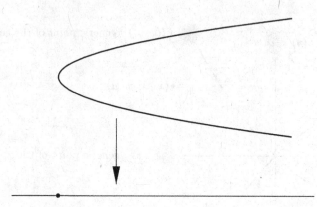

Among all schemes, those associated to affine varieties over algebraically closed fields may be characterized as spectra of rings R that are

- finitely generated
- reduced algebras
- over a field
- that is algebraically closed.

To get a sense of what more general schemes look like, and what they are good for, we will in the remainder of this section and the next consider what may happen if we remove these four restrictions. We will consider primarily examples in which exactly one of the hypotheses fails, since an understanding of these basic cases will enable one to understand the general case; we will occasionally mention more complex examples in exercises.

II.1.2   Local Schemes

Our first collection of examples of schemes other than varieties is provided by the spectra of local rings, called *local schemes*. The examples we will consider here are spectra of rings that are reduced algebras over an algebraically closed field but not, in general, finitely generated. Local schemes are for the most part technical tools in the study of other, more geometric schemes; they are often used to focus attention on the local structure of an affine scheme. The extra points we have added to classical varieties show up even more strikingly in the following examples, where in each case there is only one closed point. It would, of course, be a mistake to try to picture these schemes as geometric objects with just one point. Rather, they should be seen as germs of varieties. The phenomenon of having only one closed point is not some novelty invented by algebraists but is already present if one considers such a familiar object as the germ of a point x on a complex analytic manifold; here one pictures a "sufficiently small" neighborhood of x, in which, for example, each curve through x can be plainly distinguished, even though no other definite points beside x belong to every neighborhood. We will see that the same kind of picture is valid for the spectrum of a local ring.

Consider first the localization $K[x]_{(x)}$ of the ring $K[x]$ at the maximal ideal (x), and let $X = \operatorname{Spec} K[x]_{(x)}$. The space $|X|$ has only two points: the closed point corresponding to the maximal ideal (x), and the open point corresponding to (0), which contains the point (x) in its closure. The inclusion of $K[x]$ in $K[x]_{(x)}$ induces a map $X \to \mathbb{A}^1_K$, so that we may think of X as a subscheme of $\mathbb{A}^1_K$ (though $|X|$ is neither open nor closed in $|\mathbb{A}^1_K|$). The subscheme X is "local" in that it is the intersection of all the open subsets of $\mathbb{A}^1_K$ containing the point (x); so that, for example, the regular functions on X are exactly the rational functions on $\mathbb{A}^1_K$ regular at the point (x) — that is, they are the elements of $\mathcal{O}_{\mathbb{A}^1_K}(U)$ for some neighborhood U of the point $0 = (x)$ in $\mathbb{A}^1_K$. In these senses, X is the germ of $\mathbb{A}^1_K$ at the origin.

Next, consider the scheme $X = \operatorname{Spec} R$, where $R = K[x,y]_{(x,y)}$ is the localization of $K[x,y]$ at the maximal ideal (x,y) corresponding to the point $(0,0)$. As in the previous example, we have a map $X \to \mathbb{A}^2_K$, in terms of which we can think of X as the intersection of all open subschemes of $\mathbb{A}^2_K$ containing the closed point $(0,0)$. Again, X has only one closed point; but now there are infinitely many nonclosed points, one for every irreducible curve in the plane passing through the origin. Subschemes of X are thus germs at $(0,0)$ of subschemes of $\mathbb{A}^2_K$ and X itself is the germ of $\mathbb{A}^2_K$ at the origin.

There are analogous constructions in $\mathbb{A}^n_K$, and more generally for any subscheme of $\mathbb{A}^n_K$: if $X = \operatorname{Spec} K[x,\ldots,x_n]/I \subset \mathbb{A}^n_K$ is the scheme associated to the affine variety with ideal $I \subset K[x,\ldots,x_n]$ and $\mathfrak{m} = (x_1-a_1, \ldots, x_n-a_n)$ a maximal ideal corresponding to a closed point of X, we can consider the scheme $\operatorname{Spec} K[x_1,\ldots,x_n]_{\mathfrak{m}}/I_{\mathfrak{m}}$ as a germ of a neighborhood of $[\mathfrak{m}]$ in X. While we can talk about germs of functions on a space at a point in many contexts, in scheme theory the germ is again a scheme in its own right.

For some purposes, the local schemes introduced in this way are not local enough; the local ring of a scheme at a point still contains a lot of information about the global structure of the scheme. For example, the germs of a nonsingular variety X at various closed points will not in general be isomorphic schemes[1], although if X^{an} denotes the complex analytic variety defined by the same equations (or indeed any analytic manifold),

---

[1] This has nothing to do with schemes but is already the case for varieties over $\mathbb{C}$: for example, it is so already for the general plane curve of degree $d \geq 4$.

the germs of X^{an} at any two points are isomorphic. This phenomenon occurs essentially because the open sets used to define the germs of X are so big. To get a more local picture within the setting of schemes, we can look at the schemes associated to power series rings: for example, instead of looking at the germ $X = \operatorname{Spec} K[x,y]_{(x,y)}$ of a neighborhood of the origin in $\mathbb{A}_K^2$ above, we can consider the scheme $Y = \operatorname{Spec} K[\![x,y]\!]$. As in the previous case, this scheme has one closed point $[(x,y)]$ and one generic point $[(0)]$ whose closure is all of Y; in addition, it has one point for every irreducible power series $\sum a_{i,j} x^i y^j$ in x and y. The maps

$$K[x,y] \hookrightarrow K[x,y]_{(x,y)} \hookrightarrow K[\![x,y]\!]$$

give maps $Y \to X \to \mathbb{A}_K^2$; we think of the Y as a "smaller" neighborhood of the origin than X. (Note, however, that X and Y are neither closed subschemes nor open subschemes of $\mathbb{A}_K^2$.) For example, while the curve corresponding to the prime ideal $(y^2 - x^3 - x^2)$ is irreducible in X, because the curve in $\mathbb{A}_K^2$ defined by this equation is, the preimage in Y of this curve is the (nontrivial) union of two curves in Y, as long as the characteristic of K is not 2, because $x^2 + x^3$ has the square root

$$u = x + \tfrac{1}{2}x^2 - \tfrac{1}{8}x^3 + \cdots$$

in the power series ring. Thus we can factorize $y^2 - x^3 - x^2$ as

$$y^2 - x^3 - x^2 = (y - u)(y + u),$$

so the scheme $\operatorname{Spec} K[\![x,y]\!]/(y^2 - x^3 - x^2)$ is reducible. (See the figure on the next page.)

Of course, Y must have "more" curves than X for such things to be possible. The following exercise amplifies this fact.

Exercise II-3. (a) With $u = \sqrt{x^2 + x^3}$ as above, what is the image of $[(y-u)]$ in $\operatorname{Spec} K[x,y]$? (Hint: it's a prime ideal containing $y^2 - x^3 - x^2$.)

(b) Show that the image of the point $(y - \sum_{n \geq 1} x^n/n!)$ of Y is the generic point of $\mathbb{A}_K^2$.

In general, under the map $Y \to X$ above, the inverse image of a point corresponding to an irreducible curve $C \subset \mathbb{A}_K^2$ consists of the set of analytic branches of C at the origin. (See Walker [1950] or Brieskorn and Knörrer [1986] for further discussion of branches.)

Here is yet another important example of a local scheme. One problem with the scheme Y above is that the points described in Exercise II-3(b) are extraneous from an algebraic point of view. To avoid this, we may work with the spectrum Z of the ring $H \subset K[\![x,y]\!]$ of power series that satisfy algebraic equations over $K(x,y)$, the field of rational functions. Called the *Henselization* of X, the scheme Z sits in between Y and X in the sense that we have a series of maps

$$Y \to Z \to X \to \mathbb{A}_K^2.$$

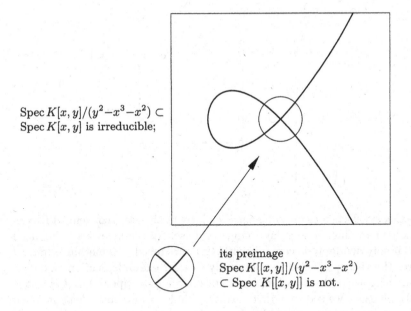

$\operatorname{Spec} K[x,y]/(y^2-x^3-x^2) \subset$
$\operatorname{Spec} K[x,y]$ is irreducible;

its preimage
$\operatorname{Spec} K[[x,y]]/(y^2-x^3-x^2)$
$\subset \operatorname{Spec} K[[x,y]]$ is not.

The usefulness of this construction is that H is the union of algebras finitely generated over K, so that Z is the inverse limit of schemes coming from ordinary varieties. Geometrically, Z is the germ of $\mathbb{A}_K^2$ in the *étale topology*, a concept we will not pursue here; see Artin [1971] for further information.

Exercise II-4. In the case $K = \mathbb{C}$, how does the spectrum of the ring of convergent power series fit into this picture?

II.2   Reduced Schemes over Non-Algebraically Closed Fields

We now consider what happens when we look at the spectrum of a finitely generated, reduced algebra over a field K that is not algebraically closed. The interest in such structures came originally from number theory, and, of course, it predates scheme theory very substantially! For example, the study of rational quadratic forms, an old subject in number theory, can be thought of as the study of varieties over the rational numbers defined by a quadratic equation. Cubic forms in three variables over the rationals still make up a very active number-theoretic research topic, now mostly pursued through the theory of elliptic curves over $\mathbb{Q}$. The basic objects themselves are varieties over $\mathbb{Q}$ (or schemes over $\mathbb{Z}$, a situation we'll return to later), but in the course of handling them, number theorists frequently make use of all the base rings shown in the following diagram, along with many intermediate fields and rings:

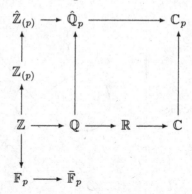

The theory of schemes provides a particularly flexible and convenient framework for handling these many changes of base. Also, a nice variety reduced mod p may suddenly become something nonreduced — something that requires the theory of schemes more fully (see for example Section II.4.4).

To start with the simplest case, consider $A_{\mathbb{R}}^1 = \operatorname{Spec} \mathbb{R}[x]$. Using the Nullstellensatz we see that there are two kinds of maximal ideals in $\mathbb{R}[x]$: those whose residue class field is $\mathbb{R}$, which have the form $(x - \lambda)$ for $\lambda \in \mathbb{R}$, and those whose residue class field is $\mathbb{C}$, which have the form $(x^2 + \mu x + \nu)$, for μ and $\nu \in \mathbb{R}$ with $\mu^2 - 4\nu < 0$. The latter type of ideals may also be written in the form $((x - z)(x - \bar{z}))$, for $z \in \mathbb{C}$ not real. A closed point of $A_{\mathbb{R}}^1$ thus corresponds either to a real number or to a conjugate pair of nonreal complex numbers. Finally, $A_{\mathbb{R}}^1$ has again a unique nonclosed point corresponding to the prime (0), whose closure is all of $A_{\mathbb{R}}^1$.

Next, we turn to the affine plane over $\mathbb{R}$, $A_{\mathbb{R}}^2 = \operatorname{Spec} \mathbb{R}[x, y]$, and consider a closed point given by a maximal ideal $\mathfrak{m}$ of $\mathbb{R}[x, y]$. Again by the Nullstellensatz the residue class field of $\mathfrak{m}$ is either $\mathbb{R}$ or $\mathbb{C}$, and the composite map

$$\mathbb{R} \to \mathbb{R}[x, y]/\mathfrak{m} \cong (\mathbb{R} \text{ or } \mathbb{C})$$

is either the identity or the inclusion of $\mathbb{R}$ in $\mathbb{C}$. Taking λ and μ to be the images of x and y in $\mathbb{C}$, we see that in the former case $\mathfrak{m} = (x - \lambda, y - \mu)$ corresponds to the ordinary point (λ, μ) in $\mathbb{R}^2$. But in the latter case $\mathfrak{m}$ corresponds to both (λ, μ) and $(\bar{\lambda}, \bar{\mu})$; put differently, the map $\mathbb{R}[x, y] \to \mathbb{C}$ sending x, y to λ, μ has the same kernel as the one sending x, y to $\bar{\lambda}, \bar{\mu}$ since they differ by the automorphism of $\mathbb{C}$ over $\mathbb{R}$.

It is not difficult to give generators for the maximal ideals described above. If $\mathbb{R}[x, y]/\mathfrak{m} \cong \mathbb{R}$, then clearly $\mathfrak{m} = (x - \lambda, y - \mu)$. In the other case, suppose first that λ is real. Then μ must satisfy an irreducible real quadratic polynomial equation $y^2 + ay + b = 0$, so $\mathfrak{m}$ contains the ideal $(x - \lambda, y^2 + ay + b)$. But this last ideal is immediately seen to be prime (for example, by factoring out $x - \lambda$ first), so $\mathfrak{m} = (x - \lambda, y^2 + ay + b)$. Of course, a similar result holds if the image of y is real.

Finally, suppose that μ and λ are both nonreal. Then $\mathfrak{m}$ contains the irreducible polynomials $f(x)$ and $g(y)$ satisfied by μ and λ, but since $g(y)$ factors as

$$g(y) = (y - \mu)(y - \bar{\mu})$$

in $\mathbb{R}[x]/(f(x)) \cong \mathbb{C}$ the ideal $(f(x), g(y))$ is not prime! The picture here, over the complex numbers, is as follows:

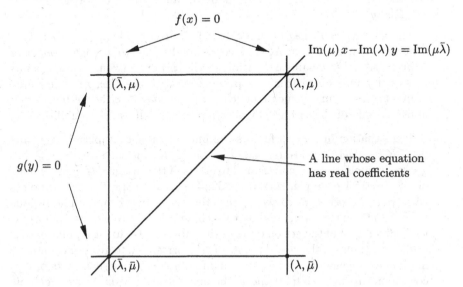

The loci defined by $f(x) = 0$ and $g(y) = 0$ are unions of two vertical and two horizontal lines, respectively, and intersect in the four points (λ, μ), $(\bar{\lambda}, \mu)$, $(\lambda, \bar{\mu})$, and $(\bar{\lambda}, \bar{\mu})$. But the polynomial

$$h(x, y) = \operatorname{Im}(\mu)\, x - \operatorname{Im}(\lambda)\, y - \operatorname{Im}(\mu \bar{\lambda})$$

defining the line joining the two points (λ, μ) and $(\bar{\lambda}, \bar{\mu})$ has real coefficients. The ideal

$$(f(x), h(x, y)) = (g(y), h(x, y)) \subset \mathbb{R}[x, y]$$

thus strictly contains the ideal $(f(x), g(y))$; and this ideal is the maximal ideal $\mathfrak{m}$ we seek, as one checks by working in

$$\mathbb{R}[x] \cong \mathbb{R}[x, y]/(h) \cong \mathbb{R}[y]$$

(for these isomorphisms, note that $(\bar{\mu} - \mu)$ and $(\bar{\lambda} - \lambda)$ are both nonzero).

In sum, then, the closed points of $\mathbb{A}_{\mathbb{R}}^2$ correspond either to points (λ, μ) of $\mathbb{A}_{\mathbb{C}}^2$ with λ and μ real, or to (unordered) pairs of points (z, w) and $(\bar{z}, \bar{w}) \in \mathbb{A}_{\mathbb{C}}^2$ with at least one of z, w not real. To put it another way, closed points of $\mathbb{A}_{\mathbb{R}}^2$ correspond to orbits of the action of complex conjugation on the points of $\mathbb{A}_{\mathbb{C}}^2$. (Note, in particular, that the closed points of $\mathbb{A}_{\mathbb{R}}^2$ are not ordered pairs of closed points of $\mathbb{A}_{\mathbb{R}}^1$!) Observe also that the residue field is

$\mathbb{R}$ at the points of $\mathbb{A}^2_{\mathbb{R}}$ corresponding to points (λ, μ) with λ, μ real, while at points of $\mathbb{A}^2_{\mathbb{R}}$ corresponding to pairs of complex conjugate points of $\mathbb{A}^2_{\mathbb{C}}$ the residue field is $\mathbb{C}$.

Exercise II-5. Show that the nonclosed points of $\mathbb{A}^2_{\mathbb{R}}$ are all either

(a) $[(0)]$, whose closure is all of $\mathbb{A}^2_{\mathbb{R}}$, or

(b) the point $[(f)]$ of $\mathbb{A}^2_{\mathbb{R}}$ corresponding to an irreducible polynomial $f \in \mathbb{R}[x, y]$.

Those of type (b) may or may not remain irreducible in $\mathbb{C}[x, y]$, so that a nonclosed point (f) in $\mathbb{A}^2_{\mathbb{R}}$ will correspond either to a single nonclosed point in $\mathbb{A}^2_{\mathbb{C}}$ (if f remains irreducible in $\mathbb{C}[x, y]$) or to two nonclosed points in $\mathbb{A}^2_{\mathbb{C}}$ (if f may be written as a product $g\bar{g}$ with $g \in \mathbb{C}[x, y]$). The closed points in the closure of such a nonclosed point may be either of both types above or only of the second. Give examples with all these possibilities.

The situation in general follows the lines of these examples: if K is any field, $\bar{K}$ its algebraic closure, and $G = \mathrm{Gal}(\bar{K}/K)$ the corresponding Galois group, the points of $\mathbb{A}^n_K$ correspond to orbits of the action of G on the points of $\mathbb{A}^n_{\bar{K}}$ (see, for example, Nagata [1962, Theorem 10.3]). The closed points correspond to orbits of closed points, the orbits being finite. The residue field at the point p corresponding to such an orbit, moreover, is isomorphic to the fixed field of the action on $\bar{K}$ of the subgroup G_p fixing a point of that orbit. For example, the closed points of $\mathbb{A}^1_{\mathbb{Q}}$ correspond to algebraic numbers modulo conjugacy; and for a prime number $q \in \mathbb{Z}$ the closed points of $\mathbb{A}^1_{\mathbb{F}_q}$ correspond to the orbits of the Frobenius automorphism of the algebraic closure of $\mathbb{F}_q = \mathbb{Z}/(q)$ (namely, 0 and the orbits of the map $a \mapsto a^q$ on the multiplicative group $\bar{K}^*$, which may be described as the inductive limit of all cyclic groups of order prime to q or as the q-torsion-free part of $\mathbb{Q}/\mathbb{Z}$).

Exercise II-6. An inclusion of fields $K \hookrightarrow L$ induces a map $\mathbb{A}^n_L \to \mathbb{A}^n_K$. Find the images in $\mathbb{A}^2_{\mathbb{Q}}$ of the following points of $\mathbb{A}^2_{\overline{\mathbb{Q}}}$ under this map.

(a) $(x - \sqrt{2}, y - \sqrt{2})$

(b) $(x - \sqrt{2}, y - \sqrt{3})$

(c) $(x - \zeta, y - \zeta^{-1})$, where ζ is a p-th root of unity, with p prime

(d) $(\sqrt{2}x - \sqrt{3}y)$

(e) $(\sqrt{2}x - \sqrt{3}y - 1)$

Where feasible, draw pictures.

Exercise II-7. We say that a subscheme $X \subset \mathbb{A}^n_K$ is *absolutely irreducible* or *geometrically irreducible* if the inverse image of X in $\mathbb{A}^n_{\bar{K}}$ is irreducible. (More generally, we say any K-scheme X is absolutely irreducible if the fiber product $X \times_{\mathrm{Spec}\,K} \mathrm{Spec}\,\bar{K}$ is irreducible.) Classify the following subschemes of $\mathbb{A}^2_{\mathbb{Q}} = \mathrm{Spec}\,\mathbb{Q}[x, y]$ as reducible, irreducible but not absolutely irreducible, or absolutely irreducible.

(a) $V(x^2 - y^2)$

(b) $V(x^2 + y^2)$

(c) $V(x^2 + y^2 - 1)$

(d) $V(x + y, \, xy - 2)$

(e) $V(x^2 - 2y^2, \, x^3 + 3y^3)$

Finally, here is an example that combines the notions of local schemes and schemes over non-algebraically closed fields. Classically, a plane curve $X \subset \mathbb{A}^2_{\mathbb{C}}$ was said to have a *node* at the origin if in some analytic neighborhood of the origin the locus of complex points of Y consisted of two smooth arcs intersecting transversely at $(0,0)$. In the language of schemes, this is the same as saying that the fiber product of X with the formal neighborhood $\operatorname{Spec} \mathbb{C}\llbracket x, y \rrbracket \to \operatorname{Spec} \mathbb{C}[x,y] = \mathbb{A}^2_{\mathbb{C}}$ is isomorphic to $\operatorname{Spec} \mathbb{C}\llbracket u, v \rrbracket/(uv)$.

Consider now a curve in the real plane $X \subset \mathbb{A}^2_{\mathbb{R}}$. We say in this case that X has a node at the origin if the corresponding complex curve

$$X \times_{\operatorname{Spec} \mathbb{R}} \operatorname{Spec} \mathbb{C} \subset \mathbb{A}^2_{\mathbb{C}}$$

does. In this case, the formal neighborhood

$$X \times_{\operatorname{Spec} \mathbb{R}[x,y]} \operatorname{Spec} \mathbb{R}\llbracket x, y \rrbracket$$

may have either one of two nonisomorphic forms: it may be isomorphic to $\operatorname{Spec} \mathbb{R}\llbracket u, v \rrbracket/(uv)$ or to $\operatorname{Spec} \mathbb{R}\llbracket u, v \rrbracket/(u^2 + v^2)$. The former is the case if the locus of real points of X (that is, the locus of points with residue field $\mathbb{R}$) looks in an analytic neighborhood of $(0,0)$ like two smooth real arcs intersecting transversely at $(0,0)$; classically, such a point was called a *crunode* of X. The latter is the case if the origin is isolated as a real point of X; this was called an *acnode* in the past.

Exercise II-8. Verify the assertions made above: specifically, show that if X is a curve in $\mathbb{A}^2_{\mathbb{C}}$ with a node at the origin, then the formal neighborhood $X \times_{\operatorname{Spec} \mathbb{C}[x,y]} \operatorname{Spec} \mathbb{C}\llbracket x, y \rrbracket$ is isomorphic to $\operatorname{Spec} \mathbb{C}\llbracket u, v \rrbracket/(uv)$; and that if $X \subset \mathbb{A}^2_{\mathbb{R}}$ is a real plane curve with a node at the origin, then the formal neighborhood $X \times_{\operatorname{Spec} \mathbb{R}[x,y]} \operatorname{Spec} \mathbb{R}\llbracket x, y \rrbracket$ has one of the two forms above. Show that there are infinitely many curves $X \subset \mathbb{A}^2_{\mathbb{Q}}$ with nodes at the origin having nonisomorphic formal neighborhoods. (As in the real case, we say that $X \subset \mathbb{A}^2_{\mathbb{Q}}$ has a node at the origin if the complex curve $X \times_{\operatorname{Spec} \mathbb{Q}} \operatorname{Spec} \mathbb{C}$ does.)

II.3 Nonreduced Schemes

We now leave the realm of objects that could be treated in the theory of varieties to look at some examples of affine schemes $\operatorname{Spec} R$ where R is a finitely generated algebra over an algebraically closed field K but may have nilpotents. The phenomena here are much less familiar, and we will spend rather more effort on them.

Schemes of this type arise already in quite simple geometric contexts: for example, the multiple points treated below occur already as intersections of two ordinary varieties and as "degenerate" fibers of maps, as in Exercise II-2. One of the most important applications of nonreduced schemes is to the theory of families of varieties: deformation theory and moduli theory. We will explain how to take limits of one-parameter families of varieties, and introduce the key notion of flatness. Finally, we will give some examples of nonreduced schemes that are interesting objects in themselves.

To start with the easiest cases, we will focus first on subschemes of affine space $\mathbb{A}_K^n$ supported at the origin — equivalently, given by ideals I whose zero locus $V(I)$ consists, as a set, just of $(0, \ldots, 0)$. (Recall that the support of a scheme is the underlying topological space.)

II.3.1 Double Points

Example II-9. The simplest such scheme is the subscheme X of $\mathbb{A}_K^1$ defined by the ideal (x^2) — that is, the scheme $\operatorname{Spec} K[x]/(x^2)$, viewed as a subscheme of $\mathbb{A}_K^1$ via the map induced by the quotient map $K[x] \to K[x]/(x^2)$. This scheme has only one point, corresponding to the ideal (x), but it differs, both as a subscheme of $\mathbb{A}_K^1$ and as an abstract scheme, from the scheme $\operatorname{Spec} K[x]/(x) = \operatorname{Spec} K$. As an abstract scheme, we can see the difference in that there exist regular functions (such as x) on X that are not equal to zero but that have value 0 at the one point of X; of course, any such function will have square 0. As a subscheme of $\mathbb{A}_K^1$, the difference is that a function $f \in K[x]$ on $\mathbb{A}_K^1$ vanishes on X if and only if both f and its first derivative vanish at 0. The data of a function on X thus consists of the values at 0 of both a function on $\mathbb{A}_K^1$ *and* its first derivative. Possibly for this reason, X is sometimes called the *first-order neighborhood* of 0 in $\mathbb{A}_K^1$.

More generally, for any n the ideal (x^n) defines a subscheme $X \subset \mathbb{A}_K^1$ with coordinate ring $K[x]/(x^n)$; a function $f(x)$ on $\mathbb{A}_K^1$ vanishes on X if and only if the value of f at 0 vanishes together with the values of the first $n-1$ derivatives of f.

Example II-10 (double points). The next step in understanding double points is to consider subschemes of $\mathbb{A}_K^2 = \operatorname{Spec} K[x,y]$ supported at the origin and isomorphic to the scheme X of Example II-9. Let $Y \subset \mathbb{A}_K^2$ be such a subscheme, $R = \mathscr{O}_Y(Y) \cong K[\varepsilon]/(\varepsilon^2)$ its coordinate ring, and

$$\varphi : K[x,y] \to R$$

the surjection defining the inclusion of Y in $\mathbb{A}_K^2$. Since the inverse image of the unique maximal ideal $\mathfrak{m}$ of R is the ideal $(x,y) \subset K[x,y]$ corresponding to. the origin, and since $\mathfrak{m}^2 = 0$ in R, the map φ vanishes on $(x, y)^2 = (x^2, xy, y^2)$ and so factors through a map

$$\bar{\varphi} : K[x, y]/(x^2, xy, y^2) \to R.$$

Equivalently, Y must be contained in the subscheme

$$\operatorname{Spec} K[x, y]/(x^2, xy, y^2).$$

But the ring $K[x, y]/(x^2, xy, y^2)$ is a three-dimensional vector space over K, whereas R is only two-dimensional. It follows that the kernel of φ will contain a nonzero homogeneous linear form $\alpha x + \beta y$, for some $\alpha, \beta \in K$. Write

$$X_{\alpha,\beta} = \operatorname{Spec} K[x, y]/(x^2, xy, y^2, \alpha x + \beta y) \hookrightarrow \mathbb{A}_K^2.$$

The subscheme $X_{\alpha,\beta}$ can be characterized either as

(i) the subscheme of $\mathbb{A}_K^2$ associated to the ideal of functions $f \in K[x, y]$ that vanish at the origin and have partial derivatives satisfying

$$\beta \frac{\partial f}{\partial x} - \alpha \frac{\partial f}{\partial y} = 0$$

there (since this implies that $f = c(\alpha x + \beta y) + \text{higher-order terms}$); or

(ii) the image of the subscheme $X \subset \mathbb{A}_K^1$ of Example II-9 under the inclusion of $\mathbb{A}_K^1$ in $\mathbb{A}_K^2$ given by $x \mapsto (\beta x, -\alpha x)$.

In the classical language, the subscheme $X_{\alpha,\beta}$ was said to consist of the point $(0, 0)$ and an "infinitely near point" in the direction specified by the line defined by $\alpha x + \beta y = 0$. We draw $X_{\alpha,\beta}$ as the small arrow in this traditional picture:

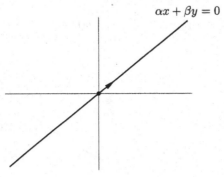

This is intended to represent a point with a distinguished one-dimensional subspace of the tangent space to the plane at that point (there is actually no distinguished tangent vector, despite the impression given by the arrow).

How do schemes such as $X_{\alpha,\beta}$ arise in practice? One way is as intersections of curves. For example, when we want to work with the intersection of a line L and a conic C that happen to be tangent, it is clearly unsatisfactory to take their intersection in the purely set-theoretic sense; a line and a conic should meet twice. Nor is it completely satisfactory to describe $C \cap L$ as their point of intersection "with multiplicity two": for example, the intersection should determine L, as it does in the non-tangent case. The satisfactory definition is that $C \cap L$ is the subscheme of $\mathbb{A}^2_K$ defined by the sum of the ideals I_C and I_L so that, for example, the line $y = 0$ and the parabola $y = x^2$ will intersect in the subscheme $X_{0,1} = \operatorname{Spec} K[x, y]/(x^2, y)$. This does indeed determine L, as the unique line in the plane containing $X_{0,1}$.

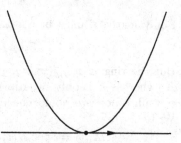

Another important way in which subschemes such as $X_{\alpha,\beta}$ arise is as limits of reduced schemes. For example, consider a pair of distinct closed points $(0,0)$ and (a, b) in the plane. Their union is the closed subscheme

$$X = \{(0,0), (a,b)\} = \operatorname{Spec} S \subset \mathbb{A}^2_K,$$

where

$$S = K[x,y]/((x, y) \cap (x - a, \ y - b))$$
$$= K[x,y]/(x^2 - ax, \ xy - bx, \ xy - ay, \ y^2 - by).$$

By the Chinese Remainder Theorem, $S \cong K \times K$; so in particular, S is a K-algebra of (vector space) dimension 2 over K.

Now suppose the point (a, b) moves toward the point $(0,0)$ along a curve $(a(t), b(t))$, with $(a(0), b(0)) = (0,0)$, where a and b are polynomials in t; we write

$$a(t) = a_1 t + a_2 t^2 + \cdots, \qquad b(t) = b_1 t + b_2 t^2 + \cdots.$$

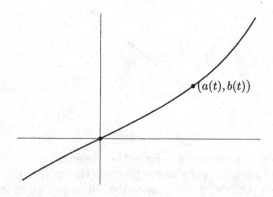

What should be the limit of $X_t = \{(0,0), (a(t), b(t))\}$ as $t \to 0$? Using schemes, we can afford the luxury of the idea that it will continue to be two points, in a suitable sense: it will be an affine scheme X whose coordinate ring is again a two-dimensional vector space over K. We may define X by taking its ideal to be the limit as $t \to 0$ of the ideal

$$I_t = (x, y) \cap (x - a(t),\, y - b(t)).$$

Of course, this only shifts the burden to describing what is the limit of a family of ideals! But this is easy: in the current case, for example, we can take their limit as codimension-2 subspaces of $K[x, y]$, viewed as a vector space over K. That this limit is again an ideal follows from the continuity of multiplication. A more delicate description is necessary in the general case, where the ideals are of infinite codimension; we will discuss this below when we come to limits of families of one-dimensional schemes and again in Section III.3.2 in the projective case.

To see what this means in practice, observe first that the generators $x^2 - a(t)x$, $xy - b(t)x$, $xy - a(t)y$, and $y^2 - b(t)y$ of the ideal I_t clearly have as their limit when $t \to 0$ the polynomials x^2, xy, xy, and y^2, so these polynomials will be in I. In addition, observe that I_t contains the linear form

$$a(t)y - b(t)x = (xy - b(t)x) - (xy - a(t)y)$$

and hence, for $t \neq 0$, also the polynomial

$$\frac{a(t)y - b(t)x}{t} = a_1 y - b_1 x + t(\dots).$$

The ideal I thus contains the limit $a_1 y - b_1 x$ of this polynomial as well; so we have $I \supset (x^2, xy, y^2, a_1 y - b_1 x)$. But the right-hand side of this expression already has codimension 2 as a vector subspace in the polynomial ring $K[x, y]$. Thus $I = (x^2, xy, y^2, a_1 y - b_1 x)$, and correspondingly

$$\lim_{t \to 0}(X_t) = X_{\alpha, \beta} \quad \text{with } \alpha = b_1,\ \beta = -a_1.$$

From this we see that X, as a subscheme of A_K^2, "remembers" the direction of approach of $(a(t), b(t))$; we think of it as consisting of the origin together with a tangent direction, along the line with equation $a_1 y - b_1 x = 0$. This line is the limit of the lines L_t joining $(0,0)$ to $(a(t), b(t))$; that is, it is the tangent line to the curve parametrized by $(a(t), b(t))$ at the origin, as shown on the right.

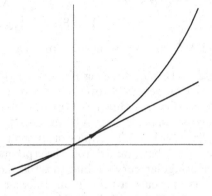

We will see how to generalize this notion of limit in Section II.3.4.

II.3.2 Multiple Points

The subschemes $X_{\alpha,\beta}$ of the preceding examples are called "double points" in the plane, the *double* referring to the vector space dimension of their coordinate rings

$$R = K[x,y]/(x^2,\, xy,\, y^2,\, \alpha x + \beta y) \cong K[t]/(t^2)$$

as K-modules. In general, if $X = \operatorname{Spec} R$ is an affine scheme and R is a finite-dimensional vector space over a field K, we define the *degree* of X relative to K, denoted $\deg_K(X)$ or simply $\deg(X)$, to be the dimension of R as a K-vector space. (Where there is unlikely to be ambiguity about the field K, we may suppress it in both the language and the notation.) In this situation we call $\operatorname{Spec} R$ a *finite K-scheme*.

We next consider examples having degree 3 or more. A number of things are different here. To begin with, all double points over an algebraically closed field K — that is, schemes of the form $\operatorname{Spec} R$, where R is a local K-algebra of vector space dimension 2 — are isomorphic, since such an R must be isomorphic to $K[x]/(x^2)$. (Proof: Let $\mathfrak{m}$ be the maximal ideal of R. Then $R/\mathfrak{m} \cong K$, since K has no finite-dimensional extension. Since R is two-dimensional, $\mathfrak{m}$ is one-dimensional. Also $\mathfrak{m}^2 = 0$ — for example, by Nakayama's Lemma — so the obvious map from $K[x]$ onto R has x^2 in the kernel and identifies R with $K[x]/(x^2)$ as required.) By contrast, this is not true of triple points: the schemes

$$\operatorname{Spec} K[x]/(x^3) \quad \text{and} \quad \operatorname{Spec} K[x,y]/(x^2,\, xy,\, y^2)$$

are readily seen to be nonisomorphic. However, any triple point is isomorphic to either of these, a fact whose proof we leave as the following exercise.

Exercise II-11. Suppose that K is algebraically closed, and let $Z = \operatorname{Spec} K[x_1,\ldots,x_n]/I \subset \mathbb{A}^n_K$ be any subscheme of dimension 0 and degree 3, supported at the origin. Show that Z is isomorphic either to $X = \operatorname{Spec} K[x]/(x^3)$ or to

$$Y = \operatorname{Spec} K[x,y]/(x^2,\, xy,\, y^2),$$

and X, Y are not isomorphic to each other.

In particular, any ring $K[x_1,\ldots,x_n]/I$ of vector space dimension 3 over K can be generated over K by two linear forms in the x_i. In geometric terms, this says that any triple point in $\mathbb{A}^n_K$ is planar — that is, lies in a linear subspace $\mathbb{A}^2_K \subset \mathbb{A}^n_K$. Inside $\mathbb{A}^2_K$ both types of triple points can be realized as limits of triples of distinct points. The ones isomorphic to X above may be obtained from three points coming together in the plane along a nonsingular curve, while those isomorphic to Y above arise when two points approach a third from different directions. The following exercises contain examples of these phenomena.

Exercise II-12. (i) Show that the subscheme of $\mathbb{A}^2_K$ given by the ideal $(y - x^2, xy)$ arises as the limit of three points on the conic curve $y = x^2$ and is isomorphic to X above, but is not contained in any line in $\mathbb{A}^2_K$.

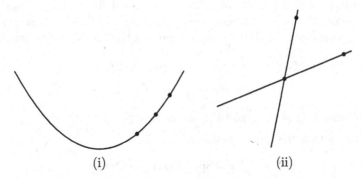

(i)   (ii)

(ii) Show that subschemes of $\mathbb{A}^2_K$ isomorphic to Y above arise when two points approach a third from different directions.

Exercise II-13. (For those familiar with the Grassmannian.) The examples above may lead one to expect that the schemes isomorphic to X are limits of those isomorphic to Y. In fact, just the opposite is the case, in the following sense. Let $\mathcal{H}$ be the set of finite subschemes of degree 3 of $\mathbb{A}^2_K$ supported at the origin; $\mathcal{H}$ naturally parametrized by a closed subscheme of the Grassmannian of codimension-3 subspaces of the six-dimensional vector space $K[x, y]/(x, y)^3$. Show that $\mathcal{H}$ is a surface, with one point corresponding to the unique subscheme $\operatorname{Spec} K[x, y]/(x^2, xy, y^2)$ isomorphic to Y and the rest corresponding to subschemes isomorphic to X. Show that the scheme $\mathcal{H}$ is isomorphic to a two-dimensional cubic cone in $\mathbb{P}^3_K$, and that the vertex is the one point corresponding to Y.

Exercise II-14. Let C be the subscheme of $\mathbb{A}^n_K$ given by the ideal

$$J = (x_2 - x_1^2, x_3 - x_1^3, \ldots).$$

A closed point in C is of the form $f(t) = (t, t^2, t^3, \ldots, t^n)$, for $t \in K$; that is, it has ideal $(x_1 - t, x_2 - t^2, \ldots)$. Consider for $t \neq 0$ the three-point subscheme

$$X_t = \{f(0), f(t), f(2t)\} \subset C.$$

(a) Show that the limit scheme as $t \to 0$ is

$$X_0 = \operatorname{Spec} K[x_1, \ldots, x_n]/(x_2 - x_1^2, x_1 x_2, x_3, x_4, \ldots, x_n)$$

and is isomorphic to the triple point $\operatorname{Spec} K[x]/(x^3)$ above.

(b) Show, however, that X_0 is not contained in the tangent line to C at the origin. Rather, the smallest linear subspace of $\mathbb{A}^n_K$ in which X_0 lies is the osculating 2-plane

$$x_3 = x_4 = \cdots = x_n = 0.$$

to C (recall that this is by definition the limit of the planes spanned by the tangent line and another point on C near the origin as the point approaches the origin), while the tangent line to C is the smallest linear subspace of $\mathbb{A}_K^n$ containing the subscheme defined by the square of the maximal ideal in the coordinate ring of X_0. Thus, in this sense, X_0 "remembers" both the tangent line and the osculating 2-plane to C.

Exercise II-15. Consider for $t \neq 0$ the subschemes

$$X_t = \{(0,0), (t,0), (0,t)\} \subset \mathbb{A}_K^2,$$

each consisting of three distinct points in $\mathbb{A}_K^2$.

(a) Show that the limit scheme as $t \to 0$ is

$$X_0 = \operatorname{Spec} K[x,y]/(x^2, xy, y^2).$$

(b) Show that the restriction of a function $f \in K[x,y]$ on $\mathbb{A}_K^2$ to X_0 determines and is determined by the values at the origin of f and its first derivatives in every direction; thus we think of it as a first-order infinitesimal neighborhood of the point $(0,0)$.

(c) Show that X_0 is contained in the union of any two distinct lines through $(0,0)$.

(d) Show that X_0 is not contained in any nonsingular curve and thus, in particular, is not the scheme-theoretic intersection of any two nonsingular curves in $\mathbb{A}_K^2$.

As we said, both types of triple point are contained in planes inside any affine space in which they are embedded. But the quadruple point $\operatorname{Spec} K[x,y,z]/(x,y,z)^2$ is not, since its maximal ideal cannot be generated by two elements. Other new phenomena occur for spatial multiple points — those not contained in the plane — and multiple points in higher-dimensional spaces. For example, not every point of degree 21 in 4-space arises as a limit of sets of 21 distinct points, as the following exercise shows. (See also Iarrobino [1985].)

Exercise II-16. Consider zero-dimensional subschemes $\Gamma \subset \mathbb{A}_K^4$ of degree 21 such that

$$V(\mathfrak{m}^3) \subset \Gamma \subset V(\mathfrak{m}^4),$$

where $\mathfrak{m}$ is the maximal ideal of the origin in $\mathbb{A}_K^4$. Show that there is an 84-dimensional family of such subschemes, and conclude that a general one is not a limit of a reduced scheme.

Exercise II-17. Classify up to isomorphism subschemes of $\mathbb{A}_K^2$ of dimension 0 and degrees 4 and 5 with support at the origin. Which are isomorphic as schemes over $\operatorname{Spec} K$?

Exercise II-18. A scheme $\operatorname{Spec} R$ supported at a point is called *curvilinear* if the maximal ideal of the (necessarily local) ring R is generated by one element; or, equivalently, if its Zariski tangent space has dimension zero or one. (The name comes from the fact that these are exactly the schemes that can be contained in a nonsingular curve.) Show that any two subschemes of $\mathbb{A}_K^2$ having degree 2 and supported at a point can be transformed into one another by a linear transformation of the plane, but that this is not possible for curvilinear schemes of length 3. (Note, however, that any two curvilinear subschemes of $\mathbb{A}_K^2$ of the same degree *can* be carried into one another by an automorphism of $\mathbb{A}_K^2$.)

Exercise II-19. (For those with some familiarity with curves.) There are infinitely many isomorphism types of degree-7 subschemes supported at the origin in 3-space and infinitely many types of degree-8 subschemes supported at the origin in the plane.

As might be expected, the behavior of nonreduced schemes over non-algebraically closed fields is more complex. The following exercise gives an example.

Exercise II-20. Classify all schemes of degree 2 and 3 over $\mathbb{R}$ supported at the origin in $\mathbb{A}_{\mathbb{R}}^2$. In particular, show that while any such scheme X whose complexification $X \times_{\operatorname{Spec} \mathbb{R}} \operatorname{Spec} \mathbb{C}$ is isomorphic to $\operatorname{Spec} \mathbb{C}[x]/(x^3)$ is itself isomorphic to $\operatorname{Spec} \mathbb{R}[x]/(x^3)$, there are exactly two nonisomorphic schemes X whose complexification is isomorphic to $\operatorname{Spec} \mathbb{C}[x,y]/(x^2, xy, y^2)$.

Degree and Multiplicity. Recall that on page 62 we defined the *degree* of a finite affine K-scheme $X = \operatorname{Spec} R$, where R is a finite-dimensional vector space over some field K, as the dimension of R over K. When K is algebraically closed, the degree of such a scheme X measures, in some sense, its nonreducedness. As the last exercise shows, however, this is not true in general: $\operatorname{Spec} \mathbb{C}$ is reduced, but has degree 2 as a scheme over $\mathbb{R}$.

There is an alternative concept, called the *multiplicity*, which measures the nonreducedness of X. Unlike the degree, which is a relative notion dependent on the specification of a base field $K \subset R$, the multiplicity is an invariant of X alone, and it is defined in a more general situation — we will define it here for any local ring R that has Krull dimension zero (equivalently, any local Artinian ring).

So let R be any zero-dimensional local ring, with maximal ideal $\mathfrak{m}$. It is possible to choose ideals of R, say

$$R \supset \mathfrak{m} = I_1 \supset I_2 \supset \cdots \supset I_{l-1} \supset I_l = 0$$

such that each successive quotient I_j/I_{j+1} is isomorphic to $R/\mathfrak{m}$ as an R-module. (For example, we could start with the coarser filtration

$$R \supset \mathfrak{m} \supset \mathfrak{m}^2 \supset \cdots \supset 0$$

and refine it by choosing arbitrary subspaces of the $R/\mathfrak{m}$-vector spaces $\mathfrak{m}^j/\mathfrak{m}^{j+1}$.) Though such a filtration is not unique, the length l is independent of the filtration chosen; we define the *multiplicity* or *length* of the ring R and of the zero-dimensional scheme X to be the number l (see for example Eisenbud [1995, Section 2.4]). Notice that in the original situation, when R is an algebra over a field K, finite-dimensional as a vector space over K, the residue field $R/\mathfrak{m} = \kappa$ is a finite extension of K and we have the relation

$$\deg_K(X) = [\kappa : K]\, \text{mult}(X).$$

For any zero-dimensional scheme X and point $p \in X$ we define the *multiplicity* of X at p, denoted $\text{mult}_p(X)$, to be the multiplicity of the local ring $\mathscr{O}_{X,p}$; if X is a finite K-scheme, the degree of X relative to K is given by

$$\deg_K(X) = \sum_{p \in X} [\kappa(p) : K]\, \text{mult}_p(X).$$

In Chapter III we will see how the notions of degree and multiplicity may each be extended to positive-dimensional schemes.

II.3.3   Embedded Points

We now consider some examples of nonreduced schemes of higher dimension; for simplicity we will restrict ourselves to the case where the underlying reduced scheme is a line. Even so, the variety of possible behaviors increases enormously; for example, we can have schemes that look like reduced schemes except at a point, or schemes that are everywhere nonreduced. In this subsection, we consider the former type. By way of terminology, we will say that a scheme $X = \text{Spec}\, K[x_1 \ldots, x_n]/I \subset \mathbb{A}_K^n$ has an *embedded component* if for some open subset $U \subset \mathbb{A}_K^n$ meeting X in a dense subset of X the closure of $X \cap U$ (as defined in Section I.2.1 above) does not equal X; or if, equivalently, the primary decomposition of the ideal I contains embedded primes (see the discussion of primary decomposition that follows). If the embedded prime is maximal — equivalently, if U may be taken to be the complement of a point — we talk about an *embedded point*; since the schemes X we will discuss below are all one-dimensional, this is all we will see.

The simplest example of a nonreduced scheme that is reduced except at one point is $X = \text{Spec}\, K[x,y]/(y^2, xy) \subset \mathbb{A}_K^2$. The ideal $I = (y^2, xy) \subset K[x,y]$ is the ideal of functions on the plane vanishing along the line $y = 0$ and in addition vanishing to order 2 at the point $(0,0)$; in algebraic terms, this means that $(y^2, xy) = (y) \cap (x,y)^2$. We can thus think of the scheme X as the line $y = 0$ with the proviso that a function f on X is defined by its restriction $f(x,0)$ to the line $y = 0$ together with the specification of its normal derivative at the point $(0,0)$ — that is, together with the number $\partial f/\partial y(0,0)$.

It is convenient to realize X as the union of the line defined by $y = 0$ with a nonreduced point — for example, the "first-order neighborhood of the origin" defined by the ideal (x^2, xy, y^2).

———————————•———————————

Such *primary decompositions* exist for any scheme: we briefly review the background from algebra. For more details see, for example, Eisenbud [1995; Atiyah and Macdonald [1969], or, for perhaps the gentlest treatment of all, Northcott [1953].

Primary Decomposition. Given any ideal I in a Noetherian ring R, we define the *associated prime ideals* of I to be the prime ideals $\mathfrak{p}$ such that $\mathfrak{p}$ is the annihilator of some element of R/I. These primes make up a finite set.

An ideal $\mathfrak{q} \subset \mathfrak{p}$ is called *primary* to $\mathfrak{p}$ if $\mathfrak{p}$ is the *radical* of $\mathfrak{q}$ (the set of elements having a power in $\mathfrak{q}$) and for any elements f, g in R with $fg \in \mathfrak{q}$ but $f \notin \mathfrak{p}$ we have $g \in \mathfrak{q}$; equivalently, $\mathfrak{q}$ is $\mathfrak{p}$-primary if $\mathfrak{p}$ is its radical and the localization map $R/\mathfrak{q} \to R_\mathfrak{p}/\mathfrak{q}R_\mathfrak{p}$ is a monomorphism.

Any ideal I may be expressed as the intersection of primary ideals. Since the intersection of ideals primary to a given prime ideal is again primary to that prime, I can even be expressed as an intersection of ideals that are primary to distinct prime ideals. If this is done in such a way that none of the primary ideals can be left out, the expression is called a *primary decomposition* of I. The primary ideals involved are called *primary components* of I.

The associated primes of I are exactly the radicals of the primary components. The primary component of I corresponding to a given associated prime is not uniquely determined by I; it is, however, so determined if the corresponding prime is minimal among the associated primes. Such primary components are called *isolated components*.

Example II-21. Taking $I = (y^2, xy)$ as above, the decomposition

$$I = (y) \cap (x, y)^2$$

already given expresses I as an intersection of primary ideals (the first is prime, the second is primary to (x, y)).

Since neither (y) nor $(x, y)^2$ can be omitted from this expression, it is a primary decomposition and the associated schemes of X (as defined below) are precisely the line X_{red} and the reduced point at the origin. The primary component associated to (x, y) in the decomposition is not unique; it could have been taken to be (x, y^2), or $(x+y, y^2)$, or indeed any of an infinite number of other such ideals, as well as their intersection (x^2, xy, y^2), or for that matter the ideal (x^n, xy, y^2) for any $n \geq 2$. Of course, the primary component (y) corresponding to (y) is unique, because X_{red} is not contained in any other associated scheme.

Despite this nonuniqueness, there is a well-defined *length* for the primary component corresponding to a given associated prime $\mathfrak{p}$, which may be computed, without choosing a primary decomposition, as the length of the largest ideal of finite length in the ring $R_{\mathfrak{p}}/IR_{\mathfrak{p}}$. Here the *length* of a module M is the maximal length l of a chain

$$M \supsetneq M_1 \supsetneq M_2 \supsetneq \cdots \supsetneq M_{l-1} \supsetneq M_l = 0$$

of submodules of M.

Exercise II-22. The length of the primary component of (xy, y^2) at the origin is 1.

It is easy to translate these matters into the geometry of schemes: any affine scheme $X = \operatorname{Spec} R$, where R is Noetherian, is the union of "primary" closed subschemes, called *primary components*, where a *primary affine scheme* is an affine scheme Y such that Y_{red} is irreducible and such that, if f, g are functions on Y,

$$\left.\begin{array}{l} fg \text{ vanishes on } Y \text{ but} \\ f \text{ does not vanish on } Y_{\mathrm{red}} \end{array}\right\} \implies g \text{ vanishes on } Y.$$

In such a *primary decomposition* of X, the components that are set-theoretically maximal — called *isolated components* — are unique. The others — called *embedded components*, because their supports are contained in larger components — are not unique. Nonetheless, the decomposition does have at least two nice uniqueness properties:

(1) The set of reduced subschemes associated to primary components in a minimal primary decomposition is unique; this is called the set of *associated schemes* to X.

(2) The "length" of the primary component associated to each of the associated schemes of X, called the *multiplicity of that associated scheme* in X, is unique.

We may use our example $X = \operatorname{Spec} K[x, y]/(y^2, xy)$ to illustrate these notions: we have already observed that X is the union of the line

$$X_{\mathrm{red}} = \operatorname{Spec} K[x, y]/(y)$$

and the multiple point

$$Y := \operatorname{Spec} K[x, y]/(x^2, xy, y^2)$$

and we have seen that this gives a primary decomposition, the multiplicity of the embedded subscheme at the origin being 1.

As we observed, we can write X in many different ways as the union of a line and a point: for example, for any $\alpha \neq 0$, we have $X = Y \cup Z$, where $Z = X_{\text{red}} = \operatorname{Spec} K[x,y]/(y)$ is the line and Y is, in the notation of Section II.3.1, the subscheme $X_{1,\alpha}$:

$$Y = \operatorname{Spec} K[x,y]/(x^2, xy, y^2, x+\alpha y).$$

Choosing two such subschemes Y, Y' gives an example of closed subschemes Y, Y' and Z in $\mathbb{A}_K^2$ such that

$$Y \cup Z = Y' \cup Z \quad \text{and} \quad Y \cap Z = Y' \cap Z, \quad \text{but} \quad Y \neq Y'.$$

In the example above, X can be described as the unique subscheme of $\mathbb{A}_K^2$ consisting of the (reduced) x-axis plus an embedded point of multiplicity 1 at the origin. But embedded points can carry geometric information, too.

Exercise II-23. Choose a linear embedding of $\mathbb{A}_K^2$ in $\mathbb{A}_K^3$, let P be the image of $\mathbb{A}_K^2$, and let X' be the image of X. Show that X' determines P as the unique plane in $\mathbb{A}_K^3$ containing X'.

It is also interesting to consider subschemes of $\mathbb{A}_K^2$ and $\mathbb{A}_K^3$ supported on a union of two given lines, with an embedded point of multiplicity 1 at the intersection of the two lines. In the plane, if we take the two lines to be the coordinate axes, such a scheme may be given as

$$X = \operatorname{Spec} K[x,y]/(x^2 y, xy^2).$$

Geometrically, this may be viewed as the union of the two lines defined by $xy = 0$ with the point $\operatorname{Spec} K[x,y]/(x^3, x^2 y, xy^2, y^3)$. In 3-space, if we take the lines to be $(x = z = 0)$ and $(y = z = 0)$, we can get such a scheme either as

$$Y_1 = \operatorname{Spec} K[x,y,z]/(z, x^2 y, xy^2)$$

or as

$$Y_2 = \operatorname{Spec} K[x,y,z]/(z^2, xz, yz, xy).$$

Y_1 is the image of the scheme X above under the embedding of $\mathbb{A}_K^2$ into $\mathbb{A}_K^3$ as the plane $z = 0$, whereas Y_2 is the union of the two lines with the subscheme of $\mathbb{A}_K^3$ defined by the square of the maximal ideal of the origin in $\mathbb{A}_K^3$.

Exercise II-24. (a) Show that $Y_1 \not\cong Y_2$.

(b) Show that $Y_1 \cong X$ is, up to isomorphism, the unique example contained in a plane of two lines meeting in a point and having an embedded point of multiplicity 1 at that intersection point.

(c) Show that Y_2 is, up to isomorphism, the unique example contained in 3-space but not in any plane of two lines meeting in a point and having an embedded point of multiplicity 1 at that intersection point.

One justification for the idea that the multiplicity of the embedded point at the origin in our scheme $X = \operatorname{Spec} K[x,y]/(xy, y^2)$ is 1 is that X is the limit as $t \to 0$ of the family of subschemes

$$X_{\mathrm{red}} \cup Y_t,$$

where Y_t is the scheme consisting of one reduced point

$$\operatorname{Spec} K[x,y]/(x, y-t) \subset \mathbb{A}_K^2.$$

This is plausible since the ideal

$$(x, y-t) \cap (y) = (xy, y^2 - ty)$$

of $X_{\mathrm{red}} \cup Y_t$ naturally seems to approach (xy, y^2) at $t \to 0$. However, the notion of limit that we introduced earlier is not quite strong enough to deal with this example, since the ideal $(x, y-t) \cap (y)$ of $X_{\mathrm{red}} \cup Y_t$ is not of finite codimension. In the next section we will rectify this, describing the general context for taking limits of schemes.

II.3.4   Flat Families of Schemes

The notion of a family of schemes is extremely general: we define a *family of schemes* to be simply a morphism $\pi : X \to B$ of schemes! The individual schemes in the family are the fibers of π over points of B. This notion includes all others that one can think of, such as a scheme defined by "equations with parameters", B being the space on which the parameters vary.

However, the notion of a family as an arbitrary morphism $\pi : X \to B$ is so general as to be virtually useless, because the fibers of the family may have nothing in common. For example, given such a family and a closed point $b \in B$, one could make a new family by replacing X by the disjoint union of $X - \pi^{-1}b$ and some other scheme Y, sending all of Y to b. Thus we must add some condition if we wish to have families of schemes that vary continuously, in some reasonable sense. What "reasonable" should mean is not obvious. It seems natural at least to ask that it include the mother of all continuously varying families, the family of projective plane curves of a given degree (see Section III.2.8). Other examples are the families of schemes defined by families of ideals of constant finite codimension in a polynomial ring, as we considered in the context of limits of multiple points.

In many geometric theories one gets the right notion of a continuously varying family by demanding some local triviality of the family; that is, locally, in some suitable sense, the family should look like the projection of a direct product to one factor. This is wrong for us on two counts. First, if we do this naively for schemes, interpreting locally as meaning locally in the Zariski topology, we get a notion that is far too restrictive to be of much use. A more sophisticated approach would be to demand this local triviality analytically; that is, to demand that if $x \in X$ and $b = \pi(x)$, then the completion of the local ring $\mathscr{O}_{X,x}$ should look like a power series ring over the completion of the local ring $\mathcal{O}_{B,b}$. This notion is quite useful (it is called *smoothness*), but it excludes, for example, the family of plane curves of a given degree, since a smooth family can't have singular fibers. Smoothness also excludes the families treated in the previous section, in which a disjoint union of distinct points approaches a multiple point; at the multiple point, the criterion is not met. Thus we must look for something more general.

The best current candidate for such a general notion is that of *flatness*. In order to motivate this definition, we consider first the more intuitive notion of limits.

Limits. The starting point for understanding the geometric content of flatness is the notion of the *limit* of a one-parameter family of schemes.

To set this up, we start with something fairly concrete: A *family of closed subschemes of a given scheme A* over a base B is a closed subscheme $X \subset B \times A$, together with the restriction to X of the projection map $B \times A \to B$; the fibers of X over $b \in B$ are then naturally closed subschemes of the fibers A_b of $B \times A$ over B.

Let B be a nonsingular, one-dimensional scheme — typically, we think of $\operatorname{Spec} R$, where $R = K[t]$, $K[t]_{(t)}$ or $K[[t]]$, but any Dedekind domain (including $\mathbb{Z}$ or $\mathbb{Z}_{(p)}$) will do. Let $0 \in B$ be any closed point, and write $B^* = B \setminus \{0\}$ for the complement of 0 in B. Let $\mathbb{A}^n_B$ and $\mathbb{A}^n_{B^*}$ be as usual affine n-space over B and B^* respectively.

We consider a closed subscheme $\mathscr{X}^* \subset \mathbb{A}^n_{B^*} = \mathbb{A}^n_{\mathbb{Z}} \times B^*$, which we view as a family of closed affine schemes parametrized by B^* — that is, for any point $b \in B^*$ we let $X_b = \pi^{-1}(b)$ be the fiber of the projection map $\pi : \mathscr{X}* \to \mathbb{A}^n_{B^*} \to B^*$, and consider these schemes X_b as the members of a family. (In case $B = \operatorname{Spec} R$ with $R = K[t]$ or $K[t]_{(t)}$ we can think of $\mathscr{X}^*$ as a "family of subschemes of $\mathbb{A}^n_K$ varying with parameter t".) We ask the basic question: what is the limit of the schemes X_b as b approaches the point 0?

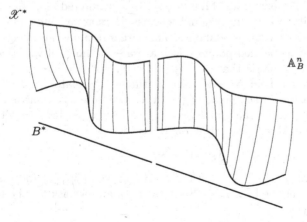

The answer — the only possible answer — is simple enough: since the limit of the schemes X_b in any reasonable sense must fit into a family with them, we take $\mathscr{X} \subset \mathbb{A}^n_B$ to be the closure $\overline{\mathscr{X}^*}$ of $\mathscr{X}^*$ in $\mathbb{A}^n_B$, and take the limit $\lim_{b \to 0} X_b$ of the schemes X_b to be the fiber X_0 of $\mathscr{X}$ over the point $0 \in B$.

To make this more concrete, if $B = \operatorname{Spec} R$ is affine and $t \in R$ a generator of the maximal ideal $\mathfrak{m} \subset R$ corresponding to the point $0 \in B$ (so that $B^* = \operatorname{Spec} R[t^{-1}]$), and $I(\mathscr{X}^*) \subset R[t^{-1}][x_1, \ldots, x_n]$ is the ideal of $\mathscr{X}^* \subset \mathbb{A}^n_{B^*}$, then the ideal of the subscheme $\mathscr{X} \subset \mathbb{A}^n_B$ is the intersection

$$I(\mathscr{X}) = I(\mathscr{X}^*) \cap R[x_1, \ldots, x_n].$$

To be even more concrete, if we take $B = \operatorname{Spec} K[t]$, the limiting scheme $X_0 \subset \mathbb{A}^n_K$ is cut out by the limits of polynomials vanishing on the schemes X_t — in other words, if we view the ideals $I(X_t) \subset K[x_1, \ldots, x_n]$ as linear subspaces of the K-vector space $K[x_1, \ldots, x_n]$ and let $V \subset K[x_1, \ldots, x_n]$ be the limiting position of the planes $I(X_t)$, the ideal $I(X_0)$ is generated by V. Thus this definition of limit generalizes the naive notion used in Section II.3.1.

For example, take $B = \operatorname{Spec} K[t]$ and $B^* = B \setminus \{0\} = \operatorname{Spec} K[t, t^{-1}]$, and let X_t be the subscheme of $\mathbb{A}^1_K$ consisting of the two points with coordinates t and $-t$ — that is, take $\mathscr{X}^* = V(x^2 - t^2) \subset \operatorname{Spec} K[t, t^{-1}][x] = \mathbb{A}^1_{B^*}$. Then the closure $\mathscr{X}$ of $\mathscr{X}^*$ in $\mathbb{A}^n_B$ is given again as $\mathscr{X} = V(x^2 - t^2) \subset \operatorname{Spec} K[t][x] = \mathbb{A}^1_B$, and the fiber X_0 of $\mathscr{X}$ over the point $0 \in B$ is simply the double point $X_0 = V(x^2) \subset \mathbb{A}^1_K$.

The notion of the limit of a family of schemes $\mathscr{X}^* \subset \mathbb{A}^n_{B^*}$ depends very much on the embedding in $\mathbb{A}^n_{B^*}$, not just on the abstract family $\mathscr{X}^* \to B^*$. Thus, in the preceding example, the schemes $\mathscr{Y}^* = V(x^2 - 1)$ and $\mathscr{Z}^* = V(x^2 - t^{-2}) \subset \operatorname{Spec} K[t, t^{-1}][x] = \mathbb{A}^1_{B^*}$ are isomorphic as B^*-schemes to the scheme $\mathscr{X}^*$, but the limit of $\mathscr{Y}^*$ is the two reduced points $V(x^2 - 1) \subset \mathbb{A}^1_K$ and that of $\mathscr{Z}^*$ is the empty set.

Examples. The examples of limits we have encountered up to now have all involved limits of zero-dimensional schemes. Here are a couple of examples involving positive-dimensional ones. They are instructive also because they illustrate how embedded points arise naturally in limits of varieties.

The first example is that of three lines through the origin in affine 3-space $\mathbb{A}^3_K$ over a field K. We take the three coordinate axes, rotate one down until it lies in the plane of the other two, and ask what is the limit of this family. Specifically, in $\mathbb{A}^3_K = \operatorname{Spec} K[x, y, z]$ we let $L = V(y, z)$ be the x-axis and $M = V(x, z)$ the y-axis, and let N_t be the line

$$N_t = V(x - y,\, z - tx).$$

For $t \neq 0$ we let $X_t = L \cup M \cup N_t$. The curves $\{X_t\}_{t \neq 0}$ form a family $\mathscr{X}^* \subset \mathbb{A}^3_{B^*}$ over the base $B^* = \operatorname{Spec} K[t, t^{-1}]$, and we ask for the limit X_0 of this family.

This is straightforward to calculate, though the answer may initially be surprising. The ideal of the union of the three coordinate axes is (xy, xz, yz), so the ideal of the scheme X_t for $t \neq 0$ is generated simply by products of linear forms:

$$I(X_t) = (Q_1, Q_2, Q_3),$$

where

$$Q_1 = z(z - tx),$$
$$Q_2 = z(z - ty),$$
$$Q_3 = (z - tx)(z - ty).$$

When we let t go to zero, we see that the ideal of the limiting scheme contains z^2, the common limit of Q_1, Q_2 and Q_3. In addition, for $t \neq 0$ the ideal $I(X_t)$ contains $Q_1 - Q_3 = tyz - t^2xy$ and $Q_2 - Q_3 = txz - t^2xy$. Thus, for $t \neq 0$ the ideal contains

$$\frac{Q_1 - Q_3}{t} = yz - txy \quad \text{and} \quad \frac{Q_2 - Q_3}{t} = xz - txy,$$

and hence the ideal of the limiting scheme X_0 contains xz and yz. Finally, the ideal of X_t contains

$$x\frac{Q_1 - Q_3}{t} - y\frac{Q_2 - Q_3}{t} = txy(x - y),$$

and hence the ideal of the limiting scheme contains $xy(x - y)$. Thus we have

$$I(X_0) \supset \left(xz, yz, z^2, xy(x-y)\right)$$

and we claim that in fact this is an equality. We will establish this in a moment, but before we do we should point out the striking fact about this: *the limit scheme X_0 of the family of schemes $\{X_t = L \cup M \cup N_t\}_{t \neq 0}$ is not simply the union $L \cup M \cup N_0$.* In fact, the ideal of the union is

$$I(L \cup M \cup N_0) = \left(z, xy(x-y)\right),$$

so that

$$I(X_0) = I(L \cup M \cup N_0) \cap (x, y, z)^2.$$

In other words, the limit scheme X_0 has an embedded point at the origin.

In fact, it's not hard to see this directly, which in turn allows us to prove the equality $I(X_0) = (xz, yz, z^2, xy(x-y))$: the schemes X_t all have three-dimensional Zariski tangent space at the origin $(0,0,0) \in \mathbb{A}^3_K$, so X_0 must as well, because if $\mathscr{X} \subset \mathbb{A}^n_B$ is any closed subscheme and $\sigma : B \to \mathscr{X}$ any section of $\mathscr{X} \to B$, the dimension of the Zariski tangent space $T_{\sigma(b)}\mathscr{X}$ is an upper-semicontinuous function of $b \in B$. This in turn implies that

$$I(X_0) \subset I(L \cup M \cup N_0) \cap (x, y, z)^2 = \left(xz, yz, z^2, xy(x-y)\right),$$

from which equality follows.

A similar example is the limit of the scheme consisting of two disjoint lines in $\mathbb{A}_K^3$ as the lines move to meet in a single point. As the following exercise shows, their limit actually has an embedded point at the point of intersection:

Exercise II-25. Let L_t be the line in $\mathbb{A}_K^3$ defined by the ideal $(y, z-t)$ and M be the line defined by (x, z); for $t \neq 0$ let X_t be their union. Show that the limit of X_t as $t \to 0$ is the scheme

$$X_0 = \operatorname{Spec} K[x, y, z]/(z^2, xz, yz, xy).$$

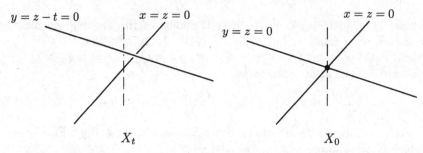

The following exercise shows that the appearance of the embedded point in the limit is no accident:

Exercise II-26. (a) Show that there does not exist a family of lines $L_t \subset \mathbb{A}_K^3$ disjoint from $M = V(x, z)$ parametrized by $B^* = \operatorname{Spec} K[t, t^{-1}]$ such that the limit of $M \cup L_t$ as $t \to 0$ is the reduced scheme

$$X = \operatorname{Spec} K[x, y, z]/(z, xy).$$

(b) Similarly, show that there does not exist a family of lines $L_t \subset \mathbb{A}_K^3$ parametrized by $B^* = \operatorname{Spec} K[t, t^{-1}]$ such that the limit of $M \cup L_t$ as $t \to 0$ is the scheme

$$X = \operatorname{Spec} K[x, y, z]/(z, x^2 y, xy^2).$$

Note that in these two examples, as well as those analyzed earlier, the limit of a union of schemes properly contains the union of their limits. We will return to this in Chapter V.

Taking the limit of a one-parameter family of subschemes of a given scheme is a fundamental operation in algebraic geometry. In the examples occurring throughout the remainder of this book, we will calculate the ideals of such limits by ad-hoc methods, as we've done here. But there is a general algorithm, best carried out by machines, for performing this computation. For example, suppose that the base $B = \operatorname{Spec} K[t]$, and we have an ideal $\mathscr{I} \subset K[t][x_1, \ldots, x_n]$ such that for $\lambda \neq 0$ the scheme $X_\lambda \subset \mathbb{A}_K^n$ is defined by the ideal

$$I_\lambda = (\mathscr{I}, t-\lambda)/(t-\lambda) \subset K[t][x_1, \ldots, x_n]/(t-\lambda) \cong K[x_1, \ldots, x_n].$$

Then we define an ideal $\mathscr{J} \subset K[t][x_1, \ldots, x_n]$ by setting

$$\mathscr{J} = \bigcup_k (\mathscr{I} : t^k);$$

that is, $\mathscr{J}$ is the ideal of polynomials $f(t, x_1, \ldots, x_n)$ such that $t^k f \in \mathscr{I}$ for some k. This can be computed using Gröbner bases; see Eisenbud [1995, Chapter XV].

Flatness. The preceding discussion suffices to describe the notion of a continuously varying family of subschemes of a fixed scheme A (such as affine or projective space) over a nonsingular one-dimensional base: we say such a family $X \subset B \times A$ is continuous if each fiber is the limit of nearby ones. This notion is still too restrictive, however: it does not suffice, for example, if the base B is nonreduced, a case that turns out to be of great utility. To extend the notion to the most general setting, Serre introduced the following notion:

Definition II-27. A module M over a ring R is *flat* if for every monomorphism of R-modules $A \to B$ the induced map $M \otimes_R A \to M \otimes_R B$ is again a monomorphism.

In particular, any free module is flat; and thus if R is a field, every module is flat. It is not hard to show that if R is a Dedekind domain, then M is flat if and only if M is torsion-free. We next make the corresponding geometric definition:

Definition II-28. A family $\pi : X \to B$ of schemes is *flat* if for every point $x \in X$ the local ring $\mathscr{O}_{X,x}$, regarded as an $\mathscr{O}_{B,\pi(x)}$-module via the map $\pi^\#$, is flat.

This notion is general enough to include the families of plane curves of given degree but restrictive enough so that the varieties in a flat family have a lot in common. It is really quite satisfactory, except for the fact that — initially, at least — it does not seem to be a very "geometric" property. In fact, however, it is the most natural — indeed, the only possible — extension of the naive notion of limits introduced above! We will establish this fact, and then go on to consider other properties of the notion of flatness; see Eisenbud [1995; Matsumura [1986; Hartshorne [1977] for good technical discussions.

To begin with, flatness expresses the quality we desire in the cases we have already considered:

Proposition II-29. *Let $B = \operatorname{Spec} R$ be a nonsingular, one-dimensional affine scheme, $0 \in B$ a closed point and $B^* = B \setminus \{0\}$. Let $\mathscr{X} \subset \mathbb{A}_B^n$ be any closed subscheme, and $\pi : \mathscr{X} \to B$ the projection. The following conditions are equivalent:*

(1) *π is flat over 0.*

(2) *The fiber $X_0 = \pi^{-1}(0)$ is the limit of the fibers $X_b = \pi^{-1}(b)$ as $b \to 0$.*

(3) *No irreducible component or embedded component of $\mathscr{X}$ is supported on X_0.*

Proof. We start with the equivalence of (2) and (3). Set $\mathscr{X}^* = \pi^{-1}(B^*) \subset \mathscr{X}$. Since $\mathscr{X} \subset \mathbb{A}^n_B$ is closed, it contains the closure of $\mathscr{X}^*$; so the fiber $X_0 = \pi^{-1}(0)$ contains the limit of the fibers $X_b = \pi^{-1}(b)$ as $b \to 0$, and to say that $X_0 = \lim_{b \to 0} X_b$ is simply to say that we have equality: $\mathscr{X} = \overline{\mathscr{X}^*}$. Conversely, X_0 properly contains the limit of the X_b if and only if $\overline{\mathscr{X}^*} \subsetneq \mathscr{X}$, that is, the expression

$$\mathscr{X} = \overline{\mathscr{X}^*} \cup X_0$$

as a union of closed subschemes is nontrivial. Thus (2) and (3) are equivalent.

To see that (1) is equivalent to (3), simply observe that $\mathscr{O}_{\mathscr{X},x}$, regarded as an $\mathscr{O}_{B,0}$-module, is flat for all $x \in X_0$ if and only if $\mathscr{O}_{\mathscr{X}}(\mathscr{X})$ is torsion-free as an R-module (see Bourbaki [1972, I.2.4, Proposition 3.ii]; because all these rings R are principal ideal domains, this also follows easily from Matsumura [1986, Theorem 7.6 and its converse on p. 50] or Eisenbud [1995, Corollary 6.3]).   □

How general is this interpretation of flatness? To begin with, since the condition of flatness is local in the domain of a morphism $\pi : \mathscr{X} \to B$, the assumption that $\mathscr{X}$ and B are affine is really no restriction at all. If we assume that $\mathscr{X}$ is of *finite type* over B, a mild extra finiteness condition described in Section III.1.1, we can further reduce to the case where $\mathscr{X}$ is a closed subscheme of $\mathbb{A}^n_B$ and π is the restriction to $\mathscr{X}$ of the projection $\mathbb{A}^n_B \to B$. All these are minor hypotheses. The serious restriction in applying the preceding result is that we take B to be nonsingular and one-dimensional. We can, however, broaden this substantially with the following lemma, which characterizes flat families of finite type over a reduced base.

Lemma II-30. *Let K be a field, B a reduced K-scheme, $b \in B$ a closed point and $\mathscr{X} \subset \mathbb{A}^n_B$ a closed subscheme. $\mathscr{X}$ is flat over b if and only if for any nonsingular, one-dimensional K-scheme B', any closed point $0 \in B'$ and any morphism $\varphi : B' \to B$ carrying 0 to b, the fiber X_b is the limit of the fibers $X_{\varphi(b')}$ as b' approaches 0 — that is, for any $\varphi : B' \to B$ carrying 0 to b, the pullback family*

$$\mathscr{X}' = \mathscr{X} \times_B B' \subset \mathbb{A}^n_{B'} \to B'$$

is flat over 0.

Proof. Since $X_{\varphi(b')} = X'_{b'}$, Proposition II-29 asserts the equivalence of the limit condition $X_b = \lim_{b' \to 0} X_{\varphi(b')}$ with the flatness of $\mathscr{X}'$ over B'. That said, one direction is clear: in general, if $\mathscr{X} \to B$ is flat and $B' \to B$ is any morphism, the fiber product $\mathscr{X} \times_B B' \to B'$ is flat; see Matsumura [1986, Chapter 2, Section 3]. For the other direction, which is much harder, see Raynaud and Gruson [1971, Cor. 4.2.10]. $\qquad\square$

When the conditions of Lemma II-30 are met, we will call the fiber X_b the *flat limit* of the nearby fibers of $\mathscr{X}$ over B.

A word of warning: while for B one-dimensional and $0 \in B$ a nonsingular point there exists a unique flat limit of a given family $\mathscr{X} \subset \mathbb{A}^n_{B^*}$ over $B^* = B \setminus \{0\}$, *two-parameter families may not admit any flat limits at all*. Consider for example the degree-2 subschemes of $\mathbb{A}^2_K$ discussed earlier. We take as our base the scheme $B = \operatorname{Spec} K[s,t] = \mathbb{A}^2_K$, with the origin as our special point $0 \in B$. For $(s,t) \neq (0,0) \in B$, we let $X_{s,t} \subset \operatorname{Spec} K[x,y] = \mathbb{A}^2_K$ be the subscheme consisting of the union of the points (x,y) and $(x-s, y-t) \in \mathbb{A}^2_K$. These subschemes form a family $\mathscr{X}^*$ over $B^* = B \setminus \{0\}$, defined by

$$\mathscr{X}^* = V\big(x(x-s),\, x(y-t),\, y(x-s),\, y(y-t)\big) \subset \mathbb{A}^2_{B^*}.$$

But we have seen that the limits of the schemes $X_{s,t}$ as (s,t) approaches the origin along lines of different slope are different double points: all supported at the origin, of course, but with different tangent lines. The fiber X_0 of the closure $\mathscr{X} = \overline{\mathscr{X}^*} \subset \mathbb{A}^2_B$ of $\mathscr{X}^*$ in $\mathbb{A}^2_B$ over the origin $0 \in B$ must therefore contain the union of these double points, that is, it must contain the "fat point" $V(x^2, xy, y^2) \subset \mathbb{A}^2_K$. It follows that the closure must be simply the subscheme

$$\mathscr{X} = V\big(x(x-s),\, x(y-t),\, y(x-s),\, y(y-t)\big) \subset \mathbb{A}^2_B,$$

whose fiber over the origin is $V(x^2, xy, y^2)$. We see in particular that *no closed subscheme of $\mathbb{A}^2_B$ containing $\mathscr{X}^*$ as an open subscheme can be flat over $0 \in B$*.

The morphism $\mathscr{X} \to B$ here is the same as the morphism $X \to Y$ of Exercise I-43(b): the scheme $\mathscr{X}$ is the union of two planes in affine four-space $\mathbb{A}^4_K$ meeting at a point, with the projection $\mathscr{X} \to B$ an isomorphism on each plane. In particular, the failure of the family $\mathscr{X}^* \subset \mathbb{A}^2_{B^*}$ to have a flat limit is very much a function of the embedding in $\mathbb{A}^2_{B^*}$: outside of the origin in $\mathbb{A}^2_B$, we could include $\mathscr{X}^*$ in the disjoint union $\mathbb{A}^2_B \coprod \mathbb{A}^2_B$ of two copies of $\mathbb{A}^2_B$ to obtain a surjective morphism $\nu : \mathscr{X} \to B$ with $\nu^{-1}(B^*) \cong \mathscr{X}^*$ as B^*-schemes. Thus the failure of this family to have a flat limit might be ascribed to our perversity in choosing a bad embedding of $\mathscr{X}^*$ in $\mathbb{A}^2_B$. The following exercise gives another classic example of a nonflat family, and one that moreover has no flat limit, irrespective of the embedding.

Exercise II-31. Consider the cone $B = V(su - t^2) \subset \operatorname{Spec} K[s,t,u] = \mathbb{A}^3_K$. Let $0 = (s,t,u) \in B$ be the origin, and let $B^* = B \setminus \{0\}$ as usual. Set $\mathscr{X} = \operatorname{Spec} K[x,y] = \mathbb{A}^2_K$, and let $\varphi : \mathscr{X} \to B$ be the map dual to the inclusion of rings

$$\varphi^\# : K[s,t,u]/(su - t^2) \longrightarrow K[x,y]$$

sending s to x^2, t to xy, and u to y^2. (Equivalently, B is simply the quotient of $\mathscr{X} = \mathbb{A}^2_K$ by the involution $(x, y) \mapsto (-x, -y)$, and φ the quotient map.) Let $\mathscr{X}^*$ be the inverse image $\varphi^{-1}(B^*) \subset \mathscr{X}$. Show that $\mathscr{X} \to B$ is not flat over 0.

In fact, the family $\mathscr{X}^* \to B^*$ has *no* flat limit, in the sense that there is no scheme $\mathscr{Y}$ and surjection $\nu : \mathscr{Y} \to B$ such that $\nu^{-1}(B^*) \cong \mathscr{X}^*$ as B^*-schemes. Nor is this really pathological: in Section IV.3.2 we'll see examples of naturally occurring families that don't admit flat limits.

Proposition II-29 and Lemma II-30 together give us a geometric interpretation of the flatness of a morphism $\varphi : \mathscr{X} \to B$, at least in case where $\mathscr{X}$ is of finite type over a base B that is reduced and over a field: it says that φ is flat at p if, under any embedding of a neighborhood of $p \in \mathscr{X}$ in affine space $\mathbb{A}^n_B$, the fiber $X_0 = \varphi^{-1}(0)$ over $0 = \varphi(p) \in B$ is (an open subset of) the limit of the fibers X_b as $b \in B$ approaches 0 along any one-parameter family. The wonderful thing about the definition of flatness in general is that it takes this basic notion and extends it, in a very natural way, to arbitrary morphisms! This is particularly remarkable (and useful) in case the base space B is a nonreduced scheme. If, for example, $B = \operatorname{Spec} K[\epsilon]/(\epsilon^2)$, it makes no sense to talk about the "fibers of $\mathscr{X} \to B$ over nearby points"; B has only one point. Nonetheless (as we will see explicitly in Chapter VI) it *does* make sense to talk about families $\mathscr{X} \to B$ of schemes parametrized by B "varying continuously"; flatness exactly captures this property. (Even in case the base B has one-dimensional Zariski tangent space, as in the example $B = \operatorname{Spec} K[\epsilon]/(\epsilon^2)$, we can't just use the criterion that no component of $\mathscr{X}$, irreducible or embedded, is supported on the inverse image of the reduced point B_{red}: for example, the morphism $\operatorname{Spec} K[x, y]/(x^2, xy, y^2) \to \operatorname{Spec} K[\epsilon]/(\epsilon^2)$ dual to the ring homomorphism $\epsilon \mapsto x$ is not flat.)

In general, if $B = \operatorname{Spec} R$ is the spectrum of a local Artinian ring R with maximal ideal $\mathfrak{m}$, $0 = V(\mathfrak{m}) = B_{\mathrm{red}} \subset B$ its unique point, a flat morphism $\varphi : \mathscr{X} \to B$ is called an "infinitesimal deformation" of the fiber $X_0 = \varphi^{-1}(0)$. Such things played an important role in the algebraization of the theory of curves on surfaces — see, for example, Mumford [1966] and the discussion in Section VI.2.3.

To conclude this section, we mention (without proof) two facts about flatness, both of which will reaffirm that flatness is indeed the correct criterion for a family $\mathscr{X} \to B$ of schemes to be "varying continuously". The first is one we mentioned at the outset: we would like families of hypersurfaces to be flat. Explicitly, if

$$f(x_1, \ldots, x_n) = \sum a_I x^I$$

is a polynomial in n variables whose coefficients a_I are regular functions on a scheme B, then the corresponding subscheme $V(f) \subset \mathbb{A}^n_B$ should be flat over B, at least away from the common zero locus $V(\{a_I\}) \subset B$ of the coefficients. In fact, more is true: the same holds for families of complete intersections $\mathscr{X} = V(f_1, \ldots, f_c) \subset \mathbb{A}_B^n$. We state this as follows:

Proposition II-32. *Let R be a local ring with maximal ideal $\mathfrak{m}$, $B = \operatorname{Spec} R$, $0 = [\mathfrak{m}] \in B$ the unique closed point of B and $\kappa = \kappa(0) = R/\mathfrak{m}$ the residue field. Let $f_1, \ldots, f_c \in R[x_1, \ldots, x_n]$ be polynomials with coefficients in R, and*

$$\mathscr{X} = V(f_1, \ldots, f_c) \subset \operatorname{Spec} R[x_1, \ldots, x_n] = \mathbb{A}_B^n.$$

If the fiber $X_0 = \pi^{-1}(0)$ of the projection $\pi : \mathscr{X} \to B$ over 0 has codimension c in $\mathbb{A}_\kappa^n$, then $\mathscr{X} \to B$ is flat.

More generally, we have the following criterion for flatness, which is extremely useful in practice.

Exercise II-33. (a) Prove that a module M over the ring $R = K[t]_{(t)}$ is flat if and only if t is a nonzerodivisor on M, that is, if and only if M is torsion-free.

(b) Let $A = R[x_1, \ldots, x_n]$ be a polynomial ring over $R = K[t]_{(t)}$, and let M be an A-module with free presentation

$$F_1 \xrightarrow{\varphi} F_0 \longrightarrow M \longrightarrow 0.$$

Consider the module $\bar{M} := M/M_t$ over the factor ring $\bar{A} := A/tA$, and let

$$\bar{F}_1 \xrightarrow{\bar{\varphi}} \bar{F}_0 \longrightarrow \bar{M} \longrightarrow 0$$

be the corresponding presentation. Show that M is flat over R if and only if every second syzygy of $\bar{M}$ over $\bar{A}$ can be lifted to a second syzygy over A in the sense that every element of the kernel of $\bar{\varphi}$ comes from an element of the kernel of φ. (Something similar is true for any local base ring R with maximal ideal $\mathfrak{m}$ if M is finitely generated over A; this is a form of the "local criterion of flatness" — see, for example, Eisenbud [1995, Section 6.4] or Matsumura [1986, p. 174].

A second thing that makes flatness a good notion is the *generic flatness theorem*, due to Grothendieck (see for example Eisenbud [1995, Section 14.2]. This says that if one has any reasonable family of schemes $X \to B$ over a reduced base, then there is an open dense subset U of B such that the restricted family $\pi^{-1}U \to U$ is flat (here "reasonable" includes, for example, any family of subschemes of a fixed affine or projective space). In some sense this vindicates our choice of flatness as the analogue of the notion of bundle in topology: it is analogous to the observation that if $f : M \to N$ is a differentiable map of compact C^∞ manifolds, then there is a dense collection of open subsets U of the target space N such that the restriction of f to each $f^{-1}(U)$ is a fiber bundle. In any event, the generic flatness theorem certainly assures us that flat families are ubiquitous in algebraic geometry.

This concludes our initial discussion of flatness. We will see other geometric interpretations of flatness when we discuss families of projective schemes in Chapter IV.

II.3.5   Multiple Lines

We now consider a nonreduced affine scheme X supported on a line and not having embedded components. We will assume that the multiplicity of the line (in the sense of the primary decomposition) is 2, and we will analyze the possibilities.

It is very easy to write down a first example: the scheme

$$X = \operatorname{Spec} K[x,y]/(y^2) \subset \mathbb{A}_K^2$$

obviously has the desired properties. It's pretty clear that there are no more examples supported on the line $y = 0$ in $\mathbb{A}_K^2$, but we can construct many in $\mathbb{A}_K^3$. A subscheme X of the sort we want will meet a general plane in $\mathbb{A}_K^3$ passing through a point of the reduced line in a double point contained in that plane. We already know that any double point may be thought of as a point plus a tangent vector at that point, and this suggests that we obtain X by choosing a normal direction at each point of the line. For example, take $L := X_{\mathrm{red}}$ to be the line $x = y = 0$, with coordinate z. Now, choose a pair of polynomials p and q in z without common zeros, and at each point $(0,0,z_0) \in L$ take the normal direction to be the one with slope $p(z_0)/q(z_0)$ in the normal plane $z = z_0$. It is easy to see that the union over all z of the double points in the given directions will be contained in the scheme $X_{p,q}$ defined by taking

$$I_{p,q} = (x^2,\, xy,\, y^2,\, p(z)x - q(z)y)$$

and

$$X_{p,q} = \operatorname{Spec} K[x,y,z]/I_{p,q}$$

The simplest nonplanar example would be one where the chosen normal directions twist just once around L—for example, the one given by the ideal

$$I_\Gamma = (x^2,\, xy,\, y^2,\, zy - x).$$

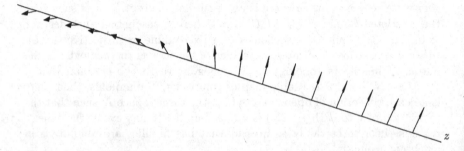

Exercise II-34. If p, q are relatively prime polynomials, then the ideal

$$(x, y)/I_{p,q}$$

in the ring

$$K[x, y, z]/I_{p,q}$$

is torsion-free of rank 1 as a $K[z]$-module; thus $X_{p,q}$ is primary, with $(X_{p,q})_{red}$ the line $\operatorname{Spec} K[x, y, z]/(x, y)$, and $X_{p,q}$ has multiplicity 2.

At first sight it looks as though these examples will possess many interesting invariants and thus, in particular, be distinct, but this is not so: we can "untwist" any of the schemes $X_{p,q}$ by an automorphism of $\mathbb{A}^3_K$ to give an isomorphism of it with the planar double line $\operatorname{Spec} K[y, z]/(y^2)$. To do this, note that since p and q have no common zeros, we may write $1 = aq + bp$ for some polynomials $a, b \in K[z]$; thus the matrix

$$\begin{pmatrix} a & b \\ p & -q \end{pmatrix}$$

has unit determinant, so the map $\mathbb{A}^3_K \to \mathbb{A}^3_K$ given by

$$(x, y, z) \mapsto (x', y', z), \quad \text{with } x' := p(z)x - q(z)y, \ y' := a(z)x + b(z)y,$$

is invertible. Again because the matrix is invertible, we have

$$(x, y) = (x', y') \quad \text{and} \quad (x^2, xy, y^2) = (x'^2, x'y', y'^2)$$

so the ideal of $X_{p,q}$ is $(x, x^2, xy, y^2) = (x, y^2)$, as required.

More generally, it turns out that there is up to isomorphism only one affine double line, in the following sense:

Exercise II-35. Prove that if A is a Noetherian K-algebra such that $X = \operatorname{Spec} A$ has no embedded components, has multiplicity 2, and satisfies $X_{red} \cong \mathbb{A}^1_K$, then X is isomorphic to $\operatorname{Spec} K[x, y]/(y^2)$.

We will see in the next chapter that this situation contrasts with the one in projective space: there are many nonisomorphic projective double lines.

II.4 Arithmetic Schemes

Our last collection of examples will be spectra of rings that are finitely generated and reduced but that do not contain any field at all. In general, the spectra of rings finitely generated over $\mathbb{Z}$ are called *arithmetic schemes*; they arise primarily in the context of number theory, although by no means all schemes of number-theoretic interest are of this type. In these examples we will see some hint of the amazing unification that schemes allow between the arithmetic and the geometric points of view.

II.4.1 Spec $\mathbb{Z}$

We start with the most obvious example, the scheme Spec $\mathbb{Z}$ itself. The prime ideals of $\mathbb{Z}$ are, of course, the ideals (p), for $p \in \mathbb{Z}$ a prime number, and the ideal (0); the former correspond to closed points of Spec $\mathbb{Z}$, with residue field $\mathbb{F}_p$, while the latter is a "generic" point, whose closure is all of Spec $\mathbb{Z}$ and whose residue field is $\mathbb{Q}$. The picture is this:

This bears a formal resemblance to an affine line $\mathbb{A}^1_K$ over a field; indeed, this similarity is just the beginning of a long sequence of analogies, and it is well to bear it in mind while looking at the following examples. However, the analogy also has its limits: while Spec $\mathbb{Z}$ behaves much like $\mathbb{A}^1_K$, for example, it is not an open dense subscheme of any scheme analogous to $\mathbb{P}^1_K$.

II.4.2 Spec *of the Ring of Integers in a Number Field*

Secondly, consider a scheme of the form Spec A, where $A \subset K$ is the ring of integers in a number field K; we will analyze the example $K = \mathbb{Q}[\sqrt{3}]$ and $A = \mathbb{Z}[\sqrt{3}]$. As in the case of Spec $\mathbb{Z}$, there are just two types of points: closed points corresponding to nonzero prime ideals in A, having finite residue fields, and a generic point corresponding to (0) with residue field K. What makes this example interesting is the map Spec $A \to$ Spec $\mathbb{Z}$ induced by the inclusion of $\mathbb{Z}$ in A. Consider, for example, the fiber over a point $[(p)] \in$ Spec $\mathbb{Z}$. This is just the set of primes in A containing the ideal $pA \subset A$, and it may behave in any one of three ways (a good basic reference for the unexplained material here is Serre [1979]):

(1) If p divides the discriminant 12 of K over $\mathbb{Q}$—that is, for $p = 2$ or 3— the ideal (p) is the square of an ideal in A: we have

$$2A = (1 + \sqrt{3})^2$$

and, of course,

$$3A = (\sqrt{3})^2.$$

The residue fields at the points $(1 + \sqrt{3})$ and $(\sqrt{3}) \in$ Spec A are the fields $\mathbb{F}_2$ and $\mathbb{F}_3$, respectively.

(2) Otherwise, if 3 is a square mod p, the prime (p) will factor into a product of distinct primes: for example

$$11A = (4 + 3\sqrt{3})(4 - 3\sqrt{3})$$

and

$$13A = (4 + \sqrt{3})(4 - \sqrt{3}).$$

The residue fields at these points will again be the prime fields, in this case $\mathbb{F}_{11}$ and $\mathbb{F}_{13}$, respectively.

(3) Finally, if $p > 3$ and 3 is not a square mod p — for example, when $p = 5$ or 7 — the ideal pA is still prime and corresponds to a single point in Spec A. In these cases, the residue field is the quadratic extension of $\mathbb{F}_p$ — for instance, $\mathbb{F}_{25}$ and $\mathbb{F}_{49}$ in the two examples.

In general, as in this example, if K is a quadratic number field, and A is the ring of algebraic integers in K, then the inclusion $\mathbb{Z} \subset A$ induces a map of schemes $\psi : \operatorname{Spec} A \to \operatorname{Spec} \mathbb{Z}$ whose fiber over each closed point $(p) \in \operatorname{Spec} \mathbb{Z}$ is one of the following:

(1) A single, nonreduced point, with coordinate ring isomorphic to $A/\mathfrak{p}^2$, whose underlying reduced point $\mathfrak{p}$ has residue field $\mathbb{F}_p$, if p *ramifies* in A — that is, if pA is the square of a prime ideal $\mathfrak{p}$ of A.

(2) The disjoint union of two reduced points, $\mathfrak{p}$ and $\mathfrak{p}'$, with residue fields $A/\mathfrak{p} = A/\mathfrak{p}' = \mathbb{F}_p$, if pA is a product of two distinct prime ideals of A.

(3) A single reduced point $\mathfrak{p}$, with residue field $A/\mathfrak{p}$ of degree 2 over $\mathbb{F}_p$, if p remains prime in A.

In every case the coordinate ring of the fiber has dimension 2 as an $\mathbb{F}_p$-algebra. That is because A is a free $\mathbb{Z}$-module of rank 2. Of interest here is the analogy between the map $\operatorname{Spec} A \to \operatorname{Spec} \mathbb{Z}$ and a branched cover of Riemann surfaces (or, more generally, of one-dimensional schemes over an algebraically closed field such as $\mathbb{C}$). Essentially, we may think of Spec A as a two-sheeted cover of Spec $\mathbb{Z}$, with branching over the "ramified" primes, just as, for example, Spec $\mathbb{C}[z]$ is a double cover of Spec $\mathbb{C}[z^2]$ branched over the origin. The one apparent difference is that over some points $(p) \in$ Spec $\mathbb{Z}$ other than ramification points we may have, instead of two distinct points with multiplicity 1, one point with multiplicity 1 but with a residue field that is a quadratic extension of the residue field $\mathbb{F}_p$ at (p). These are denoted by uniform gray dots in the picture:

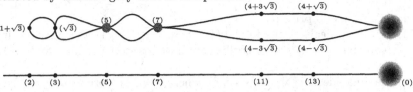

A more inclusive analogy would be with a finite map between one-dimensional schemes over a non-algebraically closed field. Consider, for example, the map

$$\operatorname{Spec} \mathbb{R}[x][y]/(y^2 - x) \to \mathbb{A}^1_\mathbb{R} = \operatorname{Spec} \mathbb{R}[x]$$

Looking just at points of $\mathbb{A}^1_\mathbb{R} = \operatorname{Spec} \mathbb{R}[x]$ with residue field $\mathbb{R}$ — that is, points of the form $(x - \lambda)$ with λ real — we have ramification over the point

(x), and for $\lambda \neq 0$ the inverse image of $(x - \lambda)$ is either two distinct points with residue field $\mathbb{R}$ (if $\lambda > 0$) or one point with residue field $\mathbb{C}$ (if $\lambda < 0$).

We may continue this analogy a little further by looking at schemes of the form $\operatorname{Spec} B$, where $B \subset A \subset K$ is an *order* in a number field — that is, a subring of the ring of integers in K having quotient field K. For example, let $A = \mathbb{Z}[\sqrt{3}]$ and consider the ring $B = \mathbb{Z}[11\sqrt{3}]$ and the associated scheme $\operatorname{Spec} B$. The map $\operatorname{Spec} A \to \operatorname{Spec} \mathbb{Z}$ described above factors through $\operatorname{Spec} B$, and indeed the map $\operatorname{Spec} A \to \operatorname{Spec} B$ is an isomorphism except that the two points $(4 + 3\sqrt{3})$ and $(4 - 3\sqrt{3}) \in \operatorname{Spec} A$ map to the same point $(11, 11\sqrt{3}) \in \operatorname{Spec} B$. We may thus picture $\operatorname{Spec} B$ as a sort of "nodal curve" — that is, the double cover $\operatorname{Spec} A$ of $\operatorname{Spec} \mathbb{Z}$ with two points identified.

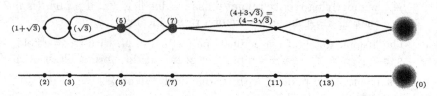

Alternatively, consider the case $A = \mathbb{Z}[\sqrt{3}]$ and $B = \mathbb{Z}[2\sqrt{3}]$. Here the map $\operatorname{Spec} A \to \operatorname{Spec} B$ is one-to-one but not an isomorphism at the point $[(1 + \sqrt{3})]$ which goes to $[(2, 2\sqrt{3})]$.

Exercise II-36. Show that the point $p = [(2, 2\sqrt{3})]$ is a "cusp" of the scheme $\operatorname{Spec} \mathbb{Z}[2\sqrt{3}]$ in the sense that it is a singular point and the desingularization $\operatorname{Spec} A \to \operatorname{Spec} B$ has fiber over p consisting of a double point.

II.4.3   Affine Spaces over Spec $\mathbb{Z}$

Our next example is of a two-dimensional scheme, $\operatorname{Spec} \mathbb{Z}[x]$; this is also denoted $\mathbb{A}^1_{\mathbb{Z}}$. The prime ideals in $\mathbb{Z}[x]$ are

(i) (0);

(ii) (p), for $p \in \mathbb{Z}$ prime;

(iii) principal ideals of the form (f), where $f \in \mathbb{Z}[x]$ is a polynomial irreducible over $\mathbb{Q}$ whose coefficients have greatest common divisor 1; and

(iv) maximal ideals of the form (p, f), where $p \in \mathbb{Z}$ is a prime and $f \in \mathbb{Z}[x]$ a monic polynomial whose reduction mod p is irreducible.

Exercise II-37. Prove this.

Of these, only the last are closed points; the first, of course, has closure all of $\mathbb{A}^1_{\mathbb{Z}}$, while the second and third types have closures we will describe below.

Probably the best way to picture $\mathbb{A}^1_{\mathbb{Z}}$ is via the map $\mathbb{A}^1_{\mathbb{Z}} \to \operatorname{Spec} \mathbb{Z}$ (again a flat map!). Under this map, points of type (ii) and (iv) above go to the corresponding points $(p) \in \operatorname{Spec}\mathbb{Z}$, while the points of types (i) and (iii) go to the generic point $(0) \in \operatorname{Spec}\mathbb{Z}$. Indeed, the fiber of this map over the point (p) is isomorphic to $\mathbb{A}^1_{\mathbb{F}_p} = \operatorname{Spec}\mathbb{F}_p[x]$, the point $(p, f) \in \mathbb{A}^1_{\mathbb{Z}}$ corresponding to the point in $\mathbb{A}^1_{\mathbb{F}_p}$ given by the set of roots of the polynomial f in the algebraic closure $\bar{\mathbb{F}}_p$ (recall that points of $\mathbb{A}^1_{\mathbb{F}_p}$ correspond to orbits of the action of the Galois group $\operatorname{Gal}(\bar{\mathbb{F}}_p/\mathbb{F}_p)$ on $\mathbb{F}_p$). Similarly, the fiber over the generic point $(0) \in \operatorname{Spec}\mathbb{Z}$ is the scheme $\mathbb{A}^1_{\mathbb{Q}} = \operatorname{Spec}\mathbb{Q}[x]$, with $(f) \in \mathbb{A}^1_{\mathbb{Z}}$ meeting $\mathbb{A}^1_{\mathbb{Q}}$ in the point corresponding to the set of roots of f in $\bar{\mathbb{Q}}$. The picture thus is as follows:

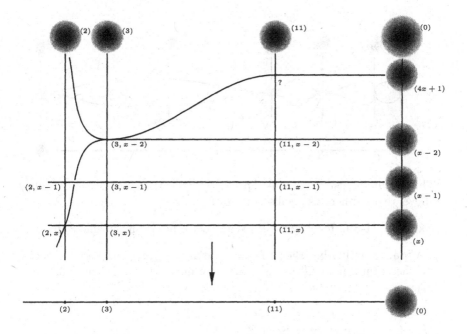

The closure of the point $(p) \in \mathbb{A}^1_{\mathbb{Z}}$ is the fiber $\mathbb{A}^1_{\mathbb{F}_p}$ over the point $(p) \in \operatorname{Spec}\mathbb{Z}$. The closures of the other nonclosed points — those of type (iii) above — are more interesting. These will consist of the point (f) itself in the fiber $\mathbb{A}^1_{\mathbb{Q}}$ over (0) together with all the points $(p, g) \in \mathbb{A}^1_{\mathbb{Z}}$, where g is a factor of f over $\bar{\mathbb{F}}_p$ — that is, in each fiber $\mathbb{A}^1_{\mathbb{F}_p}$ of $\mathbb{A}^1_{\mathbb{Z}}$, the union of the points of $\mathbb{A}^1_{\mathbb{F}_p}$ corresponding to roots of f mod p.

Exercise II-38. What is the point marked with a ? in the picture above? Why are the closures of the points $(4x+1)$ and $(x-2)$ indicated by curves meeting tangentially at the point $(3, x-2)$, while they are both transverse to the closure of (3)? (See the discussion leading up to Exercise II-44 for one answer.) Why is the closure of the point $(4x+1)$ drawn as having a vertical asymptote over the point $(2) \in \operatorname{Spec}\mathbb{Z}$?

For another example, consider the ideal generated by a simple linear polynomial, such as $(5x - 49)$. To continue the analogy between $\operatorname{Spec}\mathbb{Z}$ and the affine line over a field, we can think of the closure of this point as the graph of the function $49/5$ on $\operatorname{Spec}\mathbb{Z}$; this is a function with a simple pole at the point (5) and a double zero at (7). (This curve is tangent to the closed subscheme (x) in $\mathbb{A}^1_{\mathbb{Z}}$, as evidenced by the fact that the intersection of (x) with the subscheme $(5x - 49)$ is not just the point $(7, x)$ but a nonreduced point supported at this point.)

The closure of the point $(x^2 - 3)$ is pictured below in a slightly different style:

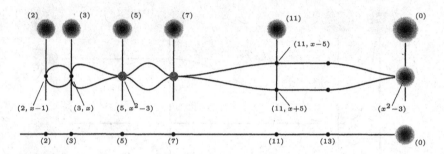

This closure is just the scheme $\operatorname{Spec}\mathbb{Z}[x]/(x^2 - 3) = \operatorname{Spec}\mathbb{Z}[\sqrt{3}]$ described above, realized here as a subscheme of $\mathbb{A}^1_{\mathbb{Z}}$.

Exercise II-39. Identify the three unlabeled points in the above diagram.

More generally, the scheme $\mathbb{A}^n_{\mathbb{Z}} = \operatorname{Spec}\mathbb{Z}[x_1, \ldots, x_n]$ can best be viewed via the natural map $\mathbb{A}^n_{\mathbb{Z}} \to \operatorname{Spec}\mathbb{Z}$, whose fibers are the schemes $\mathbb{A}^n_{\mathbb{F}_p}$ and $\mathbb{A}^n_{\mathbb{Q}}$.

II.4.4   A Conic over $\operatorname{Spec}\mathbb{Z}$

Our next example gives a hint of the depth of the unification of geometry and arithmetic achieved in scheme theory. We consider the scheme

$$\operatorname{Spec}\mathbb{Z}[x, y]/(x^2 - y^2 - 5)$$

and its morphism to $\operatorname{Spec}\mathbb{Z}$.

To begin with, the fiber of this scheme over the generic point $[(0)] \in \operatorname{Spec}\mathbb{Z}$ is the scheme $X = \operatorname{Spec}\mathbb{Q}[x, y]/(x^2 - y^2 - 5)$, which we have already described: its points are the orbits, under the action of the Galois group $G = \operatorname{Gal}(\bar{\mathbb{Q}}/\mathbb{Q})$, of the set of pairs (x, y) of elements of $\bar{\mathbb{Q}}$ satisfying $x^2 - y^2 = 5$. The fiber over (p) is similarly the subscheme of the affine plane $\mathbb{A}^2_{\mathbb{F}_p}$ over $\mathbb{F}_p$ defined by the equation $x^2 - y^2 = 5$—that is, whose points are the orbits, under the action of the Galois group $G = \operatorname{Gal}(\bar{\mathbb{F}}_p/\mathbb{F}_p)$, of the set of pairs (x, y) of elements of $\bar{\mathbb{F}}_p$ satisfying $x^2 - y^2 = 5$.

The fibers of this scheme over all primes other than 2 and 5 are nonsingular conics, as is the fiber over the generic point.

Exercise II-40. Are there plane conics over $\operatorname{Spec}\mathbb{Z}$ that are reducible but nonsingular? Classify them.

The fibers over (2) and (5) are singular, however: modulo 2, we have

$$x^2 - y^2 - 5 = (x + y + 1)^2$$

and modulo 5 we can write

$$x^2 - y^2 - 5 = (x + y)(x - y)$$

Thus the fiber over (2) is a double line, while the fiber over (5) is a union of two lines (so that in particular there are two nonclosed points mapping to the point (5), while there is only one such point mapping to each of the other points $(p) \in \operatorname{Spec}\mathbb{Z}$).

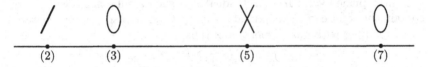

$$\text{(2)} \qquad \text{(3)} \qquad \text{(5)} \qquad \text{(7)}$$

Exercise II-41. (This assumes some knowledge of projective geometry.) The fiber of X over a point $(p) \in \operatorname{Spec}\mathbb{Z}$ such that $p \equiv 1 \bmod (4)$, $p \neq 5$, is really a hyperbola — that is, it meets the "line at infinity" in the fiber $\mathbb{A}^2_{\mathbb{F}_p}$ in two points with residue field $\mathbb{F}_p$ and is isomorphic to $\mathbb{A}^1_{\mathbb{F}_p} - \{0\}$. Thus, for example, the fiber over (p) is the curve $x^2 + y^2 - 5 = 0$; its closure in the projective plane over $\mathbb{F}_p$ has equation $X^2 + Y^2 - 5Z^2 = 0$, and so meets the line $Z = 0$ at ∞ in the two points $[1, \alpha, 0]$ where $\alpha^2 = p - 1 \bmod (p)$. Show that, by constrast, if $p \equiv 3 \bmod (4)$, the fiber is an ellipse; that is, it meets the line at ∞ in one point with residue field $\mathbb{F}_{p^2}$.

The preceding picture is very much in keeping with the geometric analogy: a surface fibered over a curve — for example, the surface $V(x^2 - y^2 - z) \subset \operatorname{Spec}K[x, y, z]$ fibered over the z-line $\operatorname{Spec}K[z]$ — will have a finite number of singular fibers, as in the classic picture shown on the right.

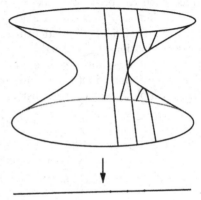

II.4.5   Double Points in $\mathbb{A}^1_\mathbb{Z}$

Next, we consider some double points over $\mathbb{Z}$. Again, let

$$X = \mathbb{A}^1_\mathbb{Z} = \operatorname{Spec}\mathbb{Z}[x].$$

If $Z \subset X$ is a closed subscheme supported at only one point, corresponding to a prime (p, f), say, we will wish to speak of the degree of Z just as we did in the case of finite subschemes over a field. In the case of schemes over a field, we defined the degree to be the dimension of $\mathscr{O}_Z(Z)$ as a vector space over K. But in the current case $\mathscr{O}_Z(Z)$ might contain no field at all — it might be $\mathbb{Z}/(p^2)$, for example. More confusing still, its residue field might not be $\mathbb{Z}/(p)$. In the case at hand the cheapest way out of this dilemma is to note that the cardinality $\#\mathscr{O}_Z(Z)$ is always of the form p^d and take the degree to be d — this is obviously the vector space dimension if $\mathscr{O}_Z(Z)$ happens to be a $\mathbb{Z}/(p)$ vector space. (A more sophisticated approach is to define the degree of a reduced closed point first as the vector space dimension over $\mathbb{Z}/(p)$ and then define the degree of Z by multiplying the degree of the reduced point by the multiplicity of Z at this point.)

Consider for example the subschemes of degree 2 supported at the point $(7, x)$. These behave in a manner analogous to subschemes of degree 2 in the affine plane over a field. The ideal I of such a subscheme will always contain the square of the maximal ideal $\mathfrak{p} = (7, x)$ and so will be generated by $\mathfrak{p}^2$ together with one element of $\mathfrak{p}$: thus,

$$I = I_{\alpha,\beta} = (49, 7x, x^2, \alpha 7 + \beta x)$$

for some $\alpha, \beta \in \mathbb{Z}$ not both divisible by 7. It will depend only on the congruence classes of α and β in $\mathbb{Z}/(7)$; and multiplying the pair (α, β) simultaneously by a unit in $\mathbb{Z}/(7)$ will not change I either. Thus for each point $[\alpha, \beta]$ of the projective line over the field of seven elements we get a double point supported at $\mathfrak{p}$.

Exercise II-42. Show that this correspondence is bijective.

The set of subschemes of degree 2 supported at $(7, x)$ may thus be identified with the projective line $\mathbb{P}^1_K$ over the field $K = \mathbb{F}_7$, much as the set of subschemes of $\mathbb{A}^2_K$ over a field K may be identified with the projective line over that field. (The identification in either case is actually with the projectivization of the *Zariski tangent space* to the ambient space at the point.) There is, however, one difference: whereas all subschemes of $\mathbb{A}^2_K$ of degree 2 supported at a point are isomorphic, the subschemes $Z_{\alpha,\beta}$ defined by $I_{\alpha,\beta}$ look different, even abstractly. We have

$$Z_{\alpha,\beta} = \operatorname{Spec}\mathbb{Z}/(49) \qquad \text{if} \quad \beta \neq 0$$

but

$$Z_{1,0} = \operatorname{Spec}(\mathbb{Z}/(7))[x]/(x^2)$$

which are not isomorphic.

Exercise II-43. Classify (a) the subschemes of degree 3 supported at the point $(7, x) \in \mathbb{A}^1_{\mathbb{Z}}$, and (b) the subschemes of degree 4 supported at the point $(2, x^2 + x + 1)$.

Exercise II-44. Referring to the diagram on page 85, use the preceding discussion to justify the fact that the curves $(4x + 1)$ and $(x - 2)$ are drawn tangent to one another, while the curves $(4x + 1)$ and (11) are drawn transverse.

Finally, here is an example of a flat family over $\operatorname{Spec}\mathbb{Z}$. Recall that in the preceding section there was a discussion of the family of pairs of lines $M \cup L_t$, where M is the line $x = z = 0$ and L_t is the line $y = z - t = 0$. The key observation there was that the flat limit of the schemes $M \cup L_t$ as t approached zero was not the scheme $M \cup L_0$ but, rather, that scheme with an embedded point at the origin.

Here is the analogous phenomenon in a family parametrized by $\operatorname{Spec}\mathbb{Z}$. Let $U = \operatorname{Spec}\mathbb{Z}[7^{-1}] = \operatorname{Spec}\mathbb{Z} - \{(7)\}$ be the complement of the point $(7) \in \operatorname{Spec}\mathbb{Z}$, and let

$$W = \mathbb{A}^3_U := \operatorname{Spec}\mathbb{Z}[7^{-1}, x, y, z] \subset \mathbb{A}^3_{\mathbb{Z}}$$

be the corresponding open subscheme of $\mathbb{A}^3_{\mathbb{Z}}$. Let $\mathscr{N}$ and $\mathscr{L}$ be the closed subschemes of $\mathbb{A}^3_{\mathbb{Z}}$ given by the ideals (x, z) and $(y, z - 7)$, respectively, and let $\mathscr{N}^* = \mathscr{N} \cap W$ and $\mathscr{L}^* = \mathscr{L} \cap W$. Let $\mathscr{X}^*$ be the union of $\mathscr{N}^*$ and $\mathscr{L}^*$ and let $\mathscr{X} \subset \mathbb{A}^3_{\mathbb{Z}}$ be the closure of $\mathscr{X}^*$ in $\mathbb{A}^3_{\mathbb{Z}}$. We may then think of $\mathscr{X}^*$ as a family of pairs of lines parametrized by U; and the fiber X_7 of $\mathscr{X}$ over $(7) \in \operatorname{Spec}\mathbb{Z}$ is the flat limit of this family "as 7 goes to 0". The fiber X_7 is, as we expect, supported on the union of the fibers $(x = z = 0)$ and $(y = z = 0)$ of $\mathscr{N}$ and $\mathscr{L}$ over (7); but the scheme X_7 is not reduced: exactly as in the picture in the preceding section, it has an embedded point at the origin.

Exercise II-45. Verify the flatness of $\mathscr{X}$ over $\operatorname{Spec}\mathbb{Z}$ and the description of X_7. Can you find analogues over $\operatorname{Spec}\mathbb{Z}$ of the other flat families discussed in the preceding section?

III
Projective Schemes

Once we have understood affine schemes, the theory of projective schemes does not really contain so much that is still novel: for the most part it differs from the classical theory of projective varieties in ways that are completely analogous to the difference between affine schemes and affine varieties.

We start by introducing two finiteness conditions, *finite* and *of finite type*. We then define and discuss *separated* and *proper* morphisms, which correspond to the attributes of Hausdorffness and compactness in most of geometry. It is partly because projective varieties and schemes have these properties that they are fundamental objects in classical algebraic geometry and in the theory of schemes.

The next part of the chapter is devoted to the introduction of projective schemes and some examples. Just as in the case of affine schemes, two approaches to projective schemes are possible: one can define projective space and then take subschemes, or one can define all projective schemes on an equal footing, starting with graded algebras. As we did in the affine case, we adopt the second possibility.

After introducing the basic definitions of projective schemes and subschemes, we describe morphisms of projective schemes, a topic that (as in the category of varieties) is more subtle than its affine counterpart. We conclude the section with some examples of projective schemes, most notably the Grassmannian.

The final section of the chapter is devoted to three invariants of projective schemes embedded in projective space that were introduced by David Hilbert: the Hilbert polynomial, Hilbert function, and free resolution. Using these, we can sometimes distinguish among similar schemes, such as the projective double lines, and we can also shed some new light on the phenomenon of flatness. Among the invariants of a projective scheme that can be defined in terms of its Hilbert polynomial is its degree; and in this connection we discuss the famous Bézout theorem.

III.1   Attributes of Morphisms

III.1.1   Finiteness Conditions

There are two finiteness conditions that play a major role in most nontrivial results about schemes. They have similar names but very different character. The first, finite type, is a straightforward condition satisfied by almost any morphism arising in a geometric contexts; it is invoked usually just to preclude infinite-dimensional fibers, or "non-geometric" schemes such as spectra of local rings. The second condition, finiteness, is by contrast a very stringent condition: it says that a morphism is proper and that all its fibers are finite (in particular, zero-dimensional).

First, we say that a morphism $\varphi : X \to Y$ of schemes is *of finite type* if for every point $y \in Y$ there is an open affine neighborhood $V = \operatorname{Spec} B \subset Y$ of y and a finite covering

$$\varphi^{-1}(V) = \bigcup_{i=1}^{n} U_i$$

of its inverse image by affine open sets $U_i \cong \operatorname{Spec} A_i$, such that the map

$$\varphi_V^{\#} : B = \mathscr{O}_Y(V) \to \mathscr{O}_X(\varphi^{-1}V) \to \mathscr{O}_X(U_i) = A_i$$

makes each A_i into a finitely generated algebra over B. Thus, for example, any subscheme X of $\mathbb{A}_K^n$ or $\mathbb{P}_K^n$ is of finite type over K (meaning the structure morphism $X \to \operatorname{Spec} K$ is of finite type), while the spectrum of a positive-dimensional local K-algebra is not.

A morphism $\varphi : X \to Y$ is called *finite* if for every point $y \in Y$ there is an open affine neighborhood $V = \operatorname{Spec} B \subset Y$ of y such that the inverse image $\varphi^{-1}(V) = \operatorname{Spec} A$ is itself affine, and if, via the pullback map

$$\varphi_V^{\#} : B = \mathscr{O}_Y(V) \to \mathscr{O}_X(\varphi^{-1}V) = A,$$

A is a finitely generated B-module. This is a far more restrictive hypothesis than being of finite type; for one thing, it immediately implies that the fibers of φ are finite, and it implies that the map $|\varphi| : |X| \to |Y|$ of underlying topological spaces is closed, that is, the image of a closed subset of X is closed in Y. Thus, for example, if $Y = \operatorname{Spec} B$ and $f \in B[x]$ is a polynomial, the morphism $\operatorname{Spec}(B[x]/(f)) \to Y$ is finite if the leading coefficient of f is a unit, but not otherwise. For all this see Eisenbud [1995, Chapter 4 and Section 9.1].

III.1.2 Properness and Separation

Many techniques of geometry yield the most complete results when applied to compact Hausdorff spaces. Although affine schemes are quasicompact in the Zariski topology, they do not share the good properties of compact spaces in other theories because the Zariski topology is not Hausdorff. For example, the image of a regular map of affine schemes $\varphi : X \to Y$ need not be closed, even though X is quasicompact.

The fact that the Zariski topology is not Hausdorff has another unpleasant consequence. Recall that in the general definition of a manifold, one starts with a topological space that is Hausdorff and admits a covering by charts of the standard form (balls in Euclidean space, say). The fact that the balls themselves are Hausdorff is not enough by itself to guarantee that the total space is. This is why the line with the doubled origin described in Exercise I-44 and shown again here

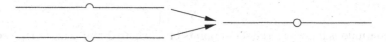

is not a manifold. However, when we work with schemes (or, for that matter, with varieties) glued together from affine schemes, we cannot afford to specify that the total space is Hausdorff because even the local pieces are not. This has the result that given two maps of schemes $\varphi, \psi : X \to Y$, the set where φ and ψ are equal may not be closed. This is illustrated in the following exercise, which is a typical case.

Exercise III-1. (a) Let Y be the line with doubled origin over a field K, defined in Exercise I-44, and let $\varphi_1, \varphi_2 : \mathbb{A}^1_K \to Y$ be the two obvious inclusions. Show that the locus where φ_1 and φ_2 agree (simply as continuous maps of topological spaces) is not closed.

(b) Now let $X = Y \times_K Y$ and let φ and ψ be the two projection maps from X to Y. Show that the set of points at which φ and ψ agree is not closed (note that this is just the diagonal, defined below). Show that the same is true for the set of closed points at which φ and ψ agree, so this is not a pathology special to schemes but occurs already in the category of varieties.

Such a pathology cannot happen, however, if X is an affine scheme; nor, it turns out, can it happen when X is a projective scheme. The desirable property that these schemes have, which is one of the most important consequences of the Hausdorff property for manifolds, is expressed by saying that X is *separated* as a scheme over K. In general, given any map $\alpha : X \to S$ of schemes, we define the *diagonal* subscheme $\Delta \subset X \times_S X$ to be the subscheme defined locally on $X \times_S X$ for each affine open

$$X \supset \operatorname{Spec} A \xrightarrow{\ \alpha|\operatorname{Spec} A\ } \operatorname{Spec} B \subset S$$

by the ideal I generated by all elements of the form

$$a \otimes 1 - 1 \otimes a \in A \otimes_B A.$$

We then say that α is *separated*, or that X is *separated as a scheme over* S, if Δ is a closed subscheme of $X \times_S X$.

Exercise III-2. Let $Y \to S$ be any map of topological spaces, and let

$$\Delta \subset Y \times_S Y$$

be the diagonal. Show that if Δ is a closed set, then for any commutative diagram

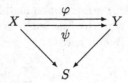

of continuous functions the set of points of X where φ and ψ agree is closed. Now prove a similar lemma for regular maps of schemes: show that there is a naturally defined (that is, maximal) closed subscheme on which φ and ψ agree.

Exercise III-3. Let X be a scheme separated over S. Show that (closed or open) subschemes of X are again separated over S.

Exercise III-4. Note that from the very definition of the diagonal it follows that affine morphisms are separated.

We shall see below that projective schemes, to be defined shortly, are also separated, so at least these features of the properties of Hausdorff spaces are valid for them as well.

In the case of classical affine varieties — even things as simple as plane curves — it was realized early in the previous century that the simplest way to get something that would behave like a compact object — would, in fact, be compact in the classical topology, in the case of varieties over the complex numbers — was to take the closure of an affine variety in projective space. It turns out that if $\varphi : X \to Y$ is a map of projective varieties, then indeed φ maps closed subvarieties of X to closed subvarieties of Y. Somewhat more generally, if we take the product of such a map with an arbitrary variety Z, to get

$$\psi := \varphi \times 1_Z : X \times Z \to Y \times Z$$

then ψ maps closed subvarieties of $X \times Z$ onto closed subvarieties of $Y \times Z$. It turns out that *this, with the separation property, is the central property of projective varieties that makes them so useful.* But it is a property satisfied by a slightly larger class of varieties than the projective ones, and it is a property that is sometimes easier to verify than projectivity, so it is of great importance to make a general definition.

If $\alpha : X \to S$ is a map of schemes of finite type, we will say that α is *proper*, or that X is *proper over* S, if α is separated and for all maps $T \to S$, the projection map of the fibered product

$$X \times_S T \to T$$

carries closed subsets onto closed subsets. As usual, if $S = \operatorname{Spec} R$ is a ring, we shall often say "proper over R" when we mean "proper over $\operatorname{Spec} R$".

The additional property given here, besides that of separation, is sometimes expressed by saying that α is *universally closed*. The name *proper* comes from an old geometric usage: a map $\alpha : M \to N$ of Hausdorff spaces is called proper if the preimage of every compact set is compact. This is a kind of relative compactness for the map α. It is related to our notion by the property expressed in the following exercise.

Exercise III-5. Let $\mathscr{C}$ be the category of locally compact Hausdorff spaces which have countable bases for their topologies. Show that a map $f : X \to Y$ in $\mathscr{C}$ is universally closed if and only if it is proper in the sense that for all compact subsets C of Y the subset $f^{-1}(C)$ is compact.

This notion of properness turns out to be the key property in algebraic geometry whether of schemes or of varieties — it plays the role played by "compact and Hausdorff" in other geometric theories. The projective schemes that we will introduce below are simply the most common examples of schemes proper over a given scheme B. We will not prove this central result here; it is not terribly difficult, but it would take us too far afield. See, for example, Hartshorne [1977, Theorem II.4.9] for a proof.

A finite morphism $\varphi : X \to Y$ is necessarily proper; see Eisenbud [1995, Section 4.4].

III.2   Proj of a Graded Ring

III.2.1   The Construction of Proj S

By far the most important examples of schemes that are not affine are the schemes *projective* over an affine scheme $\operatorname{Spec} A$, where A is an arbitrary commutative ring. (For simplicity we usually say that such a scheme is projective over A instead of over $\operatorname{Spec} A$.) Such a scheme is obtained from a graded A-algebra by a process very much analogous to the construction of a projective variety from its homogeneous coordinate ring. One can also define schemes projective over an arbitrary base scheme B by starting with a sheaf of graded $\mathscr{O}_B$-algebras, and this generalization has important applications. But most of the theory quickly reduces to the case where B is affine, and we will stick with that level of generality here.

To describe this construction, we start with a positively graded A-algebra having A as the degree 0 part, that is, an A-algebra S with a *grading*

$$S = \bigoplus_{\nu=0}^{\infty} S_\nu \qquad \text{(as A-modules)}$$

such that

$$S_\nu S_\mu \subset S_{\nu+\mu} \quad \text{and} \quad S_0 = A.$$

An element of S is called *homogeneous of degree ν* if it lies in S_ν. We will define an A-scheme $X = \operatorname{Proj} S$ from S. The schemes projective over A are by definition the schemes of the form $\operatorname{Proj} S$, where S is a finitely generated A-algebra. The algebra S is called the *homogeneous coordinate ring* of X, though (like the homogeneous coordinate ring of a projective variety) it is in fact not determined by X.

In case S is the polynomial ring

$$S = A[x_0, \ldots, x_r]$$

over A, with grading defined by giving the elements of A degree 0 and giving each variable degree 1, the resulting scheme $\operatorname{Proj} S$ is called *projective r-space over A* and is written $\mathbb{P}_A^r$. (The following exercises will make it clear that this is the same scheme $\mathbb{P}_A^n$ as defined in Chapter I.) In case $A = K$ is a field, the scheme $\mathbb{P}_K^r$ bears the same relation to the variety called projective space over K as the scheme $\mathbb{A}_K^r$ bears to the variety called affine r-space.

We will suppose for simplicity that, as in the case of the polynomial ring, the algebra S is generated over A by its elements of degree 1, and we leave the general case as an exercise. (In a different direction, most of what we say below also holds if S is not assumed to be finitely generated over A, but this generalization is less frequently used.)

$\operatorname{Proj} S$ may be defined as follows: we write

$$S_+ = \bigoplus_{\nu=1}^{\infty} S_\nu$$

for the ideal generated by homogeneous elements of strictly positive degree in S. We say that an ideal is *homogeneous* if it is generated by homogeneous elements. The underlying topological space $|\operatorname{Proj} S|$ is the set of homogeneous prime ideals in the ring S that do not contain S_+ (these are sometimes called *relevant* prime ideals, and S_+ is thus called the *irrelevant ideal*). The topology of $|\operatorname{Proj} S|$ is defined by taking the closed sets to be the sets of the form

$$V(I) := \{\mathfrak{p} \mid \mathfrak{p} \text{ is a relevant prime of } S \text{ and } \mathfrak{p} \supset I\}$$

for some homogeneous ideal I of S.

We will give $|\operatorname{Proj} S|$ the structure of a scheme by specifying this structure on each of a basis of open sets. To do this, let f be any homogeneous element of S of degree 1, and let U be the open set

$$|\text{Proj}\,S| - V(f)$$

of homogeneous primes of S not containing f (and thus not containing S_+). The points of U may be identified with the homogeneous primes of $S[f^{-1}]$. On the other hand, these homogeneous primes correspond to all the primes of the ring of elements of degree 0 in $S[f^{-1}]$, which is denoted by $S[f^{-1}]_0$; see Exercise III-6(a). Thus we may identify U with the topological space $\text{Spec}\,S[f^{-1}]_0$ and give it the corresponding structure of an affine scheme. We will write $(\text{Proj}\,S)_f$ for this open affine subscheme of $\text{Proj}\,S$. If $x_0, x_1, \ldots$ are elements of degree 1 generating an ideal whose radical is the irrelevant ideal S_+; then the open sets

$$(\text{Proj}\,S)_{x_i} := \text{Proj}\,S - V(x_i)$$

form an affine open cover of $\text{Proj}\,S$.

If g is another degree 1 element of S, then the overlap $(\text{Proj}\,S)_f \cap (\text{Proj}\,S)_g$ is the open affine subset of $(\text{Proj}\,S)_f$ given by the spectrum of

$$S[f^{-1}]_0[(g/f)^{-1}] = S[f^{-1}, g^{-1}]_0.$$

Since this expression is symmetric in f and g, we get a natural identification

$$((\text{Proj}\,S)_f)_{(g/f)} = ((\text{Proj}\,S)_g)_{(f/g)}.$$

As in the discussion of gluing in Section I.2.4, this makes $\text{Proj}\,S$ into a scheme.

The scheme $X = \text{Proj}\,S$ has a natural structure map to $\text{Spec}\,S_0$ defined by the map $S_0 \to \mathcal{O}_X(X)$. One case is so important that it deserves a definition: If $B = \text{Spec}\,A$ is an affine scheme, then a morphism $X \to B$ is *projective* if it is the structure map $\text{Proj}\,S \to \text{Spec}\,S_0$ for a graded ring S such that $S_0 = A$ and S is generated over A by finitely many elements. We will soon be in a position to generalize this to arbitrary schemes B.

In the rest of this section and the next we present some basic facts about projective schemes and their closed subschemes. Since these facts and their proofs are quite parallel to things from the theory of varieties, we present them as exercises.

Exercise III-6. (a) For any homogeneous ideal I of S and homogeneous element f of degree 1, the intersection

$$(I \cdot S[f^{-1}]) \cap S[f^{-1}]_0$$

is generated by elements obtained by choosing a set of homogeneous generators of I and multiplying them by the appropriate (negative) powers of f (see Exercise III-10 for the generalization where f has arbitrary degree). Thus the homogeneous primes of $S[f^{-1}]$ are in one-to-one correspondence with all the primes (no homogeneity condition) of the ring of elements of degree 0 in $S[f^{-1}]$; the correspondence is given by taking a prime $\mathfrak{p}$ of $S[f^{-1}]$ to $\mathfrak{q} = \mathfrak{p} \cap S[f^{-1}]_0$ and taking the prime $\mathfrak{q}$ of $S[f^{-1}]_0$ to $\mathfrak{q}S[f^{-1}]$.

(b) Let $S = A[x_0, \ldots, x_r]$ be the polynomial ring, and let U be the open affine set $(\mathbb{P}_A^r)_{x_i}$, of $\mathbb{P}_A^r = \operatorname{Proj} S$. By definition,

$$U = \operatorname{Spec} S[x_i^{-1}]_0$$

Show that

$$S[x_i^{-1}]_0 = A[x_0', \ldots, x_r']$$

the polynomial ring with generators $x_j' = x_j/x_i$. (Note that $x_i' = 1$, so that this is a polynomial ring in r variables.) Thus

$$(\mathbb{P}_A^r)_{x_i} = \mathbb{A}_A^r$$

so projective r-space has an open affine cover by $r + 1$ copies of affine r-space, as described in Chapter I.

(c) Consider the map $\alpha : S \to S[x_i^{-1}]_0$ obtained by mapping x_i to 1 and x_j to x_j' for $j \neq i$. Show from part (a) that if I is a homogeneous ideal of S, then

$$I' := I \cdot S[x_i^{-1}] \cap S[x_i^{-1}]_0 = \alpha(I)' \cdot S[x_i^{-1}]_0.$$

The process of making I' from I is called *dehomogenization*. Describe, as in the classical case, the inverse process, homogenization.

Exercise III-7. If I is a homogeneous ideal of the graded ring S, then we have an inclusion of underlying sets

$$|\operatorname{Proj} S/I| \subset |\operatorname{Proj} S|.$$

Show that the intersection of this subset with an open affine $(\operatorname{Proj} S)_f$ is a closed subset of $(\operatorname{Proj} S)_f$, and that the corresponding subscheme is isomorphic to $(\operatorname{Proj} S/I)_f$, so that $\operatorname{Proj} S/I$ can be realized as a closed subscheme of $\operatorname{Proj} S$. Every finitely generated A-algebra generated in degree 1 is a factor ring, by a homogeneous ideal, of the polynomial ring $A[x_0, \ldots, x_r]$ for some r, so we see that *every projective scheme over A is a closed subscheme of a projective space over A*. We will see in more detail the correspondence between ideals in the ring S and closed subschemes of $\operatorname{Proj} S$ in Exercises III-15 and III-16.

Exercise III-8. Show that $\mathbb{P}_A^r$ is the disjoint union of the open set $\mathbb{A}_A^r$ and the closed set $\mathbb{P}_A^{r-1}$. In particular, $\mathbb{P}_A^0 = \operatorname{Spec} A$. Thus, for example, we may picture $\mathbb{P}_\mathbb{Z}^1$ as the union of the affine line $\mathbb{A}_\mathbb{Z}^r$ over $\mathbb{Z}$ (as pictured in Chapter II) with a "point at ∞" isomorphic to $\operatorname{Spec}\mathbb{Z}$, as follows:

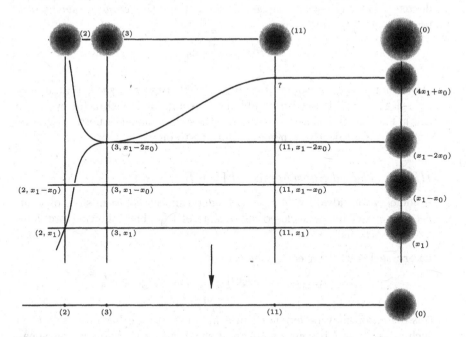

Exercise III-9. Add to this diagram pictures of the closures of the points $(4x_1 - 5x_0)$, $(2x_1 - 5x_0)$, and (5) (compare with the diagram of $\mathbb{A}^2_{\mathbb{Z}}$ in Section II.4.3). *Note:* The curve $(4x_1 - 5x_0)$ should be drawn tangent to the "point at ∞" (x_0), while the curve $(2x, -5x_0)$ should not — informally, we could say this is because the function $5/4$ has a double pole at (2), while $5/2$ has only a simple pole there. (See also the discussion in Exercise II-38.)

Exercise III-10. With notation as above, let h be a homogeneous element of S of any strictly positive degree. The set

$$(\operatorname{Proj}S)_h := \operatorname{Proj}S - V(h)$$

is as above the set of homogeneous primes of S not containing h. Show that this set is again in one-to-one correspondence with the set of primes of $S[h^{-1}]_0$ and that in fact there is an isomorphism of $\operatorname{Spec}S[h^{-1}]_0$ with an open (affine) subscheme of $\operatorname{Proj}S$. Show also that a collection

$$\{(\operatorname{Proj}S)_h\}_{h \in H}$$

of such open affines is an open cover of $\operatorname{Proj}S$ if and only if the elements of H generate an ideal whose radical equals S_+.

Exercise III-11. Extend the definition of $\operatorname{Proj}S$ to the case where S is not necessarily generated by elements of degree 1, and show that $\operatorname{Proj}S$ is a projective scheme.

Exercise III-12. Let S be a graded ring, not necessarily generated in degree 1. For any positive integer d, define the d-th *Veronese subring* of S to be the graded ring

$$S^{(d)} = \bigoplus_{\nu=0}^{\infty} S_{d\nu}$$

Show that $\operatorname{Proj} S$ is isomorphic to $\operatorname{Proj} S^{(d)}$. However, show that if $S = A[x, y]$, then $S^{(d)}$ is not isomorphic to S as a graded algebra (or even as a ring). Thus, as in the case of varieties, the correspondence between graded algebras and projective schemes is not one-to-one.

III.2.2   Closed Subschemes of $\operatorname{Proj} R$

A homogeneous ideal $I \subset A[x_0, \ldots, x_r]$ determines a coherent sheaf of ideals $\tilde{I} \subset \mathscr{O}_{\mathbb{P}_A^r}$, and hence a closed subscheme of $\mathbb{P}_A^r$. The following problems develop these facts.

Exercise III-13. For each open set

$$U_i = (\mathbb{P}_A^r)_{x_i} = \operatorname{Spec} A[x_0, \ldots, x_r, x_i^{-1}]_0 \cong \mathbb{A}_A^r,$$

let $\tilde{I}(U_i)$ be the ideal $I \cdot A[x_0, \ldots, x_r, x_i^{-1}] \cap A[x_0, \ldots, x_r, x_i^{-1}]_0$. Show that this definition may be extended in a unique way to other open sets U in such a way that $\tilde{I}$ becomes a coherent sheaf of ideals. We may thus speak of the closed subscheme $V(\tilde{I})$ of $\mathbb{P}_A^r$ *associated to a homogeneous ideal I.*

Exercise III-14. Conversely, given a closed subscheme X in $\mathbb{P}_A^r$, we may define a homogeneous ideal $I(X) \subset A[x_0, \ldots, x_r]$ to be the ideal generated by all homogeneous polynomials $p(x_0, \ldots, x_r)$ such that for every i setting the i-th variable equal to 1 gives rise to an element

$$p(x_0, \ldots, 1, \ldots, x_r) \in \mathscr{I}_X(U_i) \subset A[x_0, \ldots, x_r, x_i^{-1}]_0$$

Show that if $I = I(X)$, then $\tilde{I} = \mathscr{I}_X$.

Note that with Exercise III-7 this shows that every closed subscheme of a projective scheme is projective: if $I \subset S = A[x_0, \ldots, x_r]$ is a homogeneous ideal, then $V(\tilde{I}) \subset \mathbb{P}_A^r$ is isomorphic to the scheme $\operatorname{Proj} S/I$.

Exercise III-15. The correspondence between subschemes and ideals is not, as it was in the case of affine schemes, one to one. For example, show that in $\mathbb{P}_K^1$ with K a field, the ideals $I = (x_0)$ and $I' = (x_0^2, x_0 x_1)$ both define the same reduced, one-point subscheme. More generally, show that if $I \subset S = K[x_1, \ldots, x_r]$ is any homogeneous ideal, and for any integer n_0 we define an ideal $I' \subset I$ by

$$I' = \bigoplus_{n \geq n_0} I_n$$

then I and I' define the same subscheme of $\mathbb{P}_K^r$.

Exercise III-16. To deal with this, define the *saturation* of a homogeneous ideal $J \subset S := A[x_0, \ldots, x_r]$ to be the ideal

$$I = \{F \in S : F \cdot S_n \subset J \text{ for some } n\}$$

and say that a homogeneous ideal is *saturated* if it equals its saturation. Show that there is a bijective correspondence between subschemes of $\mathbb{P}_A^r$ and saturated ideals.

Exercise III-17. Show that the isomorphism of Exercise III-12 defines an isomorphism between projective space $\mathbb{P}_A^r$ and a closed subscheme of $\mathbb{P}_A^{N-1}$, where $N = \dim_A(A[x_0, \ldots, x_r]_d)$. (This is just the scheme-theoretic version of the Veronese map.)

Exercise III-18. Show that if R is a graded ring finitely generated over a ring $A = R_0$ (not necessarily generated by its graded part of degree 1), $\operatorname{Proj} R$ is isomorphic to a closed subscheme of some projective space $\mathbb{P}_A^r$.

We conclude with a definition and a basic theorem.

Definition III-19. A morphism $\varphi : X \to Y$ of schemes is said to be *projective* if it is the composition of a closed embedding $X \to \mathbb{P}_Y^n$ with the structure morphism $\mathbb{P}_Y^n \to Y$.

Note that if $Y = \operatorname{Spec} A$ is affine, this amounts to saying that X is of the form $\operatorname{Proj} S$ for some finitely generated A-algebra S. The basic fact about projective morphisms is the one stated above:

Theorem III-20. *Projective morphisms are proper.*

For a proof see Hartshorne [1977, Theorem II.4.9].

Exercise III-21. Show that a finite morphism $\varphi : X \to Y$ is proper, and locally projective in the sense that Y can be covered by open sets $U \subset Y$ such that the restriction $\varphi : V = \varphi^{-1}U \to U$ is projective. (We have adopted the definition of projective morphism given in Hartshorne [1977, Section II.4]; what is here called locally projective is called projective by Grothendieck [1961, EGA II, 5.5].)

III.2.3 Global Proj

Proj of a Sheaf of Graded $\mathcal{O}_X$-Algebras. The construction of Proj of a graded ring S gives rise to a scheme $X = \operatorname{Proj} S$ together with a structure morphism $X \to B = \operatorname{Spec}(S_0)$. Because the association of $\operatorname{Proj} S$ to S is functorial, there is a more general construction that gives rise to schemes X with structure morphisms $X \to B$ to arbitrary schemes B, and that specializes to the construction Proj when B is affine: all we have to do for general B is replace the graded S_0-algebra S with a sheaf of algebras over $\mathcal{O}_B$.

To carry this out, let B be any scheme. By a *quasicoherent sheaf of graded $\mathcal{O}_B$-algebras* we will mean a quasicoherent sheaf $\mathscr{F}$ of algebras on B, and a grading

$$\mathcal{F} = \bigoplus_{\nu=0}^{\infty} \mathcal{F}_{\nu}$$

such that $\mathcal{F}_{\nu}\mathcal{F}_{\mu} \subset \mathcal{F}_{\nu+\mu}$, and $\mathcal{F}_0 = \mathcal{O}_B$. Thus, for every affine open subset $U \subset B$ with coordinate ring $A = \mathcal{O}_B(U)$, the ring $\mathcal{F}(U)$ will be a graded A-algebra with 0-th graded piece $\mathcal{F}(U)_0 = A$.

Given such a sheaf $\mathcal{F}$, for each affine open subset $U \subset B$ we will let $X_U \to U$ be the scheme $X_U = \operatorname{Proj}\mathcal{F}(U)$ with the structure morphism $\operatorname{Proj}\mathcal{F}(U) \to \operatorname{Spec}(A) = U$. For every inclusion $U \subset V$ of open subsets of B, the restriction map $\mathcal{F}(V) \to \mathcal{F}(U)$ is a homomorphism of graded rings whose 0-th graded piece is the restriction map $\mathcal{O}_B(V) \to \mathcal{O}_B(U)$, and so induces a map $X_U \to X_V$ commuting with the structure morphisms $X_U \to U$ and $X_V \to V$ and the inclusion $U \hookrightarrow V$. We may thus glue together the schemes X_U to arrive at a scheme X with structure morphism $X \to B$; X is denoted $\operatorname{Proj}\mathcal{F}$; and the construction of X is called *global Proj*.

As in the case of ordinary Proj, in most situations it will be the case that the sheaf of algebras $\mathcal{F}$ is generated by its first graded piece $\mathcal{F}_1$, and that $\mathcal{F}_1$ is coherent (or, somewhat more generally, for some $d > 0$ the Veronese subsheaf

$$\mathcal{F}^{(d)} = \bigoplus_{\nu=0}^{\infty} \mathcal{F}_{d\nu}$$

is generated by $\mathcal{F}_d$, and $\mathcal{F}_d$ is coherent). Under these hypotheses it follows, again as in the case of ordinary Proj, that the morphism $\operatorname{Proj}\mathcal{F} \to B$ is proper.

The simplest example of global Proj gives us yet another construction of projective space over an arbitrary scheme S. Recall that projective space $\mathbb{P}^n_S$ over an arbitrary scheme S was defined initially in Chapter I via the gluing construction: if S is covered by affine schemes $U_\alpha = \operatorname{Spec} R_\alpha$, we define projective space $\mathbb{P}^n_S$ to be the union of the projective spaces $\mathbb{P}^n_{U_\alpha}$, with the gluing maps induced by the identity maps on $U_\alpha \cap U_\beta$. We can also define it as a product:

$$\mathbb{P}^n_S = \mathbb{P}^n_{\mathbb{Z}} \times_{\operatorname{Spec}\mathbb{Z}} S.$$

Finally, we can realize it as the global Proj of the symmetric algebra of the free sheaf of rank $n + 1$ on S:

Exercise III-22. Let S be any scheme. Show that projective space $\mathbb{P}^n_S$ over S may be constructed as a global Proj:

$$\mathbb{P}^n_S = \operatorname{Proj}\left(\operatorname{Sym}(\mathcal{O}_S^{\oplus n+1})\right).$$

In particular, we can realize products of projective spaces over a given scheme S either as fibered products, or via global Proj: if we denote by

$\mathscr{O}_{\mathbb{P}_S^n}[X_0, \ldots, X_m]$ the sheaf of graded $\mathscr{O}_{\mathbb{P}_S^n}$-algebras $\mathrm{Sym}(\mathscr{O}_{\mathbb{P}_S^n}^{\oplus m+1})$, then

$$\mathbb{P}_S^n \times_S \mathbb{P}_S^m = \mathrm{Proj}\, \mathscr{O}_{\mathbb{P}_S^n}[X_0, \ldots, X_m].$$

A third way is via the *Segre embedding*:

Exercise III-23. Let S be any scheme. Show that

$$\mathbb{P}_S^n \times_S \mathbb{P}_S^m \cong V\big(\{X_{i,j}X_{k,l} - X_{i,l}X_{k,j}\}\big) \subset \mathrm{Proj}\, \mathscr{O}_S[\{X_{i,j}\}_{0\le i\le n; 0\le j\le m}]$$
$$= \mathbb{P}_S^{(n+1)(m+1)-1}.$$

This in turn gives us a way of describing subschemes of such a product, at least locally over the base:

Exercise III-24. Let $S = \mathrm{Spec}\, R$ be any affine scheme. Show that any closed subscheme

$$X \subset \mathrm{Proj}\, R[x_0, \ldots, x_n] \times_S \mathrm{Proj}\, R[y_0, \ldots, y_m] = \mathbb{P}_S^n \times_S \mathbb{P}_S^m$$

may be given as the zero locus of a collection $\{F_\alpha(x_0, \ldots, x_n; y_0, \ldots, y_m)\}$ of bihomogeneous polynomials F_α in the two sets of variables $(x_0, \ldots, x_n)$ and $(y_0, \ldots, y_m)$. In particular, show that the ideal of 2×2 minors of the matrix $\begin{pmatrix} x_0 & x_1 & \cdots & x_n \\ y_0 & y_1 & \cdots & y_n \end{pmatrix}$ defines the diagonal subset in $\mathbb{P}_S^n \times_S \mathbb{P}_S^n$. Deduce that any projective morphism is separated.

A more serious application of the global Proj construction is the definition of the *blow-up* of a scheme X along a closed subscheme $Y \subset X$; we will discuss this in full in Chapter V. Another common use of global Proj is the construction of the projectivization of a vector bundle, which we now describe.

The Projectivization $\mathbb{P}(\mathscr{E})$ of a Coherent Sheaf $\mathscr{E}$. We saw in Exercise III-22 that projective space $\mathbb{P}_S^n$ over a scheme S is $\mathrm{Proj}(\mathrm{Sym}(\mathscr{O}_S^{\oplus n+1}))$. We make a similar construction for any coherent sheaf $\mathscr{E}$, and define the *projectivization* $\mathbb{P}(\mathscr{E})$ of $\mathscr{E}$ to be the B-scheme

$$\mathbb{P}(\mathscr{E}) = \mathrm{Proj}(\mathrm{Sym}\,\mathscr{E}) \longrightarrow B.$$

To review the simplest case, let V be an n-dimensional vector space over a field K, regarded as a vector bundle over the one-point scheme $\mathrm{Spec}\, K$. The projectivization of V is a projective space of dimension n over K. The projectivization of V^* is called the the *dual projective space* to $\mathbb{P}V$. The K-valued points of $\mathbb{P}(V)$ correspond to one-dimensional quotients of V or equivalently to hyperplanes in V. The K-valued points of $\mathbb{P}(V^*)$ correspond to one-dimensional subspaces of V; this is what was classically called $\mathbb{P}^n$.

More generally, if $\mathscr{E}$ is a locally free sheaf of rank $n + 1$, then $\mathbb{P}(\mathscr{E})$ is a *projective bundle* over B, in the sense that for sufficiently small affine open subsets $U \subset B$ the inverse image of U in $\mathbb{P}(\mathscr{E})$ is isomorphic to projective space $\mathbb{P}_U^n$ as U-scheme. (When $\mathscr{E}$ is not locally free, it is less clear what the resulting scheme $\mathbb{P}(\mathscr{E})$ will look like.) When B is a variety over an algebraically closed field K and $\mathscr{E}$ the sheaf of sections of a vector bundle E on B, the K-valued points of $\mathbb{P}(\mathscr{E}^*)$ correspond to one-dimensional subspaces of fibers of $\mathscr{E}$, while the K-valued points of $\mathbb{P}(\mathscr{E})$ correspond to one-dimensional quotients of fibers of $\mathscr{E}$, or equivalently to hyperplanes in these fibers.

Note that any closed subscheme $X \subset \mathbb{P}_B^n$ may be realized as $\mathrm{Proj}(\mathscr{F})$ for some quasicoherent sheaf $\mathscr{F}$ of graded $\mathcal{O}_B$-algebras. More generally, if $\mathscr{E}$ is any coherent sheaf, any closed subscheme $X \subset \mathbb{P}(\mathscr{E})$ of its projectivization may be so realized. Conversely, if $\mathscr{F}$ is any quasicoherent sheaf of graded $\mathcal{O}_B$-algebras generated by $\mathscr{F}_1$, the surjection $\mathrm{Sym}(\mathscr{F}_1) \to \mathscr{F}$ gives an embedding $X = \mathrm{Proj}\,\mathscr{F} \hookrightarrow \mathbb{P}(\mathscr{F}_1)$.

Exercise III-25. Let K be a field, $\mathbb{P}_K^2 = \mathrm{Proj}\,K[X, Y, Z]$ the projective plane over K and $(\mathbb{P}_K^2)^* = \mathrm{Proj}\,K[A, B, C]$ the dual projective plane. Let Σ be the universal line over $(\mathbb{P}_K^2)^*$, that is,

$$\Sigma = V(AX + BY + CZ) \subset \mathbb{P}_K^2 \times_K (\mathbb{P}_K^2)^*$$

viewed as a family over $(\mathbb{P}_K^2)^*$. Show that $\Sigma \to (\mathbb{P}_K^2)^*$ is the projectivization of a locally free sheaf $\mathscr{E}$ of rank 2 on $(\mathbb{P}_K^2)^*$, and describe the sheaf $\mathscr{E}$.

Exercise III-26. Let B be any scheme, $\mathscr{E}$ a locally free sheaf on B and $E = \mathrm{Spec}(\mathrm{Sym}\,\mathscr{E}^*) \to B$ the total space of the vector bundle associated to $\mathscr{E}$. Show that we can complete $E \to B$ to a bundle of projective spaces over B: specifically, show that we have an inclusion on E in the bundle $\mathbb{P}(\mathscr{E}^* \oplus \mathcal{O}_B)$ as an open subscheme, with complement a hyperplane bundle $\mathbb{P}(\mathscr{E}^*) \subset \mathbb{P}(\mathscr{E}^* \oplus \mathcal{O}_B)$.

III.2.4   Tangent Spaces and Tangent Cones

Affine and Projective Tangent Spaces. The Zariski tangent spaces to a scheme are abstract vector spaces. When a scheme X over a field K is embedded in an ambient space like affine space or projective space over K, however, we can also associate to a point p of X with residue field K a corresponding linear subvariety of that affine or projective space, called the *affine tangent space* or *projective tangent space* to X at p. In the case of an affine scheme $X = V(f_1, \ldots, f_k) \subset \mathbb{A}_K^n$ and point $p = (a_1, \ldots, a_n) \in X$, this is the subvariety given as

$$V\left(\left\{\sum_i \frac{\partial f_\alpha}{\partial x_i}(a_1, \ldots, a_n) \cdot (x_i - a_i)\right\}_{\alpha = 1, \ldots, k}\right).$$

To understand the relationship between this scheme and the Zariski tangent space, note that a vector space over a field K is not the same thing as affine space $\mathbb{A}_K^n$. But it is true that, given a vector space V of dimension n over K, we may associate to V a scheme $\overline{V}$, isomorphic to affine space $\mathbb{A}_K^n$, so that the points of $\overline{V}$ with residue field K correspond naturally to vectors in V: this is just the spectrum of the symmetric algebra of the dual vector space

$$\overline{V} = \mathrm{Spec}\,(\mathrm{Sym}(V^*))\,.$$

We will call $\overline{V}$ the *scheme associated to* the vector space V.

This said, the scheme $\overline{T_p(\mathbb{A}_K^n)}$ associated to the Zariski tangent space to affine space $\mathbb{A}_K^n$ over a field K at any K-rational point $p \in \mathbb{A}_K^n$ (that is, a closed point with residue field $\kappa(p) = K$) may be naturally identified with the affine space itself, via an identification carrying the origin in $T_p(\mathbb{A}_K^n)$ to p. Now, suppose $X \subset \mathbb{A}_K^n$ is any subscheme, and $p \in X$ any K-rational point. The differential $d\iota_p$ of the inclusion $\iota : X \hookrightarrow \mathbb{A}_K^n$ at p represents the Zariski tangent space $T_p(X)$ as a vector subspace

$$d\iota_p : T_p(X) \hookrightarrow T_p(\mathbb{A}_K^n).$$

We take the induced inclusion of schemes

$$\overline{d\iota_p} : \overline{T_p(X)} \hookrightarrow \overline{T_p(\mathbb{A}_K^n)} = \mathbb{A}_K^n$$

and compose it with the translation morphism $t_p : \mathbb{A}_K^n \to \mathbb{A}_K^n$ sending the origin to p to obtain an inclusion

$$t_p \cdot \overline{d\iota_p} : \overline{T_p(X)} \hookrightarrow \mathbb{A}_K^n \longrightarrow \mathbb{A}_K^n.$$

The image of this inclusion is an affine subspace of $\mathbb{A}_K^n$, which we will call the *affine tangent space* to X at p. Again, note that it is a scheme, not a vector space.

A similar construction will associate to a point p with residue field K on a projective scheme $X \subset \mathbb{P}_K^n$ a linear space $\mathbb{T}_p(X) \subset \mathbb{P}_K^n$. One way to do this is to choose an open subset $U \cong \mathbb{A}_K^n \subset \mathbb{P}_K^n$ containing p, and take the closure in $\mathbb{P}_K^n$ of the affine tangent space to $X \cap U$ at p. But there is a more intrinsic way. First, we write our ambient projective space $\mathbb{P}_K^n$ as the projective space $\mathbb{P}V$ associated to a vector space V, that is, as

$$\mathbb{P}_K^n = \mathrm{Proj}\,S$$

where

$$S = \mathrm{Sym}\,V^*$$

is the symmetric algebra of the dual of a vector space V. Thus, $(k+1)$-dimensional linear subspaces of $S_1 = V$ correspond to k-planes in $\mathbb{P}_K^n$. We let

$$I = I(X) \subset S$$

be the homogeneous ideal of $X \subset \mathbb{P}_K^n$, and let $\mathfrak{m} = \mathfrak{m}_p \subset S$ be the ideal of forms vanishing at the point $p \in X$. Let J be the saturation of the ideal $I + \mathfrak{m}^2 \subset S$. We define the *projective tangent space* $\mathbb{T}_p(X) \subset \mathbb{P}_K^n$ to X at p to be the subspace

$$\mathbb{T}_p(X) = V(J \cap S_1) \subset \mathbb{P}_K^n$$

of $\mathbb{P}_K^n$. By way of explanation, note that J is the ideal of the first-order neighborhood of p in X — that is, the intersection of X with the "fat point" $P \subset \mathbb{P}_K^n$ defined by the square of the ideal $\mathfrak{m}$ of p. The projective tangent space $\mathbb{T}_p(X) = V(J \cap S_1)$ is thus the span of this first-order neighborhood $V(J)$, that is, the smallest linear subspace of $\mathbb{P}_K^n$ containing $V(J)$.

Exercise III-27. Show that this definition coincides with the naive definition proposed initially.

Since the projective tangent space $\mathbb{T}_p(X)$ to a projective scheme X at a K-rational point $p \in X$ is a linear subspace of the ambient projective space $\mathbb{P}V$, it is of the form $\mathbb{T}_p(X) = \mathbb{P}W$ for some quotient vector space $V \to W \to 0$. We may ask then what the relationship is between the vector space W and the Zariski tangent space $T_p(X)$. The answer, which we will see in Section VI.2.1 is that we have an exact sequence

$$0 \longrightarrow K \longrightarrow W^* \longrightarrow T_p(X) \longrightarrow 0.$$

More precisely, if $V \to U \to 0$ is the one-dimensional quotient of V corresponding to the point $p \in X \subset \mathbb{P}V$, then the surjection $V \to U$ factors through a surjection $\varphi : W \to U$, and we have a natural identification

$$T_p(X) = \mathrm{Hom}(\mathrm{Ker}\,\varphi, U).$$

In any event, note that we do have a natural identification of the set of lines through p in $\mathbb{T}_p(X)$ with the set of lines through the origin in $T_p(X)$.

Exercise III-28. Let $X = V(F) \subset \mathbb{P}_K^n$ be the hypersurface in $\mathbb{P}_K^n$ given by the homogeneous polynomial $F(Z_0, \ldots, Z_n)$, and let $p = [a_0, \ldots, a_n] \in X$ be any point with residue field K. Show that the projective tangent space $\mathbb{T}_p(X)$ is the zero locus $V(L) \subset \mathbb{P}_K^n$ of the linear form

$$L(Z_0, \ldots, Z_n) = \sum_{k=0}^{n} \frac{\partial F}{\partial Z_i}(a_0, \ldots, a_n) \cdot Z_k.$$

Tangent Cones. A more accurate reflection of the tangential behavior of a scheme X at a point $p \in X$ is its tangent cone. To define this, let X now be an arbitrary scheme, $p \in X$ and point, $\mathcal{O}_{X,p}$ the local ring of X at p and $\mathfrak{m} = \mathfrak{m}_{X,p} \subset \mathcal{O}_{X,p}$ the maximal ideal in $\mathcal{O}_{X,p}$. We define the *tangent cone* $TC_p(X)$ to X at p to be the scheme

$$TC_p(X) = \mathrm{Spec}\left(\bigoplus_{\alpha=0}^{\infty} \mathfrak{m}^\alpha/\mathfrak{m}^{\alpha+1} \right).$$

A few observations about this construction are in order. First, we note that the graded ring $B = \bigoplus (\mathfrak{m}^\alpha/\mathfrak{m}^{\alpha+1})$ is generated by its first graded piece

$$B_1 = \mathfrak{m}/\mathfrak{m}^2 = (T_p X)^*$$

so that B is a quotient of the ring

$$A = \mathrm{Sym}((T_p X)^*).$$

We thus have an inclusion

$$TC_p(X) = \operatorname{Spec} B \hookrightarrow \operatorname{Spec} A = \overline{T_pX},$$

or in other words, the tangent cone to X at p is naturally a subscheme of the scheme associated to the Zariski tangent space T_pX to X at p.

To give a more concrete realization of the tangent cone, suppose that X is a subscheme of affine space over a field K, that is,

$$X \subset \operatorname{Spec} K[x_1, \ldots, x_n]$$

and let $I = I(X) \subset K[x_1, \ldots, x_n]$ be the ideal of X; suppose moreover that the point $p \in X$ is the origin $(x_1, \ldots, x_n) \in \mathbb{A}_K^n$. For any polynomial $f \in K[x_1, \ldots, x_n]$, write

$$f(x_1, \ldots, x_n) = f_m(x_1, \ldots, x_n) + f_{m+1}(x_1, \ldots, x_n) + \cdots$$

with $f_l(x_1, \ldots, x_n)$ homogeneous of degree l and $f_m \neq 0$; the first nonzero term $f_m(x_1, \ldots, x_n)$ is called the *leading term* of f. Then we have the following interpretation:

Exercise III-29. Show that the tangent cone

$$TC_p(X) \subset \overline{T_pX} \subset \overline{T_p(\mathbb{A}_K^n)} = \mathbb{A}_K^n$$

is the subscheme defined as the zero locus of the leading terms of all elements $f \in I$.

Returning to the general case, note that since the ring $B = \bigoplus(\mathfrak{m}^\alpha/\mathfrak{m}^{\alpha+1})$ is graded, we can also associate a geometric object to the pair (X, p) by taking $\operatorname{Proj} B$. This is a subscheme of the projective space $\mathbb{P}(T_pX) \cong \mathbb{P}_{\kappa(p)}^n$ associated to the Zariski tangent space to X at p, called the *projectivized tangent cone* to X at p and denoted $\mathbb{P}TC_p(X)$. In many ways it is more convenient to deal with, being a projective scheme and of one lower dimension than the tangent cone; it contains in general slightly less information (as exercise III-30 below will show, the tangent cone $TC_p(X)$ may have an embedded point at the origin, which the projectivized tangent cone will miss).

Even though the degree of a general subscheme of projective space will not be defined until Section III.3.1, we should mention here an important invariant of a scheme that can be defined in terms of the projectivized tangent cone to X at p: we define the *multiplicity* of X at p to be the degree of the projectivized tangent cone $\mathbb{P}TC_p(X) \subset \mathbb{P}(T_pX) \cong \mathbb{P}_{\kappa(p)}^n$. This definition represents one more example of how schemes arise naturally and are useful in the context of varieties: in the category of varieties we can still define the tangent cone (as the reduced scheme associated to our tangent cone) and projectivized tangent cone, but they do not behave well in families.

There are many naturally occurring examples of nonreduced tangent cones. For example, consider the family of plane cubic curves with equation $C_t = \operatorname{Spec} K[x, y]/(y^2 - tx^2 - x^3)$ (that is, we let $B = \mathbb{A}^1_K = \operatorname{Spec} K[t]$, and take our family to be $\mathscr{C} = V(y^2 - tx^2 - x^3) \subset \mathbb{A}^2_B \to B$). For each t, the curve C_t may be given parametrically as the image of the map

$$\mathbb{A}^1_K = \operatorname{Spec} K[\lambda] \longrightarrow \mathbb{A}^2_K = \operatorname{Spec} K[x, y]$$

given by $t \mapsto (\lambda^2 - t, \lambda^3 - t\lambda)$. For $t \neq 0$, this curve has a node at the origin — the two points $\lambda = \pm\sqrt{t}$ each map to the origin — and this is reflected in the tangent cone $\mathbb{T}_{(0,0)}C_t = V(y^2 - tx^2)$, which is the union of the two lines $y = \pm\sqrt{t}\,x$. When $t = 0$, we see that the node of the curve has degenerated to a cusp, and the tangent cone is now the double line $\mathbb{T}_{(0,0)}C_0 = V(y^2)$.

For more subtle examples, consider the curves C_1 and $C_2 \subset \mathbb{A}^3_K$ given as the images of the maps $\nu_i : \mathbb{A}^1_K \to \mathbb{A}^3_K$ given by

$$\nu_1 : t \longmapsto (t^3, t^4, t^5)$$

and

$$\nu_2 : t \longmapsto (t^3, t^5, t^7).$$

In each case, let p be the singular point of C_i.

Exercise III-30. (a) Show that the projectivized tangent cones

$$\mathbb{P}TC_p(C_i) \subset \mathbb{P}^2_K$$

to both curves C_i are curvilinear schemes of degree 3, that is, isomorphic to $\operatorname{Spec} K[s]/(s^3)$, and that they are not contained in any line in $\mathbb{P}^2_K$.

(b) Find an example of a curve $C \subset \mathbb{A}^3_K$ where the projectivized tangent cone to C at the origin is isomorphic to $\operatorname{Spec} K[s]/(s^3)$ and contained in a line.

(c) Find an example of a curve $C \subset \mathbb{A}^3_K$ where the projectivized tangent cone to C at the origin is isomorphic to $\operatorname{Spec} K[s, t]/(s^2, st, t^2)$.

(d) Find an example of a curve $C \subset \mathbb{A}^3_K$ where the projectivized tangent cone to C at the origin is contained in a line, but the Zariski tangent space $T_0(C)$ is three-dimensional.

There is another geometric characterization of the tangent cone to a scheme X at a point $p \in X$: simply put, *the tangent cone is the locus of limiting positions of lines $\overline{pq}$ joining p to points $q \neq p \in X$ as q approaches p.* To state this precisely, suppose first that a neighborhood of p in X is embedded in affine space $\mathbb{A}^n_K$ over a field K. Let $T = \overline{T}_p\mathbb{A}^n_K$ be the affine space associated to the Zariski tangent space $T_p\mathbb{A}^n_K$ to $\mathbb{A}^n_K$ at p, and consider the incidence correspondence

$$\Sigma = \{(v, q) : v \in \overline{T}_p(\overline{pq}) \subset T \times (\mathbb{A}^n_K \setminus \{p\})\}.$$

Equivalently, in terms of the identification of T with $\mathbb{A}_K^n$ itself, Σ is the subscheme of $\mathbb{A}_K^n \times (\mathbb{A}_K^n \setminus \{p\})$ given by the equations

$$y_i\big(x_j - x_j(p)\big) - y_j\big(x_i - x_i(p)\big) = 0.$$

Let $\Gamma = \pi_2^{-1}(X \setminus \{p\}) \subset T \times (X \setminus \{p\})$ be the inverse image of $X \setminus \{p\}$ in Σ, and $\overline{\Gamma}$ the closure of Γ in $T \times X$. We have then:

Proposition III-31. *The tangent cone TC_pX is the fiber of $\overline{\Gamma}$ over the point $p \in X$.*

This statement (modulo possible embedded components at the origin in TC_pX) will be proved in Chapter IV. It amounts to the statement that the projectivization of TC_pX is the exceptional divisor of the blow-up of X at p.

Proposition III-31 is very useful, for example in doing Exercises III-32-III-34 below.

Exercise III-32. Let V be the vector space of polynomials of degree n on $\mathbb{P}_K^1 = \operatorname{Proj} K[X, Y]$, that is, homogeneous polynomials of degree n in two variables X, Y, and let $\mathbb{P}V^* \cong \mathbb{P}_K^n$ be the projective space parametrizing one-dimensional subspaces of V. Let $\Delta \subset \mathbb{P}_K^n$ be the discriminant hypersurface, that is, the locus of polynomials with a repeated factor with the reduced scheme structure (we will see in Chapter V how to give equations for, and hence a natural scheme structure on, Δ). If

$$F(X, Y) = \prod (a_i X + b_i Y)^{m_i}$$

is any polynomial of degree n (with the factors $a_i X + b_i Y$ pairwise independent), what is the support of the tangent cone to Δ at the point $p = [F]$? (Hint: consider lines in $\mathbb{P}_K^n$ through the point $[F]$. How many other points of intersection with Δ will a general such line have, and which lines will have fewer?)

Exercise III-33. More generally, suppose $\Delta_m \subset \mathbb{P}_K^n$ is the locus of polynomials with an m-fold root. Again, what is the support of the tangent cone to Δ_m at a point $[F]$, where F is as above?

Exercise III-34. This is an exercise from classical geometry. Suppose $C \subset \mathbb{P}_K^n$ is a nonsingular curve. The union of the projective tangent lines to C is the support of a surface $S \subset \mathbb{P}_K^n$, called the *tangent developable* to C; this surface will be singular along C (see Harris [1995] for example). What is the support of its tangent cone at a general point $p \in C$? (Note that if we take C to be the rational normal curve in $\mathbb{P}_K^n$, that is, the image of the n-th Veronese map $\mathbb{P}_K^1 \to \mathbb{P}_K^n$, then this is a special case of exercise III-33 above.)

Exercise III-35. In each of the following, a finite group G acts on the affine plane $\mathbb{A}_K^2 = \operatorname{Spec} K[x, y]$. The quotient $\mathbb{A}_K^2/G$ (that is, $\operatorname{Spec} K[x, y]^G$)

will have a singularity at the image of the origin $(x, y) \in \mathbb{A}_K^2$. Describe the tangent cone in each case.

(a) $G = \mathbb{Z}/(3)$, acting by $(x, y) \mapsto (\zeta x, \zeta y)$, where ζ is a cube root of unity.

(b) $G = \mathbb{Z}/(3)$, acting by $(x, y) \mapsto (\zeta x, \zeta^2 y)$, where ζ is a cube root of unity.

(c) $G = \mathbb{Z}/(5)$, acting by $(x, y) \mapsto (\zeta x, \zeta y)$, where ζ is a fifth root of unity.

We will encounter tangent cones again in our discussion of blowing up: as we indicated, the projectivized tangent cone $\mathbb{P}TC_p(X)$ to a scheme X at a point $p \in X$ is the exceptional divisor in the blow-up $\mathrm{Bl}_p(X)$ of X at p. In particular, tangent cones to arithmetic schemes will come up again in this way in Section IV.2.4.

III.2.5 Morphisms to Projective Space

Just as there is a simple characterization of morphisms to an affine scheme (Theorem I-40), there is a simple way of viewing morphisms to projective space in terms of line bundles, or, equivalently, invertible sheaves, a concept we will introduce in this section. Invertible sheaves have another geometric realization in the notion of Cartier divisors, and we will describe this connection as well. See Hartshorne [1977, Chapter II] for further information.

If we understand morphisms to the scheme $\mathbb{P}_A^n$, we will understand morphisms to an arbitrary projective scheme $Y \subset \mathbb{P}_A^n$, since a morphism to Y is just a morphism to $\mathbb{P}_A^n$ that factors through Y (a sharp version of this is given in Exercise III-45); thus we will study morphisms to projective space.

To understand the situation, we first consider morphisms $\varphi : X \to \mathbb{P}_A^n = \mathrm{Proj}\, A[x_0, \ldots, x_n]$ in the category of A-schemes, where $X = \mathrm{Spec}\, K$ is the spectrum of a field. Since X has only one point, the image p of such a morphism must be contained in one of the open sets

$$U_i = (\mathbb{P}_A^n)_{x_i} = \mathrm{Spec}\, A\left[\frac{x_0}{x_i}, \ldots, \frac{x_n}{x_i}\right] \cong \mathbb{A}_A^n.$$

Thus the map corresponds to an n-tuple of scalars $(a_0, \ldots, \hat{a}_i, \ldots, a_n) \in K^n$. Of course, p may also be contained in another open set U_j; in this case $a_j \neq 0$ and the coordinates in U_j are

$$(b_0, \ldots, b_j, \ldots, b_n) = \left(\frac{a_0}{a_j}, \ldots, \frac{1}{a_j} \ldots, \frac{a_n}{a_j}\right).$$

To show the coordinates without prejudice toward one or another of the U_i, we may say that a map $\mathrm{Spec}\, K \to \mathbb{P}_A^n$ corresponds to an $(n+1)$-tuple of elements of K, not all zero, with two such $(n+1)$-tuples corresponding to the same map if and only if they differ by a scalar; the map above corresponds to the $(n+1)$-tuple $[\alpha_0 = a_0, \ldots, \alpha_i = 1, \ldots, \alpha_n = a_n]$, or, equivalently, $[\beta_0 = b_0, \ldots, \beta_j = 1, \ldots, \beta_n = b_n]$.

Having said this, we may extend exactly the same consideration to the case of a morphism $X \to \mathbb{P}_A^n$, where X is the spectrum of a local A-algebra:

Proposition III-36. *If T is a local A-algebra, the morphisms $\operatorname{Spec} T \to \mathbb{P}_A^n$ (in the category of A-schemes) are in one-to-one correspondence with the set of $(n+1)$-tuples $[\alpha_0, \dots, \alpha_n] \in T^{n+1}$ such that at least one of the α_i is a unit, modulo the equivalence relation $[\alpha_0, \dots, \alpha_n] \sim [\alpha\alpha_0, \dots, \alpha\alpha_n]$ for any unit $\alpha \in T$.*

Proof. Write $\mathbb{P}_A^n = \operatorname{Proj} A[x_0, \dots, x_n]$. Given an $(n+1)$-tuple $[\alpha_0, \dots, \alpha_n]$ with α_i a unit, we map $\operatorname{Spec} T$ to $U_i = (\mathbb{P}_A^n)_{x_i} \subset \mathbb{P}_A^n$ via the map corresponding to the A-algebra homomorphism

$$\left[\frac{x_0}{x_i}, \dots, \frac{x_n}{x_i}\right] \longrightarrow T,$$

$$\frac{x_j}{x_i} \longmapsto \frac{\alpha_j}{\alpha_i}.$$

Conversely, given a morphism $\varphi : \operatorname{Spec} T \to \mathbb{P}_A^n$ of A-schemes, let $p \in \operatorname{Spec} T$ be the unique closed point, and suppose that $\varphi(p) \in U_i$. The preimage $\varphi^{-1}(U_i)$ is an open subset of $\operatorname{Spec} T$ containing p, and hence in all of $\operatorname{Spec} T$; in other words, $\varphi(X) \subset U_i$. The map φ is thus given by a map of A-algebras

$$\left[\frac{x_0}{x_i}, \dots, \frac{x_n}{x_i}\right] \to T$$

and we may associate to φ the $(n+1)$-tuple

$$\left[\alpha_0 = \frac{x_0}{x_i}, \dots, \alpha_i = 1, \dots, \alpha_n = \frac{x_n}{x_i}\right].$$

(If the image $\varphi(x)$ is also contained in U_j, we arrive at the $(n+1)$-tuple

$$\left[\beta_0 = \frac{x_0}{x_j}, \dots, \beta_j = 1, \dots, \alpha_n = \frac{x_n}{x_j}\right],$$

which equals $[\alpha\alpha_0, \dots, \alpha\alpha_n]$ for $\alpha = x_i/x_j$.) $\square$

To generalize this further, to affine rings or schemes, we seek a construction that, locally, reduces to the one above. To this end, we may regard the $(n+1)$-tuple $(\alpha_0, \dots, \alpha_n)$ of the proposition as giving a module homomorphism

$$\alpha : T^{n+1} \to T.$$

To say that α is surjective is equivalent to saying that any of the α_i is a unit in T. And two such maps are equivalent if they differ by composition with an automorphism of the module T (that is, multiplication by a unit). Equivalently, the kernel is a rank-n summand of T^{n+1}.

It turns out that this last sentence generalizes to describe A-morphisms from an A-scheme X to $\mathbb{P}^n$: they correspond to subsheaves $\mathscr{E} \subset \mathscr{O}_X^{n+1}$ of rank n that are locally direct summands of $\mathscr{O}_X^{n+1}$; or, equivalently, to maps

$$\mathscr{O}_X^{n+1} \to P \to 0,$$

where P is a sheaf locally isomorphic to $\mathscr{O}_X$ (such a sheaf is called *invertible*, a term that will be explained in the following discussion), modulo units of $\mathscr{O}_X$ acting as automorphisms of P.

Theorem III-37. *For any scheme X, we have natural bijections*

$$\operatorname{Mor}(X, \mathbb{P}^n_{\mathbb{Z}})$$

$$= \{\text{subsheaves } K \subset \mathscr{O}_X^{n+1} \text{ that locally are summands of rank } n\}$$

$$= \frac{\{\text{invertible sheaves } P \text{ on } X, \text{ together with an epimorphism } \mathscr{O}_X^{n+1} \to P\}}{\{\text{units of } \mathscr{O}_X(X) \text{ acting as automorphisms of } P\}}.$$

Here "natural" means that for any morphism $\varphi : X \to Y$ of schemes, the map $\operatorname{Mor}(Y, \mathbb{P}^n_{\mathbb{Z}}) \to \operatorname{Mor}(X, \mathbb{P}^n_{\mathbb{Z}})$ given by composition with φ commutes with pullback of invertible sheaves and epimorphisms; in other words, we have an isomorphism of functors from the category of schemes to the category of sets.

Of course, if $X \to B$ is a B-scheme, we will be interested in describing the morphisms of X to $\mathbb{P}^n_B$ over B. This turns out to involve no new ideas: somewhat surprisingly, for any B-scheme $X \to B$, a B-morphism $X \to \mathbb{P}^n_B$ is exactly the same thing as a morphism $X \to \mathbb{P}^n_{\mathbb{Z}}$! The point is, since $\mathbb{P}^n_B$ is the product of $\mathbb{P}^n_{\mathbb{Z}}$ with B, a morphism of any scheme X to $\mathbb{P}^n_B$ is uniquely determined by the data of a morphism $X \to B$ and a morphism $X \to \mathbb{P}^n_{\mathbb{Z}}$.

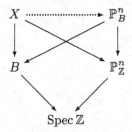

Thus, after specifying a structure morphism $\varphi : X \to B$ we get a bijection

$$\operatorname{Mor}(X, \mathbb{P}^n_{\mathbb{Z}}) \leftrightarrow \operatorname{Mor}_B(X, \mathbb{P}^n_B).$$

We now proceed with the proof of Theorem III-37. Because all the terms in these equalities are defined locally on X, the theorem reduces easily to the case where X is affine, and this is the case we will actually prove below. First, we review the corresponding notions about modules. A good basic reference is Bourbaki [1972, Chap. II-5].

Recall that a module K over a ring T is *locally free* of rank m if for every maximal ideal (or, equivalently, every prime ideal) $\mathfrak{p}$ the $T_{\mathfrak{p}}$-module $K_{\mathfrak{p}}$ is free of rank m. This is the same as the sheaf-theoretic notion.

Exercise III-38. Let K be a finitely generated module over a Noetherian ring T, and let $\tilde{K}$ be the corresponding coherent sheaf over $\operatorname{Spec} T$. Show that K is a locally free module in the sense above if and only if $\tilde{K}$ is a locally free coherent sheaf in the sense that there is an affine cover of $\operatorname{Spec} T$ by basic open sets $\operatorname{Spec} T_{f_i}$ such that the restriction of $\tilde{K}$ to each of these sets is free (equivalently, each $K[f_i^{-1}]$ is free over $T_{f_i} = T[f_i^{-1}]$).

An *invertible* T-module is a finitely generated, locally free T-module of rank 1.

In commutative algebra, locally free modules are usually called *projective modules*; their characteristic property is that if P is a locally free T-module, then any epimorphism of T-modules $M \twoheadrightarrow P$ splits. It follows that if $K \subset T^{n+1}$ is a submodule, then K is a summand of T^{n+1} if and only if T^{n+1}/K is a locally free module; in particular, K is a rank n summand of T^{n+1} if and only if T^{n+1}/K is an invertible module.

Before giving the proof of Theorem III-37, we record a result that comes from an immediate application of the definitions.

Proposition III-39. *A morphism of an arbitrary scheme X to projective space $\mathbb{P}_{\mathbb{Z}}^r = \operatorname{Proj} \mathbb{Z}[x_0, \ldots, x_r]$ may be given by a collection of maps $\varphi_i : U_i \to (\mathbb{P}_{\mathbb{Z}}^r)_{x_i}$, where $\{U_i\}$ is an open cover of X, the $(\mathbb{P}_{\mathbb{Z}}^r)_{x_i} \subset \mathbb{P}_{\mathbb{Z}}^r$ are the open subsets of Exercise III-6, and the maps φ_i and φ_j induce the same map $U_i \cap U_j \to (\mathbb{P}_{\mathbb{Z}}^r)_{x_i} \cap (\mathbb{P}_{\mathbb{Z}}^r)_{x_j} = \operatorname{Spec}(\mathbb{Z}[x_0, \ldots, x_r][x_i^{-1}, x_j^{-1}])_0$.*

The heart of Theorem III-37 is the following result, which is the affine version of the first equality.

Proposition III-40. *If T is any ring, then*

$$\operatorname{Mor}(\operatorname{Spec} T, \mathbb{P}_{\mathbb{Z}}^n)$$

$$= \{K \subset T^{n+1} \mid K \text{ is locally a rank } n \text{ direct summand of } T^{n+1}\}.$$

Proof. Suppose, first, that K is a rank n free summand of T^{n+1}, and write P for the module T^{n+1}/K. This module is locally free of rank 1 and is generated by the $n+1$ images e_i of the $n+1$ generators of T^{n+1}. Let I_j be the annihilator of (P/Te_j), and let U_j be the complement of $V(I_j)$ in $\operatorname{Spec} T$, so that the U_j form an open cover of $\operatorname{Spec} T$. Regard T-modules as sheaves on $\operatorname{Spec} T$. On U_j the map $T \to P$ defined by $1 \mapsto e_j$ is an isomorphism, and identifying $P|_{U_j}$ with $T|_{U_j}$ via this map, the projection $T^{n+1}|_{U_j} \to (T^{n+1}/K)|_{U_j} = P|_{U_j} = T|_{U_j}$ has a matrix of the form $(t_{j0}, \ldots, t_{jj} = 1, \ldots, t_{jn})$, which defines an element of $T_{U_j}^{n+1}$ and thus a morphism $\operatorname{Spec} T \to \mathbb{A}_{\mathbb{Z}}^n$. These morphisms agree on overlaps as in Proposition III-39, so they define a morphism $\operatorname{Spec} T \to \mathbb{P}_{\mathbb{Z}}^n$.

Conversely, suppose that we are given a morphism ψ from $\operatorname{Spec} T$ to $\mathbb{P}_{\mathbb{Z}}^n$. Since $\mathbb{P}_{\mathbb{Z}}^n$ is covered by $n+1$ affine n-spaces, ψ is by definition associated to an open cover $\operatorname{Spec} T = \bigcup_{i=0, \ldots, n} U_i$ and for each j an element $(t_{j0}, \ldots, t_{jj} = 1, \ldots, t_{jn})$ of $T^{n+1}|_{U_j}$ such that t_{ij} is a unit on $U_i \cap U_j$ and $t_{il} = t_{ij} t_{jl}$ in $T|_{U_i \cap U_j}$ for all i, j, l. Two such T-valued points are the same if and only if the corresponding elements of $T^{n+1}|_{U_j}$ are equal for each j.

Let K_j be the kernel of the map $T^{n+1}|_{U_j} \to T|_{U_j}$ defined by the matrix $(t_{j0}, \ldots, t_{jj} = 1, \ldots, t_{jn})$, and let

$$K = \{a \in T^{n+1} : a|_{U_j} \in K_j \text{ for each } j\}.$$

To see that K is locally a rank n summand of T^{n+1}, note that any local ring of T is a local ring of one of the U_j so the result of localizing the sequence $0 \to K \to T^{n+1} \to T^{n+1}/K \to 0$ at any prime ideal $\mathfrak{p}$ is a sequence of the form

$$0 \to K_\mathfrak{p} \to T_\mathfrak{p}^{n+1} \to T_\mathfrak{p} \to 0,$$

and such a sequence must split. $\square$

Exercise III-41. The word "locally" can be omitted in the statement of the proposition, because a submodule of a finitely generated free module that is locally a direct summand is in fact a direct summand. Prove this.

To derive a version of this with invertible modules, we use the fact that $K \subset T^{n+1}$ is a direct summand of rank n if and only if T^{n+1}/K is an invertible module, and identify the set on the right-hand side of the equality in Proposition III-40 with the set of invertible quotient modules of T^{n+1}. We may separate the isomorphism class of the quotient from the surjection that makes it a quotient and look at invertible T-modules P with surjections $T^{n+1} \to P$. Two surjections $\alpha, \beta : T^{n+1} \to P$ have the same kernel if and only if there is an automorphism $\sigma : P \to P$ such that $\beta = \sigma\alpha$. But if P is an invertible T-module, then $\mathrm{Hom}_T(P, P) = T$ (reason: the natural map $\alpha : T \to \mathrm{Hom}_T(P, P)$ taking 1 to the identity is locally the same as the natural map $T \to \mathrm{Hom}_T(T, T)$, which is an isomorphism, so α is an isomorphism). Thus the automorphisms of P may be identified with units of T, and we get the following corollary:

Corollary III-42. *If T is any ring, then*

$\mathrm{Mor}(\mathrm{Spec}\, T, \mathbb{P}_\mathbb{Z}^n)$

$$= \frac{\{\text{invertible } T\text{-modules } P \text{ with an epimorphism } T^{n+1} \to P\}}{\{\text{isomorphisms}\}},$$

where an isomorphism from $\varphi : T^{n+1} \to P$ to $\varphi' : T^{n+1} \to P$ is an isomorphism $\alpha : P \to P'$ such that $\alpha\varphi = \varphi'$. Note that the set of such isomorphisms is either empty or in (non-natural) one-to-one correspondence with the units of T.

In the classical case of the variety $\mathbb{P}_K^n$ over a field K, we can specify points of $\mathbb{P}_K^n$ by giving $(n+1)$-tuples of elements of K, not all zero. (In the scheme $\mathbb{P}_K^n$, of course, there are other, nonclosed points as well.) Analogously, for any ring A, an $(n + 1)$-tuple $(a_0, \ldots, a_n)$ of elements $a_i \in A$ that generate the unit ideal defines a surjection $A^{n+1} \to A$ of A-modules and thus defines an A-valued point of $\mathbb{P}_A^n$.

Exercise III-43. (a) Show that there are bijections between the sets

$\{(n+1)$-tuples of elements of A that generate the unit ideal$\}$ and

$\left\{\begin{array}{l}\text{maps } \operatorname{Spec} A \to \mathbb{P}^n_A \text{ such that the composite } \operatorname{Spec} A \to \mathbb{P}^n_A \to \operatorname{Spec} A \\ \text{is the identity (A-valued points of } \mathbb{P}^n_A \text{ in the category of A-schemes)}\end{array}\right\}$.

(b) Show that the image of the morphism $\operatorname{Spec} A \to \mathbb{P}^n_A$ associated to an $(n+1)$-tuple $(a_0, \ldots, a_n)$ is the closed subscheme

$$V(\{a_i X_j - a_j X_i\}_{0 \le i,j \le n}).$$

If A is a domain, show that $(\{a_i X_j - a_j X_i\}_{0 \le i,j \le n})$ is a prime ideal.

If A is a domain, Exercise III-43 shows that the image is a reduced and irreducible closed subscheme of $\mathbb{P}^n_A$, and in particular corresponds to a point of $|\mathbb{P}^n_A|$. The example of the point of $\mathbb{P}^1_{\mathbb{Z}}$ corresponding to $(2,5)$ is treated in Exercise III-9 above. Note that the $\operatorname{Spec} \mathbb{Z}$-valued point $(2,5)$ is not a $\operatorname{Spec} \mathbb{Z}$-valued point of either open set in the standard affine open cover $\mathbb{P}^1_{\mathbb{Z}} = \mathbb{A}^1_{\mathbb{Z}} \cup \mathbb{A}^1_{\mathbb{Z}}$ of $\mathbb{P}^1_{\mathbb{Z}}$, even though the point $(2x_1 - 5x_0) \in |P^1_{\mathbb{Z}}|$ lies in both!

Finally, if we are working in the category of B-schemes, we may ask for a generalization of this result describing maps of a given B-scheme X to a projective bundle. To state the result, let $\mathscr{E}$ be any coherent sheaf on B. We have then:

Theorem III-44. *For any B-scheme $\varphi : X \to B$ and coherent sheaf $\mathscr{E}$ on B, there is a natural bijection*

$\operatorname{Mor}_B(X, \mathbb{P}(\mathscr{E}))$

$= \dfrac{\{\text{invertible sheaves } P \text{ on } X, \text{ together with an epimorphism } \varphi^* \mathscr{E} \to P\}}{\{\text{isomorphisms}\}},$

where isomorphism is defined as in Corollary III-42.

We will not prove this here; the proof can be carried out by locally expressing the coherent sheaf $\mathscr{E}$ as a quotient of a free sheaf $\mathscr{O}_B^{n+1}$, and characterizing the subset of morphisms from X to $\mathbb{P}^n_B$ that factor through the resulting inclusion $\mathbb{P}(\mathscr{E}) \hookrightarrow \mathbb{P}^n_B$.

Exercise III-45. (a) Suppose that $Y \subset \mathbb{P}^n_A$ is the closed subscheme defined by homogeneous equations $\{F_i\}$. If T is a local A-algebra then, as we showed above, the morphisms from $\operatorname{Spec} T$ to $\mathbb{P}^n_A$ may be identified with $n+1$-tuples of elements of T generating the unit ideal, modulo units of T. Show that the condition that such an $n+1$-tuple correspond to a map to Y is simply that it be a zero of all the polynomials F_i.

(b) The general case of a map from an affine A-scheme to a projective A-scheme can be reduced to the local one using the following fact: if T is any A-algebra a morphism $\operatorname{Spec} T \to \mathbb{P}^n_A$ factors through X if and only if for all primes $\mathfrak{p}$ of T the composite morphisms $\operatorname{Spec} T_{\mathfrak{p}} \to \operatorname{Spec} T \to \mathbb{P}^n_A$; factor through X. Prove this.

Having characterized morphisms of schemes to projective spaces, it is instructive to look back to other geometric theories for similar characterizations. Recall that in topology the space $\mathbb{P}^n_{\mathbb{C}}$ of n-dimensional subspaces of a complex $(n+1)$-dimensional vector space $\mathbb{C}^{n+1}$ is the classifying space for subbundles of rank n of a trivial bundle of rank $n+1$ (and similarly for $\mathbb{P}^n_{\mathbb{R}}$). This means that for all spaces X, maps $X \to \mathbb{P}^n_{\mathbb{C}}$ correspond to the rank n subbundles of the trivial bundle on X. The correspondence is easy to describe: a rank n subbundle $\mathscr{J}$ of the trivial bundle $\mathscr{V} = \mathbb{C}^{n+1} \times X$ on X corresponds to the map $X \to \mathbb{P}^n_{\mathbb{C}}$ that sends a point $p \in X$ to the point of $\mathbb{P}^n_{\mathbb{C}}$ corresponding to the space

$$\mathscr{J}_p \subset \mathscr{V}_p = \mathbb{C}^n \times \{p\} = \mathbb{C}^n.$$

There are other equivalent descriptions, which may be more familiar, in terms of the rank 1 quotient bundle $\mathscr{V}/\mathscr{J}$ or the subbundle $(\mathscr{V}/\mathscr{J})^* \subset \mathscr{V}^*$ of rank 1.

Analogous results hold in the category of complex analytic spaces and maps and in the category of algebraic varieties and regular maps (taking the subbundles to be complex analytic, or algebraic, respectively). In this section we give a corresponding result for schemes. The main difference is that in algebraic geometry, it is traditional to replace vector bundles on Y by their sheaves of sections.

To see what these sheaves should look like, consider first that if $\mathscr{E}$ is a trivial vector bundle of rank 1 on a scheme X, then a section of $\mathscr{E}$ is the same as a function on X, so the sheaf of sections of $\mathscr{E}$ should be $\mathcal{O}_X$. Taking direct sums, we see that the sheaf of sections of a trivial vector bundle of rank m is the coherent sheaf that is the free $\mathcal{O}_X$-module $\mathcal{O}_X^m$. In general, since vector bundles are by definition locally trivial, their sheaves of sections are locally free sheaves of $\mathcal{O}_X$-modules of finite rank — locally free coherent sheaves. It is not hard to go in the other direction as well and to derive from a locally free coherent sheaf a vector bundle.

Given this equivalence between vector bundles and locally free coherent sheaves, why work with locally free sheaves? The reason is similar to the reason for working with schemes instead of varieties even if one is primarily interested in varieties: locally free coherent sheaves live naturally in the larger category of coherent sheaves, and working in the larger category gives us flexibility. Standard constructions in the smaller category (such as taking the fibers of a morphism of schemes, or taking the cokernel of a homomorphism of locally free sheaves) are most naturally interpreted in the larger category.

Like the line bundles to which they correspond, locally free sheaves of rank 1 play an especially important role and have a special name: they are called *invertible sheaves*. The terminology comes from number theory: an invertible module over a domain T is a finitely generated submodule I of the quotient field such that, for some other finitely generated submodule J of the quotient field (called its inverse) we have $IJ = T$, the unit ideal.

Over the scheme $\operatorname{Spec} T$, the corresponding sheaf is an invertible sheaf. More generally, given any invertible sheaf $\mathscr{I}$ over an arbitrary scheme X, the sheaf $\mathscr{I}^* = \operatorname{Hom}(\mathscr{I}, \mathscr{O}_X)$ is again invertible and the natural map $\mathscr{I} \otimes \mathscr{I}^* \to \mathscr{O}_X$ is an isomorphism (check locally). For this reason $\mathscr{I}^*$ is called the *inverse* of $\mathscr{I}$.

Knowing that invertible sheaves correspond to line bundles does not at first seem to help connect them to geometry. Just as in classical algebraic geometry, however, morphisms of a scheme X to projective space can in fact be characterized in geometric terms using the related notion of an *effective Cartier divisor*. This is defined to be a subscheme $D \subset X$ such that at every point $x \in X$ the ideal of D in the local ring $\mathscr{O}_{X,x}$ is principal and generated by a nonzerodivisor. In other words, a subscheme D is an effective Cartier divisor if and only if its ideal sheaf $\mathscr{I}_D$ is invertible. Following tradition, we define the *invertible sheaf $\mathscr{O}_X(D)$ associated to D* to be the inverse

$$\mathscr{O}_X(D) = \mathscr{I}_D^*.$$

The invertible sheaves form a group $\operatorname{Pic} X$ under the tensor product operation, and under reasonable circumstances — for example, for subschemes of projective space over a field — every invertible sheaf can be written as $\mathscr{O}_X(D) \otimes \mathscr{I}_E = \mathscr{O}_X(D) \otimes \mathscr{O}_X(E)^*$ for some effective Cartier divisors D, E.

Note the unfortunate but essentially unambiguous notation: if U is an open set of X then $\mathscr{O}(U)$ denotes the ring of sections of the sheaf $\mathscr{O}_X$ defined over U, while if D is a Cartier divisor $\mathscr{O}(D)$ denotes the sheaf above. Of course we could also manufacture such monstrosities as $\mathscr{O}_X(D)(U)\dots$

We may tighten the connection between invertible sheaves and effective Cartier divisors as follows: If D is an effective Cartier divisor then the inclusion $\mathscr{I}_D \hookrightarrow \mathscr{O}_X$ is a global section of $\operatorname{Hom}(\mathscr{I}_D, \mathscr{O}_X) = \mathscr{O}_X(D)$. This section is *regular* in the sense that for every open set $U \subset X$ no nonzero element of $\mathscr{O}_X(U)$ annihilates the restriction of this section to U (Reason: the image of $\mathscr{I}_D(U)$ in $\mathscr{O}_X(U)$ contains a nonzerodivisor.) Thus an effective Cartier divisor gives rise to an invertible sheaf with a global section.

Conversely, given an invertible sheaf $\mathscr{L}$ and a global section σ, we define the zero locus of that section to be the support of the quotient $\mathscr{L}/\mathscr{O}_X\sigma$. To understand what this means, choose a covering of X by open sets U such that $\mathscr{L}|_U \cong \mathscr{O}_U$. The zero locus of σ in U is then the zero locus of the corresponding element of $\mathscr{O}_U$. If the section is regular, it follows that the zero locus is an effective Cartier divisor. Note that another global section differing from σ by a unit in $\mathscr{O}_X(X)$ would give the same Cartier divisor. We thus have a bijection

{effective Cartier divisors}

$\updownarrow$

{invertible sheaves with choice of global section modulo units}.

The reader might wonder about the significance of "effective". An effective Cartier divisor D may be defined by giving a nonzerodivisor in

$f_U \in \mathscr{O}_X(U)$ for each set U of an open covering of X, and f_U is defined up to a unit of $\mathscr{O}_X(U)$. Thus D gives rise to a unique global section of the sheaf of invertible rational functions modulo invertible regular functions — that is, the sheaf $\mathscr{M}_X^*/\mathscr{O}_X^*$, where $*$ denotes the sheaf of multiplicative units, and $\mathscr{M}_X$ is the sheafification of the presheaf whose value on an open set U is the localization $\mathscr{O}_X(U)[S_U^{-1}]$ of the ring $\mathscr{O}_X(U)$ at the multiplicatively closed set S_U of elements that become nonzerodivisors in $\mathscr{O}_{X,x}$ for every $x \in U$. (All that one usually needs to know about this slightly baroque definition is that if $U = \operatorname{Spec} A$ for a Noetherian ring A, then $\mathscr{M}_X(U)$ is the result of inverting all nonzerodivisors in A. More general cases are subtle; see for example Kleiman [1979] for information.) We define a Cartier divisor in general to be an arbitrary section of the sheaf $\mathscr{M}_X^*/\mathscr{O}_X^*$. The Cartier divisors on X form a group called $\operatorname{Div} X$, and the association $D \mapsto \mathscr{O}_X(D)$ defines a homomorphism $\operatorname{Div} X \to \operatorname{Pic} X$.

The effective Cartier divisors form a monoid in $\operatorname{Div} X$; again, in reasonable circumstances such as for subschemes of a projective space over a field, the monoid of effective Cartier divisors generates $\operatorname{Div} X$, and in the freest possible way: $\operatorname{Div} X$ may also be realized as the Grothendieck group of the monoid. The effective Cartier divisors are then just the Cartier divisors that "effectively" define subschemes.

III.2.6 Graded Modules and Sheaves

The attentive reader may have noticed that Theorem III-37 implies the existence of a distinguished invertible sheaf on $\mathbb{P}_{\mathbb{Z}}^n$, namely, the one corresponding to the identity map. In this section we will give descriptions of this sheaf, which plays a fundamental role in projective geometry.

We begin with a general method for constructing sheaves on schemes of the form $\operatorname{Proj} \mathscr{A}$ analogous to the construction of sheaves on $\operatorname{Spec} A$ from modules over A. Let B be a scheme, and let $\mathscr{A} = \mathscr{A}_0 \oplus \mathscr{A}_1 \oplus \cdots$ be a quasicoherent sheaf of graded $\mathscr{O}_B$-algebras. Let $\mathbb{P} = \operatorname{Proj} \mathscr{A}$. Let $\mathscr{M}$ be a quasicoherent sheaf on B which has the additional structure of a sheaf of graded $\mathscr{A}$-modules; that is, we have a direct sum decomposition $\mathscr{M} = \cdots \oplus \mathscr{M}_i \oplus \mathscr{M}_{i+1} \oplus \cdots$ and there are maps $\mathscr{A}_i \otimes \mathscr{M}_j \to \mathscr{M}_{i+j}$ satisfying the usual axioms (associativity, identity, ...). We may associate to $\mathscr{M}$ a quasicoherent sheaf $\widetilde{\mathscr{M}}$ on $\mathbb{P}$ as follows: Let U be an affine subset of B, and consider the graded ring $\mathscr{A}(U)$. For each homogeneous element f of $\mathscr{A}(U)$ we have an affine open set $\mathbb{P}_{U,f} := (\operatorname{Proj} \mathscr{A}(U))_f = \operatorname{Spec}(\mathscr{A}(U)_f)_0$ of $\mathbb{P}$; the schemes $\mathbb{P}_{U,f}$ form an affine open cover of $\mathbb{P}$. The sections $\mathscr{M}(U)$ over U form a graded module over the graded ring $\mathscr{A}(U)$. Let $\mathscr{M}_{U,f}$ be the $(\mathscr{A}(U)_f)_0$-module $\mathscr{M}_{U,f} = (\mathscr{M}(U) \otimes_{\mathscr{A}(U)} \mathscr{A}(U)[f^{-1}])_0$, and let $\widetilde{\mathscr{M}}_{U,f}$ the corresponding sheaf on the affine scheme $\mathbb{P}_{U,f}$. These patch together to define a quasicoherent sheaf on $\mathbb{P}$ that we denote by $\widetilde{\mathscr{M}}$.

In fact, every quasicoherent sheaf on $\operatorname{Proj} \mathscr{A}$ corresponds to a sheaf of graded $\mathscr{A}$ modules in this way. However, unlike the correspondence between modules over a ring and quasicoherent sheaves over Spec of that ring, the correspondence is not bijective. For example, as the reader can easily check, the sheaf associated to a module $\mathcal{M}$ is the same as the sheaf associated to the truncated module $\mathcal{M}' = \oplus_{n \geq n_0} \mathcal{M}_n$ for any n_0. But in good cases this is the only kind of failure: for example, the association $\mathcal{M} \mapsto \tilde{\mathcal{M}}$ gives a bijection

{sheaves of finitely generated graded $\mathcal{A}$-modules up to truncation}

$\updownarrow$

{quasicoherent sheaves on $\mathbb{P}$}.

To start with the simplest example, if we take $\mathcal{M} = \mathcal{A}$ we get the structure sheaf $\mathcal{O}_{\mathbb{P}}$. Much more interesting is the deceptively simple modification obtained by shifting the grading by 1. In general, if $\mathcal{M} = \bigoplus_i \mathcal{M}_i$ then we define the n-th twist $\mathcal{M}(n)$ of $\mathcal{M}$ to be the same module but with degrees shifted by n, that is

$$\mathcal{M}(n)_i = \mathcal{M}_{n+i}.$$

We define $\mathcal{O}_{\mathbb{P}}(n)$ to be the sheaf $\widetilde{\mathcal{A}(n)}$ associated to the sheaf of graded modules $\mathcal{A}(n)$. The most important of these is $\mathcal{O}_{\mathbb{P}}(1)$, called the *tautological sheaf* on $\mathbb{P}$.

Exercise III-46. Assume the algebra $\mathcal{A}$ is generated in degree 1. Show that all the sheaves $\mathcal{O}_{\mathbb{P}}(n)$ are invertible. Show that $\mathcal{O}_{\mathbb{P}}(n) \otimes \mathcal{O}_{\mathbb{P}}(m) = \mathcal{O}_{\mathbb{P}}(n + m)$, and in particular $\mathcal{O}_{\mathbb{P}}(n)^{-1} = \mathcal{O}_{\mathbb{P}}(n)^* = \mathcal{O}_{\mathbb{P}}(-n)$.

Exercise III-47. Let $\pi : \mathrm{Proj}\,\mathcal{A} \to \mathrm{Spec}\,\mathcal{A}_0$ be the structure map. Show that for any quasicoherent sheaf $\mathcal{N}$ on $\mathrm{Spec}\,\mathcal{A}_0$ the pullback $\pi^*(\mathcal{N})$ is the sheaf associated to $\mathcal{A} \otimes_{\mathcal{A}_0} \mathcal{N}$.

Exercise III-48. Let K be a field, and consider the projective space $\mathbb{P}^n_K = \mathrm{Proj}\,K[x_0, \ldots, x_n]$. Let H be a hyperplane. Show that H is a Cartier divisor on $\mathbb{P}^n_K$ and that the associated invertible sheaf is $\mathcal{O}_{\mathbb{P}^n_K}(1)$.

III.2.7 Grassmannians

Grassmannians exist in the category of schemes, and behave very much like Grassmannians in classical algebraic geometry. More precisely, there is, for any scheme S and positive integers n and $k < n$, a scheme $G_S(k, n)$ called the *Grassmannian* over S; the construction is functorial in S, in the sense that for any morphism $T \to S$, the Grassmannian $G_T(k, n)$ is the fiber product $G_S(k, n) \times_S T$. (In particular, there is a scheme $G_{\mathbb{Z}}(k, n)$ — the Grassmannian over $\mathrm{Spec}\,\mathbb{Z}$ — such that any Grassmannian may be realized as $G_S(k, n) = G_{\mathbb{Z}}(k, n) \times S$.) Moreover, in case $S = \mathrm{Spec}\,K$ is the spectrum of an algebraically closed field the scheme $G_S(k, n)$ is the scheme associated to the classical Grassmann variety $G(k, n)$ over K. In fact, the constructions, which we will describe briefly below, are themselves exactly analogous to the standard constructions of the Grassmannian in the classical context. Rather, as in the case of projective space, what is new and different about the Grassmannian as a scheme are the subschemes of it; we will illustrate this with our discussion of *Fano schemes* below.

We will begin with the constructions of the Grassmannian $G_S(k, n)$ for $S = \operatorname{Spec} A$ an affine scheme (this is also called the Grassmannian *over A* and denoted $G_A(k, n)$). At the end, we will observe that the construction is natural, in the sense that for any morphism $T \to S$ of affine schemes we have

$$G_T(k, n) = G_S(k, n) \times_S T.$$

It will follow that we can construct Grassmannians over arbitrary schemes S by gluing together the Grassmannians $G_{U_\alpha}(k, n)$ over a collection of affine open subsets $U_\alpha \subset S$ covering S. Alternatively, we can simply carry out the construction of the Grassmannian $G_{\mathbb{Z}}(k, n)$ over $\operatorname{Spec} \mathbb{Z}$, and then for any scheme S simply define $G_S(k, n) = S \times G_{\mathbb{Z}}(k, n)$.

In the classical setting, there are two ways of constructing the Grassmannian $G_K(k, n)$ as a variety over a field K. Abstractly, we may describe $G_K(k, n)$ as a union of open sets, each isomorphic to affine space $\mathbb{A}_K^{k(n-k)}$. Alternatively, we may describe it at one stroke as the closed subvariety of projective space $\mathbb{P}_K^N$ given by the Plücker equations. Each of these constructions has an immediate extension to the category of schemes, and they do yield the same object. Moreover, there is in the language of schemes a third way to characterize Grassmannians: as Hilbert schemes, or more precisely as the schemes representing the functors of families of linear subspaces of a fixed vector space. We will discuss this third construction in Section VI.2.1. This is in many ways the optimal characterization of the Grassmannian: it avoids the extraneous introduction of coordinates, gives us immediately a description of morphisms of an arbitrary scheme Z to $G_K(k, n)$, and gives us a natural definition of equations for subschemes of the Grassmannian such as Fano schemes and more general Hilbert schemes.

We will start by reviewing the gluing construction of the Grassmannian as a variety over a field. We begin by realizing the set of k-dimensional linear subspaces Λ of the n-dimensional vector space K^n over a field K as the set of $k \times n$ matrices M of rank k, modulo multiplication on the left by invertible $k \times k$ matrices. For each subset $I \subset \{1, 2, \ldots, n\}$ of cardinality k we can multiply any matrix M whose I-th minor is nonzero by the inverse of its I-th submatrix M_I, to obtain a matrix M' with I-th submatrix equal to the identity. In this way, we may identify the subset $U_I \subset G_K(k, n)$ of planes Λ complementary to the subspace of K^n spanned by the basis vectors $\{e_i\}_{i \notin I}$ with the affine space $\mathbb{A}_K^{k(n-k)}$ whose coordinates are the remaining entries of M'. We thus have the following recipe for the variety $G_K(k, n)$:

Let $W \cong \mathbb{A}_K^{kn}$ be the space of $k \times n$ matrices, and for each subset $I \subset \{1, 2, \ldots, n\}$ of cardinality k, let $W_I \subset W$ be the closed subset of matrices with I-th submatrix equal to the identity. For each I and $J \neq I$, let $W_{I,J} \subset W_I$ be the open subset of matrices whose J^{th} minor is nonzero; let $\varphi_{I,J} : W_{I,J} \to W_{J,I}$ be the isomorphism given by multiplication on the left by $M_J \cdot M_I^{-1}$. We then define the Grassmannian $G_K(k,n)$ as an abstract variety to be the union of the affine spaces $W_I \cong \mathbb{A}_K^{k(n-k)}$ modulo the identifications of $W_{I,J}$ with $W_{J,I}$ given by $\varphi_{I,J}$.

This recipe applies perfectly well to define the Grassmannian $G_S(k,n)$ over any affine scheme $S = \operatorname{Spec} A$, using the gluing construction of Section I.2.4: let

$$W = \operatorname{Spec} A[\ldots, x_{i,j}, \ldots] \cong \mathbb{A}_S^{kn},$$

and for each subset $I = (i_1, \ldots, i_k) \subset \{1, 2, \ldots, n\}$ let $W_I \subset W$ be the closed subscheme corresponding to matrices whose I-th $k \times k$ submatrix is the identity; that is, the zero locus of the ideal $(\ldots, x_{\alpha, i_\beta} - \delta_{\alpha, \beta}, \ldots)$. For each I and $J \neq I$, we define exactly as before open subschemes $W_{I,J} = (W_I)_{\det M_j} \subset W_I$ and isomorphisms $\varphi_{I,J} : W_{I,J} \to W_{J,I}$; and we then define the Grassmannian $G_S(k,n)$ to be the S-scheme obtained by gluing the affine spaces $W_I \cong \mathbb{A}_S^{k(n-k)}$ along the $\varphi_{I,J}$.

An alternative construction of the Grassmannian $G_S(k,n)$ is as a subscheme of projective space $\mathbb{P}_S^N$, where $N = \binom{n}{k} - 1$, given by the Plücker equations. Again, if we are simply careful about transcribing the classical construction, it works in this new setting as well.

To set it up, start with the polynomial ring $A[\ldots, X_I, \ldots]$ in $\binom{n}{k}$ variables over A, where the variables are labeled by subsets $I = (i_1 < \cdots < i_k) \subset \{1, \ldots, n\}$. We may think of the variables X_I as corresponding to the maximal minors of a $k \times n$ matrix M. If we specify further that the first $k \times k$ submatrix of M is the identity—that is, M is of the form (I_k, B) where B is a $k \times (n-k)$ matrix—then these are in turn up to sign the minors of all sizes of the matrix B. For example, the (i, l)-th entry of B is the I-th minor of M, where $I = (1, 2, \ldots, \hat{i}, \ldots, k, k+l)$; the $((i,j), (l,m))$-th minor of B is the I-th minor of M, where $I = (1, 2, \ldots, \hat{i}, \ldots, \hat{j}, \ldots, k, k+l, k+m)$; and so on.

In terms of this description of the coordinates X_I as the minors of all sizes of a $k \times (n-k)$ matrix A, the *Plücker relations* are simply homogeneous polynomials in the variables X_I obtained by expanding the determinants of these submatrices in terms of products of complementary minors of complementary submatrices. For example, Cramer's rule translates into an expression of the determinant of an $l \times l$ submatrix of A as a sum of products of entries and determinants of $(l-1) \times (l-1)$ submatrices; in particular, on the basis of the identification made above we have the relation

$$-X_{(1,2,\ldots,\hat{i},\ldots,\hat{j},\ldots,k,k+l,k+m)} X_{(1,2,\ldots,k)}$$
$$= X_{(1,2,\ldots,\hat{i},\ldots,k,k+l)} X_{(1,2,\ldots,\hat{j},\ldots,k,k+m)} - X_{(1,2,\ldots,\hat{i},\ldots,k,k+m)} X_{(1,2,\ldots,\hat{j},\ldots,k,k+l)}.$$

We take the *Plücker ideal* $J \subset A[\ldots, X_I, \ldots]$ to be the ideal generated by the Plücker relations.

Another, more intrinsic way to describe the ideal J is simply this: we let φ be the map

$$A[\ldots, X_I, \ldots] \longrightarrow A[x_{1,1}, \ldots, x_{k,n}]$$

$$X_I \longmapsto \begin{vmatrix} x_{1,i_1} & \cdots & x_{1,i_k} \\ \vdots & & \vdots \\ x_{k,i_1} & \cdots & x_{k,i_k} \end{vmatrix}$$

sending each generator X_I of $A[\ldots, X_I, \ldots]$ to the corresponding minor of the matrix $(x_{i,j})$, and we let $J = \operatorname{Ker} \varphi$. In either case, we define the Grassmannian $G_S(k, n)$ to be the projective scheme

$$G_S(k, n) = \operatorname{Proj} A[\ldots, X_I, \ldots]/J \subset \operatorname{Proj} A[\ldots, X_I, \ldots] = \mathbb{P}_S^{\binom{n}{k}-1}.$$

Exercise III-49. Show that the two constructions yield the same scheme $G_S(k, n)$.

This description of $G_S(k, n)$ allows us to describe intrinsically the Grassmannian $G(k, V)$ of subspaces of an n-dimensional vector space V over a field K, and hence more generally to define the Grassmannian $G(k, \mathscr{E})$ of k-dimensional subspaces of a locally free sheaf $\mathscr{E}$ over a given base scheme S. In the more general setting, we take the map of sheaves

$$\mathscr{E}^{\otimes k} = \mathscr{E} \otimes \mathscr{E} \otimes \cdots \otimes \mathscr{E} \longrightarrow \wedge^k \mathscr{E}$$

given simply by $\sigma_1 \otimes \cdots \otimes \sigma_k \mapsto \sigma_1 \wedge \cdots \wedge \sigma_k$, and let φ be the induced map on symmetric algebras

$$\varphi : \operatorname{Sym}(\wedge^k \mathscr{E})^* \longrightarrow \operatorname{Sym}(\mathscr{E}^{\otimes k})^*.$$

We then define $G(k, \mathscr{E})$ to be the subscheme of $\mathbb{P}(\mathscr{E}^*) = \operatorname{Proj} \operatorname{Sym}(\wedge^k \mathscr{E})^*$ given by the ideal sheaf $\operatorname{Ker}(\varphi)$.

One notational convention: since the Grassmannian arises sometimes in the context of linear subspaces of a vector space, and sometimes in the context of subspaces of a projective space, we will adopt the convention that $G_S(k, n)$ is the scheme described above, and $\mathbb{G}_S(k, n) = G_S(k+1, n+1)$.

III.2.8 Universal Hypersurfaces

Definition III-50. Let S be any scheme. By a *hypersurface* of degree d in $\mathbb{P}_S^n$ we mean a closed subscheme $X \subset \mathbb{P}_S^n$ given locally over S as the zero locus of a homogeneous polynomial of degree d: that is, for every point $p \in S$ there is an affine neighborhood $U = \operatorname{Spec} A$ of p in S and elements $\{a_I \in A\}$ such that the a_I generate the unit ideal in A, and

$$X \cap \mathbb{P}_U^n = V\left(\sum a_I x_0^{i_0} \ldots x_n^{i_n}\right) \subset \mathbb{P}_U^n = \operatorname{Proj} A[x_0, \ldots, x_n].$$

A hypersurface $X \subset \mathbb{P}_S^n$ is flat over S (the condition that the a_I generate the unit ideal in A means that they have no common zeros in S, so that the dimensions of the fibers of $X \to S$ are everywhere $n - 1$), and of pure codimension 1 in $\mathbb{P}_S^n$. Note that the fibers of $X \to S$ have no embedded points.

By a *plane curve* over a scheme S we will mean a hypersurface in $\mathbb{P}_S^2$.

We can now introduce a fundamental object in algebraic geometry: the *universal family* of hypersurfaces of degree d in $\mathbb{P}_S^n$. This is very straightforward to define: for any positive d and n, we set $N = \binom{d+n}{n} - 1$, and let

$$\mathbb{P}_S^N = \operatorname{Proj} \mathcal{O}_S[\{a_I\}]$$

be projective space of dimension N over S, with homogeneous coordinates a_I indexed by monomials of degree d in $n + 1$ variables $(x_0, \ldots, x_n)$. We then introduce the subscheme $\mathscr{X} = \mathscr{X}_{d,n} \subset \mathbb{P}_S^N \times_S \mathbb{P}_S^n$ given by the single bihomogeneous polynomial

$$\mathscr{X} = V\left(\sum_I a_I x^I \right).$$

The scheme $\mathscr{X} \subset \mathbb{P}_S^N \times_S \mathbb{P}_S^n$, viewed as a family of closed subschemes of $\mathbb{P}_S^n$ parametrized by $\mathbb{P}_S^N$, is called the *universal hypersurface* of degree d in $\mathbb{P}_S^n$. By Proposition II-32, $\mathscr{X}$ is flat over $\mathbb{P}_S^N$.

Note that if $S = \operatorname{Spec} K$ is the spectrum of an algebraically closed field, then every hypersurface $X \subset \mathbb{P}_K^n$ of degree d is a fiber of $\mathscr{X} \to \mathbb{P}_K^N$. In fact, much more is true: as we will see in Chapter VI, if B is any S-scheme, and $\mathscr{Y} \subset \mathbb{P}_B^n$ is any closed subscheme, flat over B, whose fibers are hypersurfaces of degree d, then there is a unique morphism $\varphi : B \to \mathbb{P}_S^N$ of S-schemes such that $\mathscr{Y} = \mathscr{X} \times_{\mathbb{P}_S^n} B$. (This is the meaning of the term "universal".)

Universal hypersurfaces are fundamental objects in algebraic geometry, and arise in a number of contexts. We will see many examples of these objects, or variants of them, in the following chapter, and will describe them in more detail in Section V.1.2 and the following discussions of resultants and discriminants. We will present here a few of the simpler examples and related constructions.

We start with some notation and terminology. First, we will assume throughout that S is irreducible (with generic point Q), so that $\mathbb{P}_S^N$ is irreducible as well (for the most part, we can think of S as the spectrum of a field K, though there will be occasions when it will be handy to be able to take $S = \operatorname{Spec} \mathbb{Z}$). Let $P \in \mathbb{P}_S^N$ be the generic point, and $L = \kappa(P)$ its residue field, that is, the function field in N variables over the function field $K = \kappa(Q)$ of S. Let $X_P \subset \mathbb{P}_L^n$ be the fiber of $\mathscr{X} \to \mathbb{P}_S^N$ over the generic point $P = \operatorname{Spec} L$; X_P is sometimes called the *generic hypersurface* of degree d. We will as usual write $\mathscr{X}(L) = X_L(L)$ for the set of L-valued points of $\mathscr{X}$, or equivalently the L-rational points of X_P. Geometrically, these are sections of $\mathscr{X} \to \mathbb{P}_S^N$ defined over some open subset $U \subset \mathbb{P}_S^N$; algebraically, they are simply solutions $x_i = f_i(a)$ of the equation $\sum_I a_I x^I$, with the x_i rational functions of the a_I.

We start with a basic fact:

Exercise III-51. Show that $\mathscr{X}$ is irreducible, and smooth as an S-scheme. (Hint: consider the projection $\mathscr{X} \subset \mathbb{P}_S^N \times_S \mathbb{P}_S^n \to \mathbb{P}_S^n$.) Deduce in particular that $X_L \subset \mathbb{P}_L^n$ is smooth as an L-scheme.

Now for some examples:

Exercise III-52. If $d = 1$, so that $\mathbb{P}_S^N = (\mathbb{P}_S^n)^*$, the scheme $\mathscr{X} \subset \mathbb{P}_S^n \times_S (\mathbb{P}_S^n)^*$ is called, naturally enough, the *universal hyperplane*. Show that it is a projective bundle over $\mathbb{P}_S^n$.

Exercise III-53. Suppose now that d is arbitrary and $n = 1$, so that $N = d$ and the scheme $\mathscr{X} \subset \mathbb{P}_S^d \times_S \mathbb{P}_S^1$ is finite of degree d over $\mathbb{P}_S^d$. Show that the generic fiber X_L is a single reduced point R, with residue field an extension of degree d of the function field L of $\mathbb{P}_K^d$.

The last exercise is a little harder.

Exercise III-54. Suppose now that S is the spectrum of a field K, and take $d = n = 2$. Show that $\mathscr{X}(L) \neq \varnothing$.

Hint: Show that we can reduce to the inverse image of the subspace of $\mathbb{P}_K^N = \mathbb{P}_K^5$ corresponding to polynomials $aX^2 + bY^2 + cZ^2$; or just see the argument for Proposition IV-84.

It is in fact the case for all n and d that $\mathscr{X}(L) \neq \varnothing$ if and only if $d = 1$, as can be seen by an application of the Lefschetz Hyperplane Theorem to $\mathscr{X} \subset \mathbb{P}_K^N \times_K \mathbb{P}_K^n$.

III.3 Invariants of Projective Schemes

In this section we assume that K is a field and work with K-schemes, except when explicit mention is made to the contrary.

Suppose that we are given a scheme in a projective space; how can we find invariants of it? The simplest idea is to ask: how many independent forms of degree d vanish on it? Putting the answers together, for various d, we get what used to be called the postulation of the scheme (presumably because one was then interested in schemes for which one postulated certain values for these numbers). Nowadays, it is usual to discuss this information in the equivalent form of the Hilbert function. We will discuss here several variations of the method of Hilbert functions, which yield a wide range of invariants. Some of the invariants that we produce actually depend only on the abstract scheme and not on the given projective embedding, while others depend on the data associated to the embedding; and we will comment on these matters along the way. The approach we follow is the original one used by Hilbert [1890], rather than that of Samuel, which is more commonly adopted (see, for example, Hartshorne [1977, Chapter I]). Hilbert's method requires slightly more technique but yields a stronger and more easily understood result.

We begin by defining the basic invariants. In the last part of the chapter we will exhibit a number of simple geometric examples showing what sort of information the invariants contain.

III.3.1 Hilbert Functions and Hilbert Polynomials

To begin with, suppose that we are given a closed subscheme $X \subset \mathbb{P}_K^r$ described by a saturated ideal $I = I(X) \subset S = K[x_0, \ldots, x_r]$ defined as in Example III-14. Suppose that the homogeneous polynomials $F_1, \ldots, F_n$ generate I. Write $R = S/I(X)$ for the homogeneous coordinate ring of X, and write R_ν for the homogeneous component of degree ν.

The basic idea is to associate to $X \subset \mathbb{P}_K^r$ a function

$$H(X, \cdot) : \mathbb{N} \to \mathbb{N}$$

called the *Hilbert function of* X and defined by

$$H(X, \nu) = \dim_K R_\nu.$$

More generally, if M is any finitely generated graded S-module, we define its Hilbert function to be $H(M, \nu) := \dim_K M_\nu$. The fundamental result is as follows.

Theorem III-55 (Hilbert). *There exists a unique polynomial $P(X, \nu)$ in ν such that $H(X, \nu) = P(X, \nu)$ for all sufficiently large ν. More generally, for any finitely generated graded S-module M there exists a unique polynomial $P(M, \nu)$ such that $H(M, \nu) = P(M, \nu)$ for all sufficiently large ν.*

We will indicate below how this may be proved (along the lines of Hilbert's original proof [1890]).

The polynomial $P(X, \nu)$ is called the *Hilbert polynomial* of X. As in the classical case of varieties, it carries basic information about the scheme X. For example, we will see that its degree is the dimension of X, and in case X is of dimension 0, its (constant) value is the degree of X. More generally, we define the *degree* of any n-dimensional subscheme X of projective space over a field K to be $n!$ times the leading coefficient of the Hilbert polynomial of X; this allows us to extend to the larger class of subschemes $X \subset \mathbb{P}_K^r$ the classical notion of degree for varieties.

III.3.2 Flatness II: Families of Projective Schemes

Another aspect of the significance of the Hilbert polynomial is that it gives us a geometric interpretation of the notion of flatness.

Proposition III-56. *A family $\mathscr{X} \subset \mathbb{P}^r_B$ of closed subschemes of a projective space over a reduced connected base B is flat if and only if all fibers have the same Hilbert polynomial.*

A proof of this in the general case would take us too far afield, but the result is easy when the base is $B = \operatorname{Spec} K[t]_{(t)}$.

Proof when $B = \operatorname{Spec} K[t]_{(t)}$. A closed subscheme $X \subset \mathbb{P}^r_K \times B$ is given by an ideal I in

$$K[t]_{(t)}[x_0, \ldots, x_r]$$

which is homogeneous in $x_0, \ldots, x_r$. Thus each graded piece of the homogeneous coordinate ring

$$R = K[t]_{(t)}[x_0, \ldots, x_r]/I$$

is a module over $K[t]_{(t)}$.

As we know, the family $X \to B$ is flat if and only if each local ring $\mathcal{O}_{X,x}$ is $K[t]_{(t)}$-torsion-free. This is the same as saying that the torsion submodule of R goes to zero if we invert any of the x_i. It follows that the torsion submodule is killed by a power of the ideal $(x_0, \ldots, x_r)$ and thus meets only finitely many graded components of R. But if R_ν is a graded component of R, then since $K[t]_{(t)}$ is a principal ideal ring and R_ν is finitely generated as a $K[t]_{(t)}$-module, R_ν is torsion-free if and only if it is free. Further, R_ν is free if the number of generators it requires, which by Nakayama's Lemma is

$$\dim_K R_\nu \otimes_{K[t]_{(t)}} K$$

is equal to its rank

$$\dim_{K(t)} R_\nu \otimes_{K[t]_{(t)}} K(t)$$

that is, if and only if the value of the $H(X_{(0)}, \nu)$ is equal to the value of $H(X_{(t)}, \nu)$, where $X_{(0)}$ and $X_{(t)}$ are the fibers of the family X over the two points (0) and (t) of B. (By the same argument, the Hilbert function itself is constant if and only if the family of affine cones $\operatorname{Spec} R$ is a flat family over B.) $\square$

This proposition shows that flat limits of closed subschemes of projective space behave better than flat limits in general. For example, though it is certainly possible that the flat limit of nonempty subschemes of an affine scheme may be empty, the proposition shows that this is not possible for flat limits of nonempty subschemes of a projective space. This, together with the existence and uniqueness of flat limits of closed subschemes in a one-parameter family (Sections II.3.4 and II.3.4), gives one approach to proving that projective schemes are proper, using the "valuative criterion." For all this, see, for example, Hartshorne [1977, Chap. II].

Of course $H(X, \nu)$ contains more information than $P(X, \nu)$, but it may appear that $P(X, \nu)$, as a polynomial with only finitely many coefficients, is easier to manipulate than the whole Hilbert function. Actually, the Hilbert function has a finite expression too, in terms of binomial coefficients. To see this, we will introduce a still finer set of invariants, the *graded Betti numbers of the free resolution of R*, in terms of which both the Hilbert function and the Hilbert polynomial can be written conveniently. (The real advantage that the Hilbert polynomial has over the Hilbert function is that the information it contains depends a little less — in a sense we will make precise — on the details of the embedding of X.)

III.3.3   Free Resolutions

We will write $S(-b)$ for the graded, free module of rank 1 with generator in degree b; the apparently unfortunate choice of sign is recompensed by the convenient and eminently memorable formula

$$S(-b)_\nu = S_{\nu-b}.$$

We can resolve R, or indeed any graded S-module, by using graded, free modules, which are direct sums of copies of modules of the form $S(-b)$. Here is how.

Suppose that $F_1, \ldots, F_n$ is a minimal set of homogeneous generators for M. We will write b_{0j} for the degree of F_j. We define an epimorphism

$$\varphi_0 : E_0 := \bigoplus_{j=1}^{n} S(-b_{0j}) \to M$$

by sending the generator of $S(-b_{0j})$ to $F_j \in M$. Let $M^{(1)}$ be the kernel of φ_0. If $M^{(1)} \neq 0$, we repeat the process above with $M^{(1)}$ in place of M (which could be called $M^{(0)}$); choosing a minimal set of homogeneous elements $e_i^{(1)}$ of E_0 that generate $M^{(1)}$ with degrees b_{1i}, we map a graded free module with generators of degrees b_{1i} onto M_1, by a map

$$\varphi_1 : E_1 := \bigoplus_{j=1}^{m} S(-b_{1j}) \to E_0$$

sending the i-th generator of E_1 to $e_i^{(1)}$. Continuing in this way, we obtain a resolution

$$\mathbb{E} : \cdots \longrightarrow E_i \xrightarrow{\varphi_i} E_{i-1} \longrightarrow \cdots \xrightarrow{\varphi_1} E_0,$$

with

$$E_i = \bigoplus_j S(-b_{ij}).$$

Of course, the process stops if some φ_i is a monomorphism. Hilbert's fundamental discovery was that this always occurs if S is a polynomial ring.

Theorem III-57 (Hilbert's syzygy theorem). *Let* $S = K[x_0, \ldots, x_r]$. *In any minimal free resolution as above,* φ_i *is a monomorphism for some* $i \leq r + 1$, *the number of variables; in particular, any graded S-module has a finite, graded, free resolution.*

We will not prove this here; see Hilbert [1890] or, for a modern account, Eisenbud [1995, Section 1.10, Chap. 19] or Matsumura [1986, Theorem 19.5]. The syzygy theorem allows us to prove Theorem III-55.

Proof of Theorem III-55. The Hilbert function of the module $S(-b)$ is easy to write down. Since

$$S(-b)_\nu = S_{\nu-b}$$

has a basis consisting of all monomials of degree $\nu - b$ in $r + 1$ variables, we see that

$$H(S(-b), \nu) = \binom{r + \nu - b}{r},$$

where the binomial coefficient is to be interpreted as 0 when the bottom is larger than the top. For $\nu \geq b - r$ this agrees with the polynomial

$$P(S(-b), \nu) = \frac{(r + \nu - b)(r + \nu - b - 1) \cdots (\nu - b)}{r(r - 1) \cdots 1}$$

so we see that $H(X, \nu)$ is a polynomial for large ν.

From a finite, free resolution for M as an S-module

$$\mathbb{E} : 0 \longrightarrow E_{r+1} \xrightarrow{\varphi_{r+1}} E_r \longrightarrow \cdots \longrightarrow E_1 \longrightarrow M \longrightarrow 0,$$

with

$$E_i = \bigoplus_j S(-b_{ij}),$$

we see that the Hilbert function of M can be written in the form

$$H(M, \nu) = \sum_{i=0}^r (-1)^i H(E_i, \nu) = \sum_{i=0}^r (-1)^i \sum_j H(S(-b_{ij}), \nu).$$

Since we have already shown that each $H(S(-b_{ij}), \nu)$ is a polynomial for large ν, we see that $H(M, \nu)$ is a polynomial for large ν, as required. This proves Theorem III-55. $\square$

The Hilbert function and polynomial are clearly invariants of $X \subset \mathbb{P}_K^r$, but it is perhaps not obvious that the *graded Betti numbers* b_{ij} are too. This follows from Nakayama's Lemma; see, for example, Eisenbud [1995, Chap. 19] or Matsumura [1986, Section 19] for a discussion of minimal free resolutions over a local ring that translates immediately to the graded case.

We have thus three progressively weaker sets of invariants of a projective scheme: the graded Betti numbers, the Hilbert function, and the Hilbert polynomial. To orient the reader, we will list some facts about them that we will not prove here and that will not be used in an essential way. Then we will give some examples.

(1)   As we have already mentioned, the degree d of the polynomial $P(X,\nu)$ is the dimension of X.

(2)   The leading term is of the form

$$\frac{\delta(X)}{d!}\nu^d$$

and $\delta(X)$ is called the *degree* of X. It may be identified with the length of the subscheme in which X meets a general plane in $\mathbb{P}_K^r$ of dimension $r - d$. (See, for example, Hartshorne [1977, Chapter I, 7.3 and 7.7].) This follows from the observation, proved below (Proposition III-59), that the Hilbert polynomial of a zero-dimensional subscheme of degree δ in $\mathbb{P}_K^n$ is the constant polynomial δ together with the fact that if Y is a general hyperplane section of X, then the Hilbert polynomial of Y is the first difference function of the Hilbert polynomial of X — that is,

$$P(Y,\nu) = P(X,\nu) - P(X,\nu-1).$$

(3)   In terms of the description given in Section III.2.5 of maps to projective space, the Hilbert polynomial $P(X,\nu)$ of a subscheme $X \subset \mathbb{P}_K^r$ depends only on the invertible sheaf $\mathscr{L}$ corresponding to the embedding $X \hookrightarrow \mathbb{P}_K^r$, and not on the particular epimorphism $\mathscr{O}_X^{r+1} \to \mathscr{L}$ In fact, for readers familiar with cohomology of coherent sheaves, $P(X,\nu)$ is equal, for all ν, to the alternating sum of dimensions of cohomology groups

$$\chi(\mathscr{L}^{\otimes\nu}) = \sum (-1)^i \dim_K H^i(X, \mathscr{L}^{\otimes\nu}).$$

In particular, $P(X,0) = \chi(\mathscr{O}_X) = \sum(-1)^i \dim_K H^i(\mathscr{O}_X)$ is a number depending on X and not on the embedding! In case X is a nonsingular curve over the complex numbers — that is, a Riemann surface — the number

$$\dim_K H^1(\mathscr{O}_X) = g = 1 - P(X,0)$$

is the *genus* of X, and $1 - P(X,0)$ turns out to be the right notion of genus for any one-dimensional scheme. It is called the *arithmetic genus* of the scheme. In the case where the dimension d of X is greater than one, it was at first felt that the normal case was the case where $H^i(\mathscr{O}_X) = 0$ for $1 < i < d$ (and this cohomology group always vanishes for $i > d$), so the arithmetic genus of X was by analogy defined as $1 + (-1)^d P(X,0)$.

(4)   The set of all varieties in $\mathbb{P}_K^r$ with Hilbert polynomial equal to a given polynomial turns out to be itself naturally the set of K-valued points of a projective scheme, called the *Hilbert scheme* associated with the given polynomial. For example, any subscheme $X \subset \mathbb{P}_K^r$ with Hilbert polynomial $P(\nu) = \binom{\kappa+\nu}{k}$ (that is, the Hilbert polynomial of a k-plane) is in fact a k-plane; and the Hilbert scheme of all such subschemes turns out to be the

Grassmannian $\mathbb{G}(k,r) = G(k{+}1, r{+}1)$. There are, however, not many other cases in which these Hilbert schemes have been understood geometrically! We will return to this construction in Sections VI.2.2 and VI.2.2 of the final chapter.

Exercise III-58. Let A be a Noetherian ring and $\mathscr{X}$ a closed subscheme of $\mathbb{P}^n_A$, regarded as a family of schemes over $\operatorname{Spec} A$. Since the fiber X_p of $\mathscr{X}$ over a point $p \in \operatorname{Spec} A$ is a closed subscheme of $\mathbb{P}^n_{\kappa(p)}$, it has a Hilbert function $H(X_p, \nu)$. Show that the function $H(X_p, \nu)$, regarded as a function in p, is upper semicontinuous in the Zariski topology on $\operatorname{Spec} A$; that is, for any ν and any number m,

$$\{p \in \operatorname{Spec} A \mid H(X_p, \nu) \geq m\}$$

is a closed subset of $\operatorname{Spec} A$.

We extend the definition of the Hilbert polynomial to the case of a subscheme $X \subset \mathbb{P}^r_S$ of projective space over an arbitrary irreducible base S by defining the polynomial $P(X, \nu)$ to be the Hilbert polynomial of the fiber of X over the generic point of S. This doesn't involve anything new — by the generic flatness theorem of Section II.3.4 combined with Proposition III-56, or by Exercise III-58, X will be flat over an open dense subset $U \subset S_{\mathrm{red}}$, and $P(X, \nu)$ is simply the common Hilbert polynomial of the fibers of X_U over U — but it's convenient terminology.

(5) In many ways the invariant provided by the graded Betti numbers is the most subtle of all, and until very recently nothing was known of its geometric significance beyond that of the Hilbert function and polynomial. Now, however, we know in a few cases (and conjecture in a few more) how they reflect some subtle aspects of the intrinsic geometry of X. See, for example, Green [1984; Green and Lazarsfeld [1985] for more information.

III.3.4 Examples

Points in the Plane. Already for the case of zero-dimensional subschemes in the plane we get different information from the Hilbert polynomial, Hilbert function, and graded Betti numbers.

First of all, we have stated above that the Hilbert polynomial of a subscheme $X \subset \mathbb{P}^r_K$ is a polynomial whose degree is equal to the dimension of X; so when X is zero-dimensional, the Hilbert polynomial is a constant. We can easily prove this and somewhat more in the case of points.

Proposition III-59. *The Hilbert function of a 0-dimensional subscheme of degree δ in $\mathbb{P}^r_K$ satisfies*

$$H(X, \nu) \leq \delta$$

for all ν, with equality for large ν. Thus $P(X, \nu) \equiv \delta$.

Proof. We must show that the codimension in $K[x_0, \ldots, x_r]$ of the set of homogeneous forms of degree ν that vanish on X —that is, codim $I(X)_\nu$— is less than or equal to δ, with equality for large ν. The reason is that vanishing at a point is one linear condition on the coefficients of a polynomial, and thus vanishing at X should be δ linear conditions; for large ν we will show that these conditions are always linearly independent.

To make this precise, we pass to an affine open set. Changing coordinates, we may suppose that X is contained in the affine open set $x_r \neq 0$, so that a form F of degree ν belongs to $I(X)$ if and only if $F(x_0, \ldots, x_{r-1}, 1)$ belongs to the ideal $J \subset K[x_0, \ldots, x_{r-1}]$ of X in the affine open set $x_r \neq 0$. To say that X is of length δ means that J is of codimension δ in $K[x_0, \ldots, x_{r-1}]$ and thus of codimension less than or equal to δ in the space of those polynomials that can be written as $F(x_0, \ldots, x_{r-1}, 1)$ for F of degree ν —these are simply the polynomials in $K[x_0, \ldots, x_{r-1}]$ of degree less than or equal to ν. This shows at once that $H(X, \nu) \leq \delta$ for all ν, with equality if J has codimension δ in the space of polynomials of degree less than or equal to ν. But J will have codimension δ in the space of polynomials of degree less than or equal to ν as soon as a set of representatives for $K[x_0, \ldots, x_{r-1}]/J$ can be chosen from among the polynomials of degree less than or equal to ν, which is certainly true for all large ν. $\qquad\square$

If $X \subset \mathbb{P}_K^r$ is nonempty, $I(X)$ contains nothing of degree 0 (we are working over a field!), so $H(X, 0) = 1$. Thus the proposition provides easy examples where $P(X, 0) \neq H(X, 0)$.

We can easily exhibit a family of subschemes of $\mathbb{P}_K^2$ with constant Hilbert polynomial but varying Hilbert function. To construct such a family $\mathscr{X} \subset \mathbb{P}_K^2 \times \operatorname{Spec} K[t]$, for example, we can take the "constant" points P and Q given by $(x_2 = x_1 + x_0 = 0)$ and $(x_2 = x_1 - x_0 = 0)$, and the variable point R given by $(x_1 = x_2 - tx_0 = 0)$, and let $\mathscr{X}$ be the (disjoint) union of P, Q, and R in $\mathbb{P}_K^2 \times \operatorname{Spec} K[t]$.

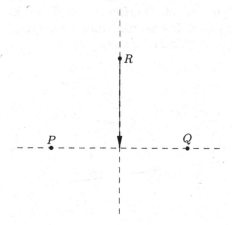

We regard $\mathscr{X}$ as a flat family over $\operatorname{Spec} K[t]$ by means of the projection to the second factor, whose fibers $X_{(0)}$ and X_λ over the generic point and over every closed point $(t - \lambda)$ (as schemes over $K(t)$ and K, respectively) have Hilbert polynomials $P(X_{(0)}, \nu) = P(X_\lambda, \nu) \equiv 3$; but while the Hilbert function $H(X_\lambda, 1) = 3$ for $\lambda = 0$, we have $H(X_0, 1) = 2$.

Exercise III-60. Let $\tilde{X}_\lambda \subset \mathbb{A}_K^3$ be the cone over the fiber X_λ of the family $\mathscr{X}$ above. Show that there does not exist a flat family $\tilde{\mathscr{X}} \subset \mathbb{A}_K^3 \times_K \operatorname{Spec} K[t]$ whose fiber over each point $(t - \lambda)$ is $\tilde{X}_\lambda$. (There does exist such a family over the complement of the origin in $\operatorname{Spec} K[t]$, however.) What is the flat limit of the cones $\tilde{X}_\lambda$ as λ approaches 0? (See the example in Section II.3.4.)

Now consider the case where X is a set of four distinct points in the plane $\mathbb{P}_K^2$. We already know that $P(X, \nu) \equiv 4$. We will treat separately the cases where all the points or all but one of the points lie on a line.

(1) *X is contained in a line.* Suppose, first, that the points lie on a line L, with equation $l = 0$, say. The only line containing X is L, so

$$H(X, 1) = H(\mathbb{P}_K^2, 1) - 1 = 2.$$

If $q = 0$ is the equation of a conic containing X, then q restricts to a form of degree 2 on L, vanishing at the four points of X, so q must vanish identically on L. Thus $q = 0$ is the union of L and one other line, and the set of equations of conics containing X is the three-dimensional space of multiples of l by linear forms. This gives

$$H(X, 2) = H(\mathbb{P}_K^2, 2) - 3 = 3.$$

Starting with $\nu = 3$, however, vanishing at the four points imposes four independent conditions on forms of degree ν, so $H(X, \nu) = 4$. To prove this, it is enough, for each 3 point subset X' of X, to find a curve of degree ν that contains X' but not the fourth point of X. We may do this with a curve consisting of ν straight lines, three of these passing through one each of the points of X' and the rest far away from X:

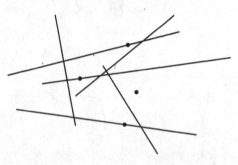

To compute the minimal free resolution in this and the next examples, we will use a result of Hilbert, which was generalized and extended to the local case by Lindsay Burch.

Theorem III-61. *If I is the homogeneous ideal of a zero-dimensional subscheme $X \subset \mathbb{P}_K^2$, then any minimal free resolution of the homogeneous coordinate ring S/I has the form*

$$0 \longrightarrow \sum_{j=1}^{n-1} S(-b_{2j}) \xrightarrow{A} \sum_{j=1}^{n} S(-b_{1j}) \longrightarrow S.$$

Further, the j-th generator of I — that is, the image of $S(-b_{1j})$ in S — is up to a nonzero scalar the determinant of the matrix A with the j-th row deleted.

For a proof, see Eisenbud [1995, Section 20.4], for example.

We will make use of this to compute minimal generators of the ideal $I(X)$ through the following corollary.

Corollary III-62. *If I is the homogeneous ideal of a zero-dimensional subscheme $X \subset \mathbb{P}_K^2$, and if I contains an element of degree e, then I can be generated by $e + 1$ elements.*

Proof. If the minimal number of generators of I is g, then I is generated by $(g-1) \times (g-1)$ determinants of a matrix A whose entries are in the graded maximal ideal of S and are thus forms of positive degree. Consequently, no element of I has degree less than $g - 1$, and we have $g \leq e + 1$, as claimed. $\square$

By the theorem, knowing the degrees of the entries of the matrix A is equivalent to knowing the graded Betti numbers in this case: the b_{1j} are just the degrees of the minors of A, and b_{2j} is the sum of b_{1j} plus the degree of the ij-th entry of A.

Applying this to the example at hand, we see that since X lies on a line, $I(X)$ may be generated by two elements, which may, of course, be taken to be L and a form of smallest possible degree in I that is not divisible by L.

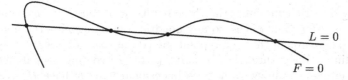

$L = 0$

$F = 0$

As we have noted, this smallest possible degree is 4, and we may, for example, take F to be the equation of a quartic consisting of four lines, each through one point of X.

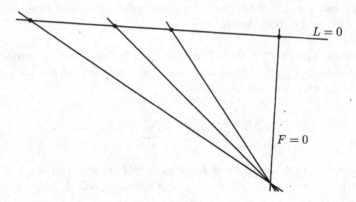

Since L and P have no common factor, we see that the minimal free resolution of $S/I(X)$ has the form

$$0 \longrightarrow S(-5) \longrightarrow S(-4) \oplus S(-1) \longrightarrow S,$$

giving the expression for the Hilbert function

$$H(X,\nu) = \binom{\nu+2}{2} - \binom{\nu+1}{2} - \binom{\nu-2}{2} + \binom{\nu-3}{2}.$$

(2) *All but one of the points of X lie on a line.* Next, consider the case where only three of the four points lie on the line L. Now there is no linear form in $I(X)$, so $H(X,1) = 3$.

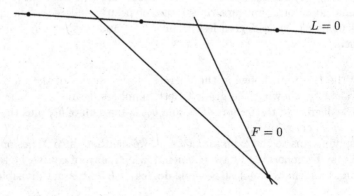

Any quadric containing the three points on L must, by the same argument as before, contain L; so any quadric containing X is the union of L and a line through the fourth point. Since the space of linear forms corresponding to lines through the fourth point is two-dimensional, the space of quadrics containing X is two-dimensional and we have $H(X,2) = 4$. Following the same argument as before, we show that $H(X,\nu) = 4$ for all larger ν, so this is the case for all $\nu \geq 2$.

As for the resolution, we see by the corollary above that $I(X)$ requires at most three generators. But $I(X)$ is not generated by the two independent quadrics it contains, since these have a common factor; thus it is minimally generated by these two quadrics and another generator, an element of smallest possible degree not contained in the ideal generated by the two quadrics or, equivalently, vanishing on a curve not containing L. It is easy to see that there is a cubic curve with the desired properties; it may be taken, for example, to be the union of three lines, each passing through one of the points of L and one passing, in addition, through the fourth point. Since the minimal generators of $I(X)$ have degrees 2, 2, 3, the matrix A must be a 2×3 matrix whose entries have degrees as given in the following diagram (up to a rearrangement of the rows and columns):

$$\begin{pmatrix} 1 & 2 \\ 1 & 2 \\ 0 & 1 \end{pmatrix}.$$

(Of course, the entry of degree 0 must actually be 0, since all the entries must be in the maximal graded ideal.) Thus the minimal free resolution has the form

$$0 \longrightarrow S(-3) \oplus S(-4) \xrightarrow{A} S(-2) \oplus S(-2) \oplus S(-3) \longrightarrow S.$$

(3) *No three points of X lie on a line.* Finally, consider the case where X consists of four points, no three of which lie on a line. We claim that the Hilbert function of X is the same as in the previous case: $H(X,1) = 3, H(X,\nu) = 4$ for $\nu \geq 2$. The first of these values is obvious, since X lies on no lines. For the second, it is enough as before to note that there are quadrics (and thus a fortiori forms of higher degree) containing any subset of the four points but missing the last; these may be constructed as before as unions of lines.

Now we compute the free resolution of $S/I(X)$. Taking the two pairs of opposite sides of the quadrilateral formed by the points gives us two quadrics q_1 and q_2 without common factor in the ideal of X.

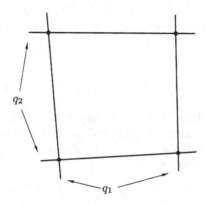

Since q_1 and q_2 are relatively prime, the free resolution of the ideal I they generate has the form

$$0 \xrightarrow{} S(-4) \xrightarrow{A} S(-2) \oplus S(-2) \xrightarrow{B} S,$$

where

$$B = (q_1 \ q_2) \quad A = \begin{pmatrix} -q_2 \\ q_1 \end{pmatrix}.$$

Computing the Hilbert function of S/I from this resolution, we see that it is the same as that of $S/I(X)$, and since $I \subset I(X)$, we must have $I = I(X)$; that is, $I(X)$ is generated by q_1 and q_2 and X is correspondingly the intersection of the two conics containing it.

Summing up, we see that all three of the examples look the same from the point of view of Hilbert polynomials; the first two examples are distinguished by their Hilbert functions; and the last two examples look the same from the point of view of Hilbert functions but are distinguished by their graded Betti numbers. It is not hard to find corresponding examples of subschemes X of length 4 where the properties distinguished are actually intrinsic properties of the schemes, not dependent on the embedding. For example, while the scheme $\operatorname{Spec} K[x]/(x^4)$ may be embedded in $\mathbb{P}^2_K$ so as to have any of the Hilbert functions and Betti numbers above (for instance, as the subschemes defined by the ideals (x_0, x_1^4), $(x_0 x_2^2 - x_1^3, x_0 x_1, x_0^2)$, and $(x_0 x_2 - x_1^2, x_0^2)$ respectively, the subscheme defined by (x_0^2, x_1^2) will always have the graded Betti numbers and Hilbert function of case 3).

Exercise III-63. Find the Hilbert polynomial, the Hilbert function, and the graded Betti numbers of all subschemes of the plane of length 3.

Examples: Double Lines in General and in $\mathbb{P}^3_K$. So far, most of our discussion of projective schemes has been parallel to the theory of varieties. We will now look at one genuinely nonclassical family of examples.

Exercise II-35 asked you to show that all affine double lines are equivalent. This is not true for projective double lines. Here are some simple examples.

Let K be a field. Consider the graded ring

$$S = K[u, v, x, y]/(x^2, xy, y^2, u^d x - v^d y)$$

and the scheme

$$X = X_d = \operatorname{Proj} S.$$

To see that X is a double line, we construct an open affine covering of X. The elements x and y are nilpotent in S, so the radical of the ideal generated by u and v is the irrelevant ideal of S, and X is covered by X_u and X_v. From the definitions we see that

$$X_u = \operatorname{Spec}(S[u^{-1}])_0.$$

To analyze the ring $(S[u^{-1}])_0$, we note that it is a factor ring of

$$K[u, v, x, y][u^{-1}]_0 = K[v', x', y'],$$

where

$$v' = \frac{v}{u}, \qquad x' = \frac{x}{u}, \qquad y' = \frac{y}{u},$$

and the kernel of the map to $(S[u^{-1}])_0$ is generated by $(x')^2$, $x'y'$, $(y')^2$, and $x' - (v')^d y'$ (see Exercise III-6). Thus

$$(S[u^{-1}])_0 \cong K[v', y']/(y')^2$$

and X_u is an affine double line. By symmetry, X_v is too, and this proves that X is a projective double line. Explicitly,

$$X_v \cong \mathrm{Spec}(S[v^{-1}])_0$$

and

$$
\begin{aligned}
(S[v^{-1}])_0 &\cong K[u'', x'', y'']/(x''^2, x''y'', y''^2, (u'')^d x'' - y'') \\
&\cong K[u'', x'', y'']/(x''^2),
\end{aligned}
$$

where

$$u'' = \frac{u}{v} = \frac{1}{v'}, \qquad x'' = \frac{x}{v}, \qquad y'' = \frac{y}{v}.$$

The simplest way to see that, in contrast to the affine case, not all double lines are isomorphic to one another, is to show that the isomorphism class of X depends on the integer d, which may be thought of as specifying how fast the double line twists around the reduced line inside it. To demonstrate this, we will show that the ring of global sections $\mathcal{O}_X(X)$ of the structure sheaf of X depends on d. To compute it, suppose first that $\sigma \in \mathcal{O}_X(X)$. The element σ restricts to an element of $\mathcal{O}_X(X_u)$, which is isomorphic to $K[v', y']/(y')^2$ by the above, so we may write

$$\sigma|_{X_u} = a(v') + b(v')y'$$

and similarly

$$\sigma|_{X_v} = f(u'') + g(u'')x''$$

for unique polynomials a, b, f, and g with coefficients in K. But on $X_u \cap X_v$ we have

$$u'' = \frac{1}{v'}$$

and

$$x'' = \frac{x}{v} = \frac{x'u}{v} = (v')^d y' \frac{u}{v} = (v')^{d-1} y'.$$

Thus $f(1/v') = a(v')$, which is only possible if f and a are constant polynomials and $f = a$. Also, $g(1/v')(v')^{d-1} = b(v')$, which is only possible if both g and b have degree less than or equal to $d - 1$ (and then each of g and b determines the other). Conversely, any element of $\mathcal{O}_X(X_u)$ of the form $a + b(v')y'$ with a a constant and b a polynomial of degree less than or equal to $d-1$ extends uniquely to a global section of $\mathcal{O}_X$, so we see that the dimension of $\mathcal{O}_X(X)$ is $d+1$. This shows that the isomorphism class of X depends on d, as claimed.

In fact, we will see below that the integer d is the negative of the arithmetic genus of X, as defined in Section III.3.3. As it turns out, every projective double line of genus $-d$, with $d \geq 0$, is isomorphic to X.

There are also double lines of positive arithmetic genus—the simplest example of which is the *double conic* $\operatorname{Proj} K[x,y,z]/(xy-z^2)^2$, which has genus 3—and even continuous families of these when the genus is greater than or equal to 7. These objects arise naturally in the study of nonsingular curves: as a nonsingular nonhyperelliptic curve degenerates to a hyperelliptic curve, a phenomenon well known in the classical theory of varieties, the canonical model of the smooth curve approaches a projective double line (see Bayer and Eisenbud [1995] and Fong [1993] for more details).

Exercise III-64. What is the ring structure of $\mathcal{O}_X(X)$ for the double line X above?

Exercise III-65. Compute $\mathcal{O}_X(X)$ for the double line

$$X = \operatorname{Proj} K[u,v,x,y]/(x^2, xy, y^2, p(u,v)x+q(u,v)y),$$

where p and q are any homogeneous polynomials of degree d without common zeros in $\mathbb{P}^1_K$. Prove that this double line is isomorphic to the double line of the example (and thus does not depend on the choice of p and q).

To calculate the Hilbert polynomial of X, observe that for each d, the ideal $I_d = (x^2, xy, y^2, u^d x - v^d y)$ contains the ideal

$$I = (x^2, xy, y^2).$$

Since S/I is a free $K[u,v]$-module on the generators 1, x, and y, we see that

$$H(S/I, \nu) = H(S/(x,y), \nu) + 2H(S/(x,y), \nu-1).$$

Further, we see easily, using this basis, that if we write $p = u^d x - v^d y$ for the fourth generator of I_d as written above, then for any homogeneous form $q = q(x,y,u,v)$ we have $qp \in I$ if and only if $q \in (x,y)$. Thus

$$H(S/I_d, \nu) = H(S/I, \nu) - H(S/(x,y), \nu-d-1).$$

But $P(S/(x,y), \nu) = \nu + 1$. Putting all these equalities together, we get

$$P(X, \nu) = 2\nu + d + 1,$$

so the Hilbert polynomial, and in particular the arithmetic genus

$$p_a(X) = 1 - P(X,0) = -d,$$

distinguishes between these double lines for different d.

Here is an exercise that will be useful for the following three examples.

Exercise III-66. Compute the Hilbert polynomials of the following sub-schemes of $\mathbb{P}^3_K$:

(a) The union of two skew lines.

(b) The union of two incident lines.

(c) The subscheme supported on the union of two incident lines with an embedded point of degree 1 at their point of intersection, not lying in the plane spanned by the two lines. Also, show that for any point p on the double line X_0 (or on any of the double lines above) there is a unique subscheme of $\mathbb{P}^3_K$ consisting of X_0 with an embedded point of degree 1 at p; and compute the Hilbert polynomial of this subscheme.

Given this exercise, we can use the notion of Hilbert polynomial to further illuminate the example of a family of pairs of skew lines tending to a pair of incident lines. Recall that in Exercise II-25 we discussed such a family and showed that the flat limit was not reduced: it was supported on the union of the incident lines but had an embedded point at their point of intersection. As the exercise above suggests, if we complete these families in $\mathbb{P}^3_K$, we see that this is necessary from the point of view of Hilbert polynomials.

Consider next a family of pairs of skew lines in $\mathbb{P}^3_K$, described as follows. First, let $L \subset \mathbb{P}^3_K$ be the constant line $x = y = 0$, and let $M \subset \mathbb{P}^3_K$ be the line $x = tv$, $y = tu$. Let Y_t be the union of these two lines. We may ask then for the flat limit of the family Y_t; or in other words, the fiber Y_0 over the origin in $\mathbb{A}^1_K$ of the union $\mathscr{Y}$ of the subschemes $\mathscr{L}$ and $\mathscr{M}$ of $\mathbb{P}^3_K \times \mathbb{A}^1_K$ given by $x = y = 0$ and $x = tv$, $y = tu$, respectively. Of course, the support of Y_0 will be the line L, but it is equally clear that it must have some nonreduced structure. In fact, the flat limit is none other than the double line X_1 above.

Exercise III-67. Verify that the flat limit Y_0 is the double line X_1. (By comparing Hilbert polynomials, it is enough to prove inclusion in one direction.)

An interesting wrinkle on this last construction is to consider a slightly different family of pairs of skew lines: we let L be as above, and let M_t be the line given by $x = tv$, $y = -t^2 u$. At first glance it might appear that the flat limit of the unions $Y_t = L \cup M_t$ will be the double line given by $(x^2 = y = 0)$, which is isomorphic to the double line X_0 above; but this cannot be, since the Hilbert polynomials are not equal. The following exercise gives the real situation.

Exercise III-68. Show that with L and M_t as above, the flat limit as $t \to 0$ of the union $L \cup M$ is the double line $x^2 = y = 0$ with an embedded point of degree 1 located at the point $[0, 0, 1, 0]$.

Exercise III-69. Let L, M, and $N_t \subset \mathbb{P}^3_K$ be the lines $u = v = 0$, $y = v = 0$, and $y + u = ty + (1-t)v = 0$, respectively; let Z_t be their union in

$\mathbb{P}^3_K$. Find the Hilbert polynomial of Z_t and the Hilbert polynomial of Z_0. What is the limit, in the sense discussed above, of the subschemes $Z_t \subset \mathbb{P}^3_K$ as $t \to 0$?

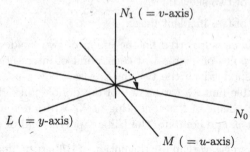

Finally, there are arithmetic analogues of each of the last three examples. For example, consider the following three families of subschemes of $\mathbb{P}^3_\mathbb{Z}$ (which are flat families over $\operatorname{Spec}\mathbb{Z}$):

(a) Let $\mathscr{L} \subset \mathbb{P}^3_\mathbb{Z}$ be the constant line $x = y = 0$, and let $\mathscr{M} \subset \mathbb{P}^3_\mathbb{Z}$ be the line $x = 7v$, $y = -7u$; let $\mathscr{U}$ be the union of these two subschemes.

(b) Let $\mathscr{L} \subset \mathbb{P}^3_\mathbb{Z}$ be the constant line $x = y = 0$, and let $\mathscr{M} \subset \mathbb{P}^3_\mathbb{Z}$ be the line $x = 7v$, $y = -49u$; let $\mathscr{U}$ be the union of these two subschemes.

(c) Let $\mathscr{L}$, $\mathscr{M}$, and $\mathscr{N} \subset \mathbb{P}^3_\mathbb{Z}$ be the subschemes defined by $u = v = 0$, $y = v = 0$, and $y + u = 7y - 6v = 0$, respectively; let $\mathscr{U}$ be their union.

Exercise III-70. For each of the subschemes $\mathscr{U} \subset \mathbb{P}^3_\mathbb{Z}$ above, find the fiber of $\mathscr{U}$ over the point $(7) \in \operatorname{Spec}\mathbb{Z}$. Compare your answer with that found in the preceding three exercises.

III.3.5 Bézout's Theorem

The most classical form of Bézout's theorem asserts that if plane curves $C, C' \subset \mathbb{P}^2_K$ defined by equations of degrees d and e meet in only finitely many points, then the number of points of intersection is at most de, with equality if the two curves meet transversely and the field K is algebraically closed. This important result has gone through many successive generalizations. In particular, the language of schemes allows us to give a version that is simultaneously simpler and more general than the original; and, while this version is not the most general possible, we will focus on this.

For the following, we will work with schemes over a field K. As in the discussion of degree, we could state Bézout's theorem for a projective scheme $X \subset \mathbb{P}^n_S$ over any base S, but this conveys no more information than Bézout's theorem for schemes over a field, applied to the fibers of X over the generic points of S. Also, note that we do not assume K is algebraically closed. We will see in Exercises III-72 through III-75 below examples over non-algebraically closed fields.

The statement of Bézout's theorem is very simple. Recall that by a *hypersurface* in projective space $\mathbb{P}^n_K = \operatorname{Proj} K[X_0, \ldots, X_n]$ over a field K we mean not any $(n-1)$-dimensional subscheme of $\mathbb{P}^n_K$ but specifically the zero locus $V(F)$ of a single homogeneous polynomial F. In particular, it will have pure dimension $n-1$ (see Eisenbud [1995], for example); and while it may be nonreduced (if F has repeated factors) it will have no embedded components. Recall also that the *degree* of an arbitrary subscheme $X \subset \mathbb{P}^n_K$ of projective space over a field K is defined in terms of its Hilbert polynomial; and that if in particular the dimension of X is zero, then its degree is simply the dimension of the space $\mathscr{O}_X(X)$ of global sections as a K-vector space.

Theorem III-71 (Bézout's Theorem for complete intersections). *Assume that $Z_1, \ldots, Z_r \subset \mathbb{P}^n_K$ are hypersurfaces of degrees $d_1, \ldots, d_r$ in projective space over a field K, and that the intersection $\Gamma = \bigcap Z_i$ has dimension $n - r$. Then*

$$\deg(\Gamma) = \prod d_i.$$

Thus, for example, if D and $E \subset \mathbb{P}^2_K$ are plane curves of degrees d and e with no common components, then the intersection $\Gamma = D \cap E$ will have degree de. As an immediate consquence, we can deduce from this the classical "$\deg(\Gamma) \leq de$" form of the theorem, together with the fact that equality holds if and only if Γ is reduced and each point of Γ has residue field K.

More generally, we can deduce from Theorem III-71 the general form of the equality statement of the classical Bézout theorem for complete intersections over an algebraically closed field, in which we express the product $\prod d_i$ of the degrees of the hypersurfaces as a linear combination of the degrees of the irreducible components Γ_i of the reduced scheme Γ_{red}, with coefficients referred to as the *multiplicity of the intersection $Z_1 \cap \ldots \cap Z_r$ along Γ_i* arising from the nonreduced structure. In this form we can further generalize the statement of Bézout's theorem to arbitrary proper intersections in projective space (that is, intersections of subschemes $X, Y \subset \mathbb{P}^n_K$ of pure codimensions k and l such that $X \cap Y$ has codimension $k + l$); but to do this we will need also to define in general the multiplicity of an intersection along one of its components. We postpone this, and the proof of Bézout's theorem for complete intersections, in order to give the reader a chance to try some examples.

Exercise III-72. Let $C \subset \mathbb{P}^2_{\mathbb{R}}$ be the conic curve given as

$$C = \operatorname{Proj} \mathbb{R}[X, Y, Z]/(X^2 + Y^2 - Z^2) \subset \operatorname{Proj} \mathbb{R}[X, Y, Z]$$

and let L_1, L_2 and L_3 be the lines given by X, $X - Z$ and $X - 2Z$ respectively. Show that no two of the schemes $C \cap L_i$ are isomorphic, but that they all have degree 2 as schemes over $\mathbb{R}$.

The following four exercises describe a situation in which Bézout's theorem over a non-algebraically closed field arises naturally: an intersection of universal curves over the product of the schemes parametrizing such curves. The situation is one that occurs frequently, and is of interest (apart from its value as an illustration of Bézout) as an example of "generic" intersections.

Exercise III-73. Let K be a field, and let

$$B = \mathbb{A}_K^{12} = \operatorname{Spec} K[a,b,c,d,e,f,g,h,i,j,k,l].$$

Consider the two conic curves $\mathscr{C}_i \subset \mathbb{P}_B^2$ given by

$$\mathscr{C}_1 = V(aX^2 + bY^2 + cZ^2 + dXY + eXZ + fYZ)$$
$$\subset \operatorname{Proj}\big(K[a,b,c,d,e,f,g,h,i,j,k,l][X,Y,Z]\big) = \mathbb{P}_B^2$$

and similarly

$$\mathscr{C}_2 = V(gX^2 + hY^2 + iZ^2 + jXY + kXZ + lYZ) \subset \mathbb{P}_B^2.$$

By considering the projection map

$$\pi_2 : \mathscr{C}_1 \cap \mathscr{C}_2 \subset \mathbb{P}_B^2 = \mathbb{P}_K^2 \times_{\operatorname{Spec} K} B \longrightarrow \mathbb{P}_K^2$$

show that $\mathscr{C}_1 \cap \mathscr{C}_2$ is an irreducible K-scheme.

Exercise III-74. (a) With $\mathscr{C}_1$ and $\mathscr{C}_2$ as above, show that the intersection $\mathscr{C}_1 \cap \mathscr{C}_2$ is generically reduced by showing that the projection

$$\pi_1 : \mathscr{C}_1 \cap \mathscr{C}_2 \subset \mathbb{P}_B^2 \longrightarrow B = \mathbb{A}_K^{12}$$

has a fiber consisting of four distinct (hence reduced and K-rational) points.

(b) Although part (a) is enough for the application in the following exercise, deduce that $\mathscr{C}_1 \cap \mathscr{C}_2$ is everywhere reduced by unmixedness of complete intersections (see Eisenbud [1995], for example). Alternatively, show that it is nonsingular by a direct tangent space calculation.

Exercise III-75. Let $L = K(a,b,c,d,e,f,g,h,i,j,k,l)$ be the field of rational functions in 12 variables over K (that is, the function field of $B = \mathbb{A}_K^{12}$). Let

$$C_1 = V(aX^2 + bY^2 + cZ^2 + dXY + eXZ + fYZ) \subset \mathbb{P}_L^2$$

and

$$C_2 = V(gX^2 + hY^2 + iZ^2 + jXY + kXZ + lYZ) \subset \mathbb{P}_L^2;$$

that is, C_1 and C_2 are the fibers of $\mathscr{C}_1$ and $\mathscr{C}_2$ over the generic point of $\mathbb{A}_K^{12}$. Deduce from the preceding two exercises that the intersection $C_1 \cap C_2$ is a single, reduced point P.

Exercise III-76. Keeping the notation of the preceding problem, show that, as Bézout predicts, the intersection $C_1 \cap C_2$ has degree 4 as a scheme over L — that is, the residue field $\kappa(P)$ of the point $P = C_1 \cap C_2$ is a quartic extension of L.

Hint: introduce affine coordinates $x = X/Z$ and $y = Y/Z$ on an open subset of $\mathbb{P}_L^2$, and express $\kappa(P)$ as

$$\kappa(P) = L[x]/(R(x))$$

where $R(x)$ is the resultant of the dehomogenized form of the defining polynomials for C_1 and C_2 with respect to x as in Section V.2.

There is an interesting sidelight to this example, which we will mention in passing. One question we may ask in this situation is, what is the Galois group of the Galois normalization of the extension $L \subset \kappa(P)$? To answer this, at least in case the ground field $K = \mathbb{C}$, we should introduce what we call the *monodromy group* of the four points of intersection of two general conics. Briefly, there is an open subset $U \subset B$ over which the fibers of the projection $\varphi : \mathscr{C}_1 \cap \mathscr{C}_2 \to B$ are reduced; and in terms of the classical topology, the restriction of the map φ to the inverse image of U is a topological covering space. As such, for any point $p \in U$ we have a monodromy action of the fundamental group $\pi_1(U, p)$ on the points of the fiber $\varphi^{-1}(p)$: to an arc $\gamma : [0, 1] \to U$ starting and ending at p and any point $q \in \varphi^{-1}(p)$ we associate the end point of the unique lifting $\tilde{\gamma} : [0, 1] \to \varphi^{-1}(U)$ of γ to $\varphi^{-1}(U)$ with $\tilde{\gamma}(0) = q$. Informally, suppose we allow two conics $C_1(t), C_2(t) \subset \mathbb{P}_\mathbb{C}^2$ to vary with a real parameter $t \in [0, 1]$, keeping them transverse at all times. As t varies, the four points of the intersection $C_1(t) \cap C_2(t)$ vary; and if the conics return to their original positions — that is, $C_i(0) = C_i(1)$ — we find that while the intersection $C_1(0) \cap C_2(0) = C_1(1) \cap C_2(1)$ the four points individually may not return to their original positions; the resulting group of permutations of the four is called the monodromy group. It turns out that the answer to our original problem — that is, the Galois group of the Galois normalization of $\kappa(P)$ over L — coincides with the monodromy group of the four points, which it is possible to see from this geometric characterization is the symmetric group on four letters.

More generally, in many enumerative problems that depend on parameters (in this example, the intersection of two conics), the universal solution turns out to be a single point P, with residue field $\kappa(P)$ a finite extension of the function field L of the scheme (in this case $B = \mathbb{A}_K^{12}$) parametrizing the problems. In this situation, we may ask, what is the Galois group of the Galois normalization of the extension $L \subset \kappa(P)$? This turns out in general to coincide with the monodromy group of the problem. For a general treatment see Harris [1979].

We will now give a proof of Bézout's theorem, and also discuss its possible generalizations. We will prove it by using the *Koszul complex* to calculate the Hilbert polynomial of Γ (and in particular its degree). The Koszul complex is fully described in Eisenbud [1995, Chapter 17]; we will sketch the construction here and simply state the properties we need.

First, we introduce the defining equations of the hypersurfaces Z_i: we write

$$Z_i = \operatorname{Proj} K[X_0, \ldots, X_n]/(F_i) \subset \operatorname{Proj} K[X_0, \ldots, X_n],$$

so that

$$\Gamma = \operatorname{Proj} K[X_0, \ldots, X_n]/(F_1, \ldots, F_r).$$

We now describe a resolution of the homogeneous coordinate ring S_Γ as follows. First, for any subset

$$I = \{i_1, i_2, \ldots, i_k\} \subset \{1, 2, \ldots, r\},$$

we will denote by $|I| = k$ the number of elements of I, and by

$$d_I = \sum_{\alpha=1}^{k} d_{i_\alpha}$$

the sum of the degrees of the corresponding polynomials. We then set

$$M_k = \bigoplus_{|I|=k} S(-d_I)$$

where as usual $S = K[X_0, \ldots, X_n]$ is the polynomial ring. As there is a unique I with $|I| = 0$, we set $M_0 = S$. We will write an element of M_k as a collection $\{G_I\}$ of polynomials, where I ranges over all multi-indices of size k; by our definition, $\{G_I\}$ will be homogeneous of degree d if $\deg(G_I) = d - d_I$ for each I.

We now define a complex

$$0 \longrightarrow M_r \longrightarrow M_{r-1} \longrightarrow \ldots \longrightarrow M_2 \longrightarrow M_1 \longrightarrow M_0 = S.$$

The map $\varphi_k : M_k \to M_{k-1}$ is given by setting $\varphi_k(\{G_I\})$ equal to the collection of polynomials $\{H_J\}$, where

$$H_J = \sum_{\alpha \notin J} \pm F_\alpha \cdot G_{J \cup \{\alpha\}}$$

and the sign depends on the number of elements of J less than α.

Notice that the image of $\varphi_1 : M_1 \to M_0 = S$ is exactly the ideal of Γ. In fact, *this sequence is a free resolution of the coordinate ring* S_Γ. This is a general phenomenon: whenever we have a collection of elements $F_1, \ldots, F_r$ in a ring S, we can form a sequence in this way, which is called the *Koszul complex*. It is a standard theorem that whenever the collection $F_1, \ldots, F_r$ is a regular sequence, then the associated Koszul complex is a resolution (see Eisenbud [1995, Chapter 17], for example). In the present circumstance, where the polynomials F_i are homogeneous, the hypothesis on the dimension of Γ together with the fact that the polynomial ring S is Cohen–Macaulay implies that the polynomials $F_1, \ldots, F_r$ form a regular sequence in S (just as in the local case), so the sequence above is a resolution.

Given the Koszul resolution, it is straightforward to describe the Hilbert polynomial $P(\Gamma, \nu)$. If we write $H(M_k, \nu)$ for the Hilbert polynomial of the module M_k (that is, $H(M_k, \nu)$ is the polynomial that agrees with the dimension of M_k in degree ν when ν is large), then from the exactness of the Koszul complex we see that

$$P(\Gamma, \nu) = \sum (-1)^k P(M_k, \nu)$$

depends only on the numbers d_i and not on the particular polynomials F_i. For convenience, we will denote the Hilbert polynomial of such a complete intersection by $P_{d_1,\ldots,d_r}(\nu)$. (Note that we don't need to write down the Koszul complex to see that complete intersections of given multidegree all have the same Hilbert polynomial; this follows directly from the flatness of families of complete intersections as stated in Proposition II-32.)

Now, simply adding up the contributions of the summands in the Kozsul complex above, we see that

$$P_{d_1,\ldots,d_r}(\nu) = \sum_{I \subset \{1,\ldots,r\}} (-1)^{|I|} \binom{n + \nu - d_I}{n}$$

where the sum ranges over all subsets of $\{1, 2, \ldots, r\}$, including the empty set and the whole set.

This in a sense the complete answer to the question of the Hilbert polynomial of Γ, but there remains the problem of reading off from it things like the degree of Γ. To do this, we use an induction on the number r to relate the functions $P_{d_1,\ldots,d_r}(\nu)$ and $P_{d_1,\ldots,d_{r-1}}(\nu)$. This is simple: in the expression above for $P_{d_1,\ldots,d_r}(\nu)$, we simply separate out those terms in which $r \in I$ and those terms in which it is not. The terms in which $r \notin I$ visibly add up to $P_{d_1,\ldots,d_{r-1}}(\nu)$; and comparing terms in which $r \in I$ to the term corresponding to $I \setminus \{r\}$, we see that these add up to $P_{d_1,\ldots,d_{r-1}}(\nu - d_r)$. Thus,

$$P_{d_1,\ldots,d_r}(\nu) = P_{d_1,\ldots,d_{r-1}}(\nu) - P_{d_1,\ldots,d_{r-1}}(\nu - d_r).$$

Now, since

$$\nu^m - (\nu - \alpha)^m = m\alpha\nu^{m-1} + O(\nu^{m-2})$$

(where, following the analysts' convention, we have written $O(\nu^{m-2})$ to denote a sum of terms of degree at most $m - 2$), we see that if $f(\nu)$ is any polynomial, written as

$$f(\nu) = c_m \nu^m + O(\nu^{m-1})$$

then

$$f(\nu) - f(\nu - \alpha) = m\alpha c_m \nu^{m-1} + O(\nu^{m-2}).$$

Since the Hilbert polynomial of projective space itself is

$$P(\mathbb{P}_K^n, \nu) = \binom{\nu + n}{n} = \frac{1}{n!}\nu^n + O(\nu^{n-1}),$$

we can write

$$P_{d_1,\ldots,d_r}(\nu) = n(n-1)\cdots(n-r+1)d_1d_2\cdots d_r\frac{1}{n!}\nu^{n-r} + O(\nu^{n-r-1})$$

$$= \frac{d_1d_2\cdots d_r}{(n-r)!}\nu^{n-r} + O(\nu^{n-r-1}).$$

Hence

$$\deg(\Gamma) = d_1d_2\cdots d_r,$$

as desired.

We could have avoided the final computation in this proof by specializing: By using the fact established at the outset that complete intersections of given multidegree all have the same Hilbert polynomial, we can just choose for each pair (i,j) with $1 \le i \le r$ and $1 \le j \le d_i$ a general linear form $L_{i,j}$ and let $Z_i = V(F_i)$ where

$$F_i = \prod_{j=1}^{d_i} L_{i,j}$$

for each i. The intersection $\Gamma = \bigcap Z_i$ is then the union of $\prod d_i$ reduced linear subspaces in $\mathbb{P}^n_K$, and so has degree $\prod d_i$; we conclude that all complete intersections of multidegree $(d_1,\ldots,d_r)$ do.

Exercise III-77. For another specialization, let $Z_i \subset \mathbb{P}^n_K$ be the subscheme defined by $F_i(X_0,\ldots,X_n) = X_i^{d_i}$. Show directly that the intersection $\bigcap Z_i$ has degree $\prod d_i$. (Hint: you can reduce to the case $r = n$.)

Multiplicity of Intersections. Bézout's theorem for complete intersections (Theorem III-71) gives the degree of a complete intersection of hypersurfaces, but in practice we are often interested in intersecting more general subvarieties or subschemes of projective space. Since we have already defined the degree of any subscheme of projective space, it seems natural to ask whether the degree of an arbitrary intersection of subschemes X, $Y \subset \mathbb{P}^n_K$ is the product of the degrees of X and Y, always assuming the intersection is proper, that is, has the expected codimension. This turns out to be false in general, although it does hold if we make some hypothesis on the singularities of the schemes being intersected: if X and Y are *locally complete intersection subschemes* of $\mathbb{P}^n_K$, or more generally *Cohen–Macaulay subschemes* of $\mathbb{P}^n_K$, we have:

Theorem III-78 (Bézout's Theorem for Cohen–Macaulay schemes). *Let X and $Y \subset \mathbb{P}^n_K$ be Cohen–Macaulay schemes of pure codimensions k and l in $\mathbb{P}^n_K$. If the intersection $X \cap Y$ has codimension $k + l$, then*

$$\deg(X \cap Y) = \deg X \deg Y.$$

Example III-79. As we indicated, the statement of Theorem III-78 fails without the hypothesis that X and Y are Cohen–Macaulay, and it's instructive to see an example of this. Perhaps the simplest occurs in $\mathbb{P}^4_K =$

$\operatorname{Proj} K[Z_0, Z_1, Z_2, Z_3, Z_4]$: we take $X = \Lambda_1 \cup \Lambda_2$ the union of the two 2-planes

$$\Lambda_1 = V(Z_1, Z_2) \quad \text{and} \quad \Lambda_2 = V(Z_3, Z_4)$$

and we take Y the two-plane

$$Y = V(Z_1 - Z_3, \, Z_2 - Z_4).$$

We have already discussed this example in Exercise I-43 and following Lemma II-30; in particular, we have seen that the scheme $X \cap Y$ of intersection is the subscheme of the plane Y defined by the square of the maximal ideal of the origin, and so has degree 3. (Alternatively, since the projective tangent space to X is all of $\mathbb{P}_K^4$, it follows that the Zariski tangent space to $X \cap Y$ is two-dimensional, from which we may see immediately that $\deg(X \cap Y) \geq 3$.) But $\deg X \, \deg Y = 2 \cdot 1 = 2$, and so Theorem III-78 cannot hold.

What is going on in this example is not mysterious. Express Y as the intersection of two general hyperplanes H_1, H_2 containing it, and reparenthesize the intersection $X \cap Y$ as

$$X \cap Y = X \cap (H_1 \cap H_2) = (X \cap H_1) \cap H_2.$$

The first time we intersect, we find that the intersection scheme $X \cap H_1$ has an embedded point at the point (Z_1, Z_2, Z_3, Z_4). The second hyperplane H_2 passes through this point, in effect picking up the extra intersection.

This example both demonstrates the need for a refined way of ascribing multiplicity to a component of the intersection of subschemes of projective space, and suggests a way to do it. Here is the idea: Suppose we are given schemes $X, Y \subset \mathbb{P}_K^n$, of pure codimensions k and l, intersecting in a scheme of codimension $k + l$. We first reduce to the case where the scheme Y is a linear subspace of projective space, as follows: choose two complementary n-dimensional linear subspaces Λ_1, $\Lambda_2 \subset \mathbb{P}_K^{2n+1}$, and an isomorphism of $\mathbb{P}_K^n$ with each. (Concretely, we can label the homogeneous coordinates of $\mathbb{P}_K^{2n+1}$ as $x_0, \ldots, x_n, y_0, \ldots, y_n$ and take the linear spaces to be given by $x_0 = \ldots = x_n = 0$ and $y_0 = \ldots = y_n = 0$.) Write X' and Y' for the images of X and $Y \subset \mathbb{P}_K^n$ under these two embeddings. Let $J \subset \mathbb{P}_K^{2n+1}$ be the subscheme defined by the equations of X, written in the variables x_i, together with the equations of Y, written in the variables y_i — in other words, the intersection of the cone over X' with vertex Λ_2 with the cone over Y' with vertex Λ_1. J is called the *join* of X' and Y'; set theoretically, it is the union of the lines joining points of X' to points of Y'. Let $\Delta \subset \mathbb{P}_K^{2n+1}$ be the subscheme defined by the equations $x_0 - y_0 = \ldots = x_n - y_n = 0$. It is clear that the scheme $X \cap Y$ is isomorphic to the scheme $J \cap \Delta$, and we will define the multiplicity of intersection of X and Y along an irreducible component $Z \subset X \cap Y$ to be the intersection multiplicity of J and Δ along the corresponding component of $J \cap \Delta$. We have thus reduced the problem of defining the multiplicity of intersection of X and Y along an irreducible component $Z \subset X \cap Y$ to the case where Y is a linear space.

We will handle this case, as suggested by the example above, by writing Y as an intersection of hyperplanes $H_1 \cap \ldots \cap H_l$ and intersecting X with the H_i one at a time. After each step we discard the embedded components of the intersection. In the end we arrive at a scheme W contained in the actual intersection $X \cap Y$, which has degree satisfying Bézout's theorem: $\deg(W) = \deg(X) \deg(Y)$. To relate this to the classical language, for each irreducible component Z of the intersection $X \cap Y$, we define the *intersection multiplicity* of X and Y along Z, denoted $\mu_Z(X \cdot Y)$, to be the length of the local ring of W at the generic point of W corresponding to the component Z. We have then:

Theorem III-80 (Bézout's Theorem with multiplicities)**.** *Let X and $Y \subset \mathbb{P}_K^n$ be schemes of pure codimensions k and l in $\mathbb{P}_K^n$. If the intersection $X \cap Y$ has codimension $k + l$, then*

$$\deg(X \cap Y) = \sum_Z \mu_Z(X \cdot Y) \deg Z_{\mathrm{red}}.$$

There are other approaches to the definition of the multiplicity $\mu_Z(X \cdot Y)$ of intersection of two schemes X and $Y \subset \mathbb{P}_K^n$ along a component $Z \subset X \cap Y$; the classical literature is full of attempts at definitions, and there is also a modern approach involving the sheaves $\mathrm{Tor}(\mathscr{O}_X, \mathscr{O}_Y)$. Most of these approaches will work as well to define intersection multiplicities of any two subschemes X, Y of a nonsingular subscheme, as long as the intersection is proper.

Beyond this, there is a still more general version of Bézout's theorem that works for arbitrary subschemes X and Y of pure codimensions k and l in a nonsingular scheme T, even when the intersection $X \cap Y$ does not have codimension $k + l$ (or even for subschemes X, Y of a possibly singular scheme T, in case one of the two is locally a complete intersection subscheme of T). In this setting, one associates multiplicities to certain subschemes, or equivalence classes of subschemes, of the actual intersection $X \cap Y$, in such as way that (in case $T = \mathbb{P}_K^n$) the degrees of these subschemes times the corresponding multiplicities add up to $\deg X \deg Y$. For this and further refinements, see Fulton [1984] and Vogel [1984].

Exercise III-81. In case the idea of taking X reducible in Example III-79 strikes the reader as cheating: show that the same phenomenon occurs if we take $X \subset \mathbb{P}_K^4$ the cone over a nonsingular rational quartic curve $C \subset \mathbb{P}_K^3$, with Y again a two-plane passing through the vertex.

Exercise III-82. To see that the failure of Theorem III-78 to hold in general cannot be remedied by replacing $\deg(X \cap Y)$ by any other invariant of the scheme $X \cap Y$ in the left hand side of the statement of Theorem III-78, find an example of a scheme $\Gamma \subset \mathbb{P}_K^n$ and subschemes $X, Y, Z, W \subset \mathbb{P}_K^n$ of the appropriate dimensions, such that $X \cap Y = Z \cap W = \Gamma$, and

$$\deg X \, \deg Y = \deg \Gamma \neq \deg Z \, \deg W.$$

III.3.6   Hilbert Series

As the final note in our discussion of Hilbert functions, Hilbert polynomials and free resolutions, we mention the *Hilbert series* of a subscheme $X \subset \mathbb{P}_K^n$, or more generally of a graded module M over the coordinate ring S of $\mathbb{P}_K^n$. This is simply a very useful vehicle for conveying the information of the Hilbert polynomial; as an illustration, we will be able to write down the Hilbert polynomial of a complete intersection in a much more transparent way.

The Hilbert series $H_M(t)$ of a module M is easy to define: if $P(M, \nu)$ is the Hilbert function of M, we let $H_M(t)$ be the Laurent series

$$H_M(t) = \sum_{\nu = -\infty}^{\infty} P(M, \nu) t^\nu.$$

We define the *Hilbert series* $H_X(t)$ of a subscheme $X \subset \mathbb{P}_K^n$ to be the Hilbert series of its coordinate ring $S_X = S/I(X)$. The first thing to note is that the Hilbert series of projective space itself is simple: we have

$$H_{\mathbb{P}_K^n}(t) = H_S(t) = \frac{1}{(1-t)^{n+1}}.$$

Similarly, the Hilbert series of any twist $S(d)$ of S is simply

$$H_{S(d)}(t) = \frac{t^d}{(1-t)^{n+1}}.$$

Given any exact sequence of graded S-modules

$$0 \longrightarrow M_r \longrightarrow M_{r-1} \longrightarrow \cdots \longrightarrow M_2 \longrightarrow M_1 \longrightarrow M_0 \longrightarrow 0,$$

we see that their Hilbert series satisfy the relation

$$\sum_{k=0}^{r} (-1)^k H_{M_k}(t) = 0.$$

Thus, if we have a free resolution of a scheme $X \subset \mathbb{P}_K^n$

$$\cdots \longrightarrow \bigoplus_{i=1}^{k_2} S(-a_{2i}) \longrightarrow \bigoplus_{i=1}^{k_1} S(-a_{1i}) \longrightarrow S \longrightarrow S_X \longrightarrow 0,$$

we see that the Hilbert series

$$H_X(t) = \frac{\sum (-1)^i t^{a_{i,j}}}{(1-t)^{n+1}}.$$

(where we adopt the convention that $k_0 = 1$ and $a_{01} = 0$). One thing we see from this is that *the Hilbert series of any subscheme of projective space is a rational function of t.*

Exercise III-83. Show that if $X \subset \mathbb{P}^n_K$ is a subscheme of dimension m, then the rational function $H_X(t)$, reduced to lowest terms, has numerator

$$\tilde{H}_X(t) = (1-t)^{m+1} H_X(t);$$

in particular, this is a polynomial in t. Show that its value at 1 is

$$\tilde{H}_X(1) = \deg(X).$$

Now suppose that $X \subset \mathbb{P}^n_K$ is a complete intersection of r hypersurfaces of degrees $d_1, \ldots, d_r$. By the Koszul resolution above, we see that the Hilbert series

$$H_X(t) = \frac{\sum (-1)^{|I|} t^{|I|}}{(1-t)^{n+1}}.$$

We can factor this, and cancel factors, writing

$$H_X(t) = \frac{\prod (1 - t^{d_i})}{(1-t)^{n+1}} = \frac{\prod (1 + t + \cdots + t^{d_i-1})}{(1-t)^{n-r+1}}.$$

Hence,

$$\tilde{H}_X(t) = (1-t)^{\dim(X)+1} H_X(t) = \prod (1 + t + \cdots + t^{d_i-1}).$$

Since the value of this polynomial at $t = 1$ is the product $\prod d_i$, Bézout's theorem follows.

IV

Classical Constructions

In this chapter, we illustrate how some geometric constructions from classical algebraic geometry are carried out in the setting of scheme theory. We will see in each case how the new language allows us to extend the range of the definitions (and of the questions we may ask about the objects); how it enables us to give precise formulations of classical problems; and in some cases how it helps us to solve them.

IV.1 Flexes of Plane Curves

In this section, we will describe the classical definition of a *flex* of a nonsingular plane curve $C \subset \mathbb{P}^2_K$ over an algebraically closed field K. We will then indicate how this definition may be extended to the setting of schemes, and show how this extension sheds light on the geometry of flexes, even in the classical case.

IV.1.1 Definitions

We need one preliminary definition. Let K be any field, let $C, D \subset \mathbb{P}^2_K$ be two plane curves without common components, and let $p \in C \cap D$ be a point of intersection. We define the *intersection multiplicity* of C and D at p, denoted $\mu_p(C \cdot D)$, to be the multiplicity of the component Γ of the scheme $C \cap D$ supported at p. Since plane curves are Cohen–Macaulay, this coincides with the notion of intersection multiplicity introduced in

Section III.3.5. Note also the relation with the notion of degree: the degree of Γ as a subscheme of $\mathbb{P}^2_K$ is the intersection multiplicity $\mu_p(C \cdot D)$ times the degree $(\kappa(p) : K)$ of the residue field as an extension of K. Thus, for example, the Bézout theorem (III-71) for plane curves asserts that

$$\deg C \deg D = \sum_{p \in C \cap D} (\kappa(p) : K) \, \mu_p(C \cdot D).$$

This said, the notion of a flex of a plane curve in classical algebraic geometry is a straightforward and geometrically reasonable one: if $C \subset \mathbb{P}^2_\mathbb{C}$ is a nonsingular plane curve of degree d over the complex numbers, a point $p \in C$ is called a *flex* if the projective tangent line $\mathbb{T}_pC \subset \mathbb{P}^2_\mathbb{C}$ (see Section III.2.4) has contact of order 3 or more with C at p; or, in modern language, if the intersection multiplicity $\mu_p(C \cdot \mathbb{T}_pC)$ of $\mathbb{T}_pC$ and C at p is at least 3. (Here, since we are working over an algebraically closed field, the intersection multiplicity coincides with the degree of the component of $\mathbb{T}_pC \cap C$ supported at p, that is, $\dim_\mathbb{C}(\mathcal{O}_{\mathbb{T}_pC \cap C, p})$.) It is a classical theorem (which we will establish below) that if C is not a line, then C has finitely many flexes, and that if they are counted with the proper multiplicity the number is $3d(d-2)$.

This simple definition was extended to singular curves — see, for example, Coolidge [1931] — though the definitions are not always precise by modern standards. There are also problems with the definition if we consider curves $C \subset \mathbb{P}^2_K$ over non-algebraically closed fields K, or over fields K of finite characteristic, or curves that contain a line or a multiple component.

What we will do here is to give a uniform definition of flexes for an arbitrary plane curve $C \subset \mathbb{P}^2_S$ over any scheme S. First recall from Section III.2.8 that by a *plane curve* of degree d over a scheme S we mean a subscheme $C \subset \mathbb{P}^2_S$ that is, locally on S, the zero locus $V(F)$ of a single homogeneous polynomial

$$F(X, Y, Z) = \sum_{i+j+k=d} a_{ijk} X^i Y^j Z^k$$

of degree d whose coefficients a_{ijk} are regular functions on S not vanishing simultaneously. Recall also that if S is affine, we can dispense with the word "locally"; that is, if $S = \operatorname{Spec} A$, a plane curve C over S is of the form

$$C = \operatorname{Proj} A[X, Y, Z]/(F)$$

for some polynomial F.

Now, given a plane curve $C \subset \mathbb{P}^2_S$ over S, we will define a closed subscheme $\mathscr{F} = \mathscr{F}_C \subset C$, which we will call the *scheme of flexes* on C. This will commute with base change $S' \to S$ (that is, if we set $C' = S' \times_S C \subset \mathbb{P}^2_{S'}$, then $\mathscr{F}_{C'} = (\pi_2)^{-1}(\mathscr{F}_C)$) and $\mathscr{F}$ will be finite and flat of degree $3d(d-2)$ over at least the open subset of S where the relative dimension of $\mathscr{F}$ is zero. The significance of this is that if we have a family of plane curves, the limits of the flexes of the general fiber are flexes of the special fiber (that is, $\mathscr{F}$ is closed), and conversely in case $\mathscr{F}$ has relative dimension zero ($\mathscr{F}$ is flat). Moreover, in the classical setting — that is, if C is a nonsingular plane curve over the spectrum $S = \operatorname{Spec} K$ of an algebraically closed field of characteristic zero — the support of $\mathscr{F}$ will be the set of flexes of C as defined classically (and hence in the general case, if $s \in S$ is any point whose residue field $\kappa(s)$ is algebraically closed of characteristic zero, the support of the fiber $\mathscr{F}_s$ of $\mathscr{F}$ over s will be the set of flexes of C_s).

To motivate our definition in the general case, we recall one of the earliest results in the classical setting: for a nonsingular plane curve $C = V(F) \subset \mathbb{P}_K^2$ over an algebraically closed field K, given as the zero locus of a polynomial $F(X, Y, Z)$, the flexes of C are the points of its intersection with its *Hessian*, the curve defined as the zero locus of the polynomial

$$H(X, Y, Z) = \begin{vmatrix} \dfrac{\partial^2 F}{\partial X^2} & \dfrac{\partial^2 F}{\partial X \partial Y} & \dfrac{\partial^2 F}{\partial X \partial Z} \\[2mm] \dfrac{\partial^2 F}{\partial Y \partial X} & \dfrac{\partial^2 F}{\partial Y^2} & \dfrac{\partial^2 F}{\partial Y \partial Z} \\[2mm] \dfrac{\partial^2 F}{\partial Z \partial X} & \dfrac{\partial^2 F}{\partial Z \partial Y} & \dfrac{\partial^2 F}{\partial Z^2} \end{vmatrix} .$$

We leave the proof of this fact as an exercise:

Exercise IV-1. Let K be an algebraically closed field of characteristic zero, $C \subset \mathbb{P}_K^2$ a plane curve and $p \in C$ a nonsingular point of C. Show that the projective tangent line $\mathbb{T}_p C$ has contact of order 3 or more with C at p if and only if $H(p) = 0$.

Hint: introduce affine coordinates

$$x = \frac{X}{Z}, \quad y = \frac{Y}{Z}$$

on the corresponding subset of $\mathbb{P}_K^2$ and use Euler's relation to see that the dehomogenization $h(x, y) = H(x, y, 1)$ of the Hessian determinant is (up to scalars)

$$h(x, y) = \begin{vmatrix} f & \dfrac{\partial f}{\partial x} & \dfrac{\partial f}{\partial y} \\[2mm] \dfrac{\partial f}{\partial x} & \dfrac{\partial^2 f}{\partial x^2} & \dfrac{\partial^2 f}{\partial x \partial y} \\[2mm] \dfrac{\partial f}{\partial y} & \dfrac{\partial^2 f}{\partial x \partial y} & \dfrac{\partial^2 f}{\partial y^2} \end{vmatrix} ,$$

where $f(x, y) = F(x, y, 1)$ is the dehomogeneization of F.

To define the scheme of flexes of an arbitrary plane curve $C \subset \mathbb{P}_S^2$ in the general setting, we simply generalize the Hessian and extend this characterization: Suppose that in some affine open subset $U = \operatorname{Spec} R \subset S$ the curve

$$C \cap \mathbb{P}_U^2 = \operatorname{Proj} R[X, Y, Z]$$

is the zero locus of the polynomial $F \in R[X, Y, Z]$. We define the *Hessian determinant* to be the polynomial

$$H(X,Y,Z) = \begin{vmatrix} \dfrac{\partial^2 F}{\partial X^2} & \dfrac{\partial^2 F}{\partial X \partial Y} & \dfrac{\partial^2 F}{\partial X \partial Z} \\[2ex] \dfrac{\partial^2 F}{\partial Y \partial X} & \dfrac{\partial^2 F}{\partial Y^2} & \dfrac{\partial^2 F}{\partial Y \partial Z} \\[2ex] \dfrac{\partial^2 F}{\partial Z \partial X} & \dfrac{\partial^2 F}{\partial Z \partial Y} & \dfrac{\partial^2 F}{\partial Z^2} \end{vmatrix}.$$

Since F, and hence H, is determined by C up to multiplication by a unit in $R = \mathcal{O}(U)$, we may define the *Hessian* C' of C to be the subscheme of $\mathbb{P}^2_S$ defined by the Hessian determinant over each affine open $U \subset S$; and we define the scheme $\mathscr{F}$ of flexes of C to be the intersection

$$\mathscr{F} = C \cap C'.$$

We see immediately that this is a closed subscheme of C and that its formation commutes with base change. In particular, for any point $s \in S$, the fiber $\mathscr{F}_s$ of $\mathscr{F}$ over s will be simply the scheme of flexes of the fiber $C_s \subset \mathbb{P}^2_{\kappa(s)}$ of C over s. As the intersection of two plane curves of degrees d and $3(d-2)$ it is finite and flat of degree $3d(d-2)$ over at least the open subset of S where the fiber dimension is zero (by Proposition II-32, families of complete intersections are flat). And, by Exercise IV-1, a nonsingular point of a curve C over an algebraically closed field of characteristic zero lies in $\mathscr{F}$ if and only if it is a flex in the classical sense.

One word of warning: our definition does not coincide with the classical one in the case of a singular curve $C \subset \mathbb{P}^2_K$: in our definition the singular points of C will always be in the support of $\mathscr{F}$. (As we will see, this is as it must be if the flexes of a family of curves are to be closed in the total space.) As for the classical formulas, we will see below how to derive them from our definition.

We can go further and relate the scheme structure of $\mathscr{F}$ at p to the geometry of C at p:

Exercise IV-2. Let $C \subset \mathbb{P}^2_K$ be as in Exercise IV-1, and $p \in C$ a nonsingular point of C. Show that the projective tangent line $\mathbb{T}_p C \subset \mathbb{P}^2_K$ to C at p has intersection multiplicity $m \geq 3$ with C at p if and only if the component Γ_p of the intersection $C \cap C'$ supported at p is isomorphic to

$$\Gamma_p \cong \operatorname{Spec} K[x]/(x^{m-2}).$$

As this exercise suggests, we define the *multiplicity* of a flex $p \in C_{\text{smooth}}$ to be the order of contact of $\mathbb{T}_p C$ with C at p minus 2. We would like to apply Bézout's theorem to deduce that a nonsingular plane curve of degree $d > 1$ over an algebraically closed field K has exactly $3d(d-2)$ flexes, counting multiplicity, but there is one further issue: we need to know that $\mathscr{F}$ is a proper subscheme of C, that is, that not every point of C is a flex! Although this seems intuitively obvious, it is actually false in positive characteristic:

Exercise IV-3. Let K be a field of characteristic p, and let $C \subset \mathbb{P}^2_K$ be the plane curve given by the polynomial $X^p Y + X Y^p - Z^{p+1}$. Show that C is nonsingular, but that every point of C is a flex.

In characteristic 0, however, our intuition is correct:

Theorem IV-4. *If $C \subset \mathbb{P}^2_K$ is any nonsingular plane curve of degree $d > 1$ over an algebraically closed field K of characteristic zero, then not every point of C is a flex (so that in particular C has exactly $3d(d-2)$ flexes, counting multiplicity).*

Proof. See for example Hartshorne [1977, Chapter IV, Exercise 2.3e] or Griffiths and Harris [1978, Chapter 2, Section 4].   □

Flexes of multiplicity $m > 1$ certainly can occur on nonsingular curves. This naturally raises the question of whether, on a general curve, all the flexes are simple (that is, have multiplicity 1). In fact, this is the case:

Exercise IV-5. Let K be an algebraically closed field. Fix an integer $d > 2$ and let $B = \mathbb{P}^N_K$ the projective space parametrizing plane curves $C \subset \mathbb{P}^2_K$ of degree d. Show that for a general point $[C] \in B$ — that is, for all points $[C]$ in a dense open set in B — all the flexes on the corresponding curve $C \subset \mathbb{P}^2_K$ are simple.

Hint: Consider the scheme of flexes $\mathscr{F}$ of the universal curve $\mathscr{C} \subset \mathbb{P}^2_B$ (as defined in Section III.2.8). Show that $\mathscr{F}$ is irreducible, and deduce that it is sufficient to exhibit a single plane curve $C \subset \mathbb{P}^2_K$ with a single simple flex.

Exercise IV-6. Suppose we want to remove the hypothesis that K is algebraically closed in Theorem IV-4 above. How should we define the multiplicity of a flex point $p \in C$ with residue field a finite extension L of K so as to preserve the conclusion that X has $3d(d-2)$ flexes?

IV.1.2 Flexes on Singular Curves

Interesting new questions arise when we consider singular curves. First of all, every singular point is a flex:

Exercise IV-7. Let $C \subset \mathbb{P}^2_K$ be a plane curve. Show that all singular points of C are flexes.

Hint: either exhibit a line through a singular point p of C with intersection multiplicity 3 or more by looking at the tangent cone to C at p (that is, expanding f around p and taking a component of the zero locus of the quadratic term); or use Exercise IV-2 and show that the Hessian vanishes at p by observing that X times the first column of the Hessian determinant, plus Y times the second column, plus Z times the third, vanishes at p.

There are two sorts of questions about flexes on singular curves. First, we can consider curves C with isolated singularities and no line components, so that the Hessian C' will still meet C in a zero-dimensional scheme Γ, and thus C will have a finite number of flexes; we ask for the number of flexes supported at nonsingular points of C. To find this number, we simply have to find the degree of the part of the scheme Γ whose support is contained in the singular locus C_{sing} and subtract this from $3d(d-2)$. It turns out that this has a nice answer in particular cases, two of which are expressed in the following exercise.

Exercise IV-8. Let $C \subset \mathbb{P}^2_K$ be irreducible and reduced, with Hessian C'. Looking ahead to Definition V-31, let $p \in C$ be an ordinary node of C ("ordinary" here means neither branch of C at p has contact of order 3 or more with its projective tangent line). Show that the component Γ_p of the intersection $C \cap C'$ supported at p has degree 6 over the residue field $\kappa(p)$ of p. Similarly, show that the component supported at a cusp p of C has degree 8. What is the degree if p is an ordinary tacnode of C? (For formal definitions of node, cusp and tacnode see Definition V-31).

Thus, over an algebraically closed field, the number of nonsingular flexes of a plane curve of degree d not containing any lines and having as singularities δ ordinary nodes and κ cusps is

$$3d(d-2) - 6\delta - 8\kappa.$$

This is an example of the classical *Plücker formulas* for plane curves.

Exercise IV-9. Verify that if C is reducible (again assuming no component of C is a line), we can get the same answer by considering the components of C individually.

IV.1.3 Curves with Multiple Components

A very different sort of question emerges when we consider curves with multiple components, for example the curve defined by a power $F = G^m$ of a polynomial $G(X, Y, Z)$. Of course, for such a curve C the scheme $\mathscr{F}_C$ of flexes is positive-dimensional, and typically not that interesting. Rather, the interesting questions arise when we consider *families of curves* specializing to such a multiple curve. We ask: in such a family, where do the flexes go?

To give just an example of such a problem, consider the case of a nonsingular quartic plane curve degenerating to a double conic in a linear family. Let K be an algebraically closed field of characteristic zero and consider a curve $\mathscr{C}$ over the scheme $B = \mathbb{A}^1_K = \operatorname{Spec} K[t]$. Suppose $U = U(X, Y, Z)$ is an irreducible quadric polynomial and $G = G(X, Y, Z)$ any quartic polynomial such that the curves $V(U)$ and $V(G) \subset \mathbb{P}^2_K$ intersect transversely.

Consider the family $\pi : \mathscr{C} \to \mathbb{A}^1_K$ of quartic plane curves given by the equation $F = U^2 + tG = 0$ — that is, the scheme

$$\pi : \mathscr{C} = \operatorname{Proj} K[t][X, Y, Z]/(U(X, Y, Z)^2 + tG(X, Y, Z))$$

$$\subset \operatorname{Proj} K[t][X, Y, Z] = \mathbb{P}^2_B \longrightarrow B.$$

Let $\mathscr{F}$ be the scheme of flexes of the curve $\mathscr{C} \subset \mathbb{P}^2_B$. To set up the problem, let $\mathscr{F}^* \subset \mathbb{P}^2_{B^*}$ be the inverse image in $\mathscr{F}$ of the punctured line $B^* = \operatorname{Spec} K[t, t^{-1}] \subset B$, and $\mathscr{F}'$ the closure of $\mathscr{F}^*$ in $\mathbb{P}^2_B$. The scheme $\mathscr{F}^*$ is finite and flat over B^*, and readily described: if $C_\mu \subset \mathbb{P}^2_K$ is the fiber of $\mathscr{C}$ over the point $(t - \mu) \in B = \mathbb{A}^1_K$, then for $\mu \neq 0$, the fiber F_μ of $\mathscr{F}$ over $(t - \mu)$ will be the $3d(d - 2) = 24$ flexes of C_μ. In other words, away from the origin $(t) \in B = \mathbb{A}^1_K$ the flexes of the curves C_μ themselves form a flat family.

Let $\mathscr{F}'$ be the closure of $\mathscr{F}^*$ in $\mathbb{P}^2_B$, and let F'_0 be the fiber of $\mathscr{F}'$ over the origin. Since B is one-dimensional and nonsingular, $\mathscr{F}'$ will be flat over all of B; it follows in particular that $F'_0 \subset C_0 \subset \mathbb{P}^2_K$ has dimension zero and degree 24 over K. We may think of F'_0 as the "limiting position" of the 24 flexes of the nearby nonsingular curves C_μ as μ approaches zero. Thus, the naive question, "where do the flexes of a plane quartic go when the quartic degenerates into a double conic?" translates into the precise problem: determine the flat limit F'_0, and in particular its support.

What makes this tricky is that the scheme F'_0 is not the fiber of $\mathscr{F}$ over the origin. Rather, $\mathscr{F}$ will have two components: one, the closure $\mathscr{F}'$ of $\mathscr{F}^*$ consisting of the "real" flexes and their limits, and the other supported on the conic $V(t, U)$ in the special fiber $\pi^{-1}((t)) = \mathbb{P}^2_K$ of $\mathbb{P}^2_B$. Thus we cannot hope to gain any clues to the answer simply by looking only at the curve C_0 (indeed, since the group of automorphisms of $\mathbb{P}^2_K$ carrying C_0 into itself acts transitively on the closed points of the conic $(C_0)_{\mathrm{red}}$, we see that the answer must depend on the family $\mathscr{C}$).

To answer the question, we first write down the ideal I of the scheme $\mathscr{F}$ (in an affine open subset $\operatorname{Spec} K[t][x, y] \cong \mathbb{A}^2_B \subset \mathbb{P}^2_B$), then the ideal $I^* = I \cdot K[t, t^{-1}][x, y]$ of $\mathscr{F}^*$, then the ideal $I' = I^* \cap K[t][x, y]$ of the closure $\mathscr{F}'$, and finally the ideal $I'_0 = (I', t)$ of the fiber F'_0 of $\mathscr{F}'$ over the origin $(t) \in B$. To illustrate how such calculations are done, we will carry out these steps in detail. (You may wish to wait to look at these details until you have a similar problem of your own to solve!)

To start, if $u(x, y) = U(X, Y, 1)$ and $g(x, y) = G(X, Y, 1)$ are the inhomogeneous forms of U and G respectively in the affine open $\operatorname{Spec} K[t][x, y] \cong \mathbb{A}^2_B \subset \mathbb{P}^2_B$, the ideal I is by definition generated by two elements, the equation $u^2 - tg$ and the affine Hessian

$$\begin{vmatrix} u^2 - tg & 2uu_x + tg_x & 2uu_y + tg_y \\ 2uu_x + tg_x & 2uu_{xx} + 2u_x^2 + tg_{xx} & 2uu_{xy} + 2u_xu_y + tg_{xy} \\ 2uu_y + tg_y & 2uu_{xy} + 2u_xu_y + tg_{xy} & 2uu_{yy} + 2u_y^2 + tg_{yy} \end{vmatrix}.$$

Thus, $I = (u^2 - tg, H)$, where

$$H = \begin{vmatrix} 0 & 2uu_x + tg_x & 2uu_y + tg_y \\ 2uu_x + tg_x & 2uu_{xx} + 2u_x^2 + tg_{xx} & 2uu_{xy} + 2u_xu_y + tg_{xy} \\ 2uu_y + tg_y & 2uu_{xy} + 2u_xu_y + tg_{xy} & 2uu_{yy} + 2u_y^2 + tg_{yy} \end{vmatrix}.$$

We may expand out H, grouping terms involving like powers of t:

$$\begin{aligned} H = {}& 8u^2 \left(-u_x^2(uu_{yy} + u_y^2) + 2u_xu_y(uu_{xy} + u_xu_y) - u_y^2(uu_x + u_x^2)\right) \\ &+ 8tu\big(-g_xu_x(uu_{yy} + u_y^2) + g_xu_y(uu_{xy} + u_xu_y) \\ &\qquad + g_yu_x(uu_{xy} + u_xu_y) - g_yu_y(uu_{xx} + u_x^2)\big) \\ &+ 4tu^2 \left(-u_x^2g_{yy} + u_xu_yg_{xy} - u_y^2g_{xx}\right) \\ &+ 2t^2 \left(-g_x^2(uu_{yy} + u_y^2) + 2g_xg_y(uu_{xy} + u_xu_y) - g_y^2(uu_{xx} + u_x^2)\right) \\ &+ 4t^2u \left(-g_xu_xg_{yy} + g_xu_yg_{xy} + g_yu_xg_{xy} - g_yu_yg_{xx}\right) \\ &+ t^3 \left(-g_x^2g_{yy} + g_xg + yg_{xy} - g_y^2g_{xx}\right). \end{aligned}$$

The first two terms on the right may be simplified, yielding the expression

$$\begin{aligned} H = {}& 8u^3 \left(-u_x^2u_{yy} + u_xu_yu_{xy} - u_y^2u_{xx}\right) \\ &+ 8tu^2 \left(-g_xu_xu_{yy} + g_xu_yu_{xy} + g_yu_xu_{xy} - g_yu_yu_{xx}\right) \\ &+ 4tu^2 \left(-u_x^2g_{yy} + u_xu_yg_{xy} - u_y^2g_{xx}\right) \\ &+ 2t^2 \left(-g_x^2(uu_{yy} + u_y^2) + 2g_xg_y(uu_{xy} + u_xu_y) - g_y^2(uu_{xx} + u_x^2)\right) \\ &+ 4t^2u \left(-g_xu_xg_{yy} + g_xu_yg_{xy} + g_yu_xg_{xy} - g_yu_yg_{xx}\right) \\ &+ t^3 \left(-g_x^2g_{yy} + g_xg + yg_{xy} - g_y^2g_{xx}\right). \end{aligned}$$

Now, modulo the other generator $u^2 - tg$ of I, we may replace u^2 by $-tg$ in this expression to arrive at a polynomial divisible by t. Thus the ideal $I^* = I(\mathscr{F}^*) = I \cdot K[t, t^{-1}][x, y] \subset K[t, t^{-1}][x, y]$, and hence the ideal $I' = I(\mathscr{F}') = I^* \cap K[t][x, y] \subset K[t][x, y]$, contain as well the element

$$\begin{aligned} H' = {}& -8ug \left(-u_x^2u_{yy} + u_xu_+yu_{xy} - u_y^2u_{xx}\right) \\ &- 8tg \left(-g_xu_xu_{yy} + g_xu_yu_{xy} + g_yu_xu_{xy} - g_yu_yu_{xx}\right) \\ &- 4tg \left(-u_x^2g_{yy} + u_xu_yg_{xy} - u_y^2g_{xx}\right) \\ &+ 2t \left(-g_x^2(uu_{yy} + u_y^2) + 2g_xg_y(uu_{xy} + u_xu_y) - g_y^2(uu_{xx} + u_x^2)\right) \\ &+ 4tu \left(-g_xu_xg_{yy} + g_xu_yg_{xy} + g_yu_xg_{xy} - g_yu_yg_{xx}\right) \\ &+ t^2 \left(-g_x^2g_{yy} + g_xg + yg_{xy} - g_y^2g_{xx}\right). \end{aligned}$$

Moreover, if we multiply this generator of I^* by u and once more replace u^2 by $-tg$, we arrive again at a polynomial divisible by t; we conclude that the ideals I^* and I' contain as well

$$J = 8g^2 \left(-u_x^2 u_{yy} + u_x u_{+} y u_{xy} - u_y^2 u_{xx}\right)$$
$$- 8gu \left(-g_x u_x u_{yy} + g_x u_y u_{xy} + g_y u_x u_{xy} - g_y u_y u_{xx}\right)$$
$$- 4gu \left(-u_x^2 g_{yy} + u_x u_y g_{xy} - u_y^2 g_{xx}\right)$$
$$+ 2u \left(-g_x^2 (u u_{yy} + u_y^2) + 2g_x g_y (u u_{xy} + u_x u_y) - g_y^2 (u u_{xx} + u_x^2)\right)$$
$$- 4tg \left(-g_x u_x g_{yy} + g_x u_y g_{xy} + g_y u_x g_{xy} - g_y u_y g_{xx}\right)$$
$$+ tu \left(-g_x^2 g_{yy} + g_x g_{+} y g_{xy} - g_y^2 g_{xx}\right).$$

To continue with this analysis, we have to use the fact that, for any homogeneous quadratic polynomial $U(X, Y, Z)$, the Hessian

$$\begin{vmatrix} \dfrac{\partial^2 U}{\partial X^2} & \dfrac{\partial^2 U}{\partial X \partial Y} & \dfrac{\partial^2 U}{\partial X \partial Z} \\[2mm] \dfrac{\partial^2 U}{\partial Y \partial X} & \dfrac{\partial^2 U}{\partial Y^2} & \dfrac{\partial^2 U}{\partial Y \partial Z} \\[2mm] \dfrac{\partial^2 U}{\partial Z \partial X} & \dfrac{\partial^2 U}{\partial Z \partial Y} & \dfrac{\partial^2 U}{\partial Z^2} \end{vmatrix}$$

is a scalar $\mu = \mu(U)$, nonzero if U is irreducible (that is, if the curve $V(U) \subset \mathbb{P}^2$ is nonsingular), and zero otherwise. It follows that

$$\begin{vmatrix} u & \dfrac{\partial u}{\partial x} & \dfrac{\partial u}{\partial y} \\[2mm] \dfrac{\partial u}{\partial x} & \dfrac{\partial^2 u}{\partial x^2} & \dfrac{\partial^2 u}{\partial x \partial y} \\[2mm] \dfrac{\partial u}{\partial y} & \dfrac{\partial^2 u}{\partial x \partial y} & \dfrac{\partial^2 u}{\partial y^2} \end{vmatrix} = -u_x^2 u_{yy} + u_x u_{+} y u_{xy} - u_y^2 u_{xx} = \lambda u + \mu,$$

for some scalar λ. Substituting this in the expression for J, we have

$$J = 8\mu g^2$$
$$+ 8\lambda g^2 u$$
$$- 8gu \left(-g_x u_x u_{yy} + g_x u_y u_{xy} + g_y u_x u_{xy} - g_y u_y u_{xx}\right)$$
$$- 4gu \left(-u_x^2 g_{yy} + u_x u_y g_{xy} - u_y^2 g_{xx}\right)$$
$$+ 2u \left(-g_x^2 (u u_{yy} + u_y^2) + 2g_x g_y (u u_{xy} + u_x u_y) - g_y^2 (u u_{xx} + u_x^2)\right)$$
$$- 4tg \left(-g_x u_x g_{yy} + g_x u_y g_{xy} + g_y u_x g_{xy} - g_y u_y g_{xx}\right)$$
$$+ tu \left(-g_x^2 g_{yy} + g_x g_{+} y g_{xy} - g_y^2 g_{xx}\right).$$

Now, we have seen that the ideal $I' \supset (u^2 + tg, H', J)$. Restricting to the fiber over the origin in B—that is, setting $t = 0$—we see that the ideal $I'_0 = (I', t)$ of the fiber F'_0 of $\mathscr{F}'$ contains

$$u^2 + tg \equiv u^2 \quad \mathrm{mod}\ (t),$$
$$H' \equiv ug \quad \mathrm{mod}\ (t, u^2),$$

and

$$J \equiv 8\mu g^2 - 2\nu u \quad \mathrm{mod}\ (t, u^2, ug),$$

where

$$\nu = g_x^2(uu_{yy} + u_y^2) - 2g_x g_y(uu_{xy} + u_x u_y) + g_y^2(uu_{xx} + u_x^2).$$

We see from this that $I_0' \supset (t, u^2, ug, 8\mu g^2 + 2\nu u)$. Now, we may write

$$\nu = g_x^2(uu_{yy} + u_y^2) - 2g_x g_y(uu_{xy} + u_x u_y) + g_y^2(uu_{xx} + u_x^2)$$
$$\equiv (g_x u_y - g_y u_x)^2 \quad \mathrm{mod}\ (u)$$
$$\equiv \begin{vmatrix} g_x & g_y \\ u_x & u_y \end{vmatrix}^2.$$

In particular, given that $V(U)$ and $V(G)$ intersect transversely, ν cannot be zero at a point where $u = g = 0$. We may thus recognize the ideal $(t, u^2, ug, 8\mu g^2 + 2\nu u)$ as the ideal of a subscheme of the special fiber C_0, supported at the eight points $t = U = G = 0$ of intersection of the conic $U = 0$ and the quartic $G = 0$ in the plane $t = 0$ and having degree 3 at each point. Since $8 \times 3 = 24$, the fiber F_0' cannot be any smaller than this, and so we must have equality, that is,

$$I_0' = (t,\ u^2,\ ug,\ 8\mu g^2 + 2\nu u).$$

In other words:

Proposition IV-10. *The scheme F_0' is supported at the eight points $t = U = G = 0$ of intersection of the conic $V(U)$ and the quartic $V(G)$ in the plane $V(t)$. At each point, it consists of a curvilinear scheme of degree 3, tangent to, but not contained in, the conic $V(U)$.*

One aspect of this answer is that any closed point of the reduced curve $(C_0)_{\mathrm{red}}$ could be a limit of flexes of nonsingular curves for a suitable family of curves C_μ tending to C_0. This is a general phenomenon; in fact, every point of a multiple component of a curve is a limit of flexes of nearby nonsingular curves.

The phenomenon described in this example is fairly general. The following exercises give two generalizations.

Exercise IV-11. Let K be as before an algebraically closed field of characteristic zero and $B = \mathbb{A}_K^1 = \mathrm{Spec}\,K[t]$. Let $F = V(U)$ be a nonsingular conic, and $D = V(G)$ and $E = V(H)$ nonsingular plane curves of degrees d and $d - 4$ respectively intersecting C transversely, such that $F \cap D \cap E = \varnothing$, and the points of $E \cap F$ are not flexes of E. Consider the family $\pi : \mathscr{C} \to B$ of plane curves of degree d given by the equation $F = U^2 H + tG = 0$ — that is, the scheme $\mathscr{C} = \mathrm{Proj}\,K[t][X, Y, Z]/(U^2 H + tG) \subset \mathbb{P}_B^2$. Describe the limiting position of the flexes of the fiber C_λ over the point $(t - \lambda) \in B$ as λ goes to zero. In particular, show that of the $3d(d - 2)$ flexes of C_λ, 3 approach each of the $2d$ points $U = G = t = 0$; 9 approach each of the

$2(d-4)$ points $U = H = t = 0$, while the remaining $3(d-4)(d-6)$ approach flexes of the curve $H = t = 0$.

Exercise IV-12. With K and B as above, suppose now that $F = V(U)$ is a nonsingular plane curve of degree e and $X = V(G)$ a nonsingular plane curve of degree $d = 2e$ intersecting F transversely. Consider the family $\pi : \mathscr{C} \to B$ of plane curves of degree d given by the equation $F = U^2 + tG = 0$, and once more describe the limiting position of the flexes of the fiber C_λ as λ goes to zero. In particular, show that of the $3d(d-2)$ flexes of C_λ, 3 approach each of the de points $U = G = t = 0$ and 2 approach each of the $3e(e-2)$ flexes of $G = t = 0$.

There is one case, other than that of a curve with multiple components, in which a plane curve over a field of characteristic 0 may have a positive-dimensional scheme of flexes: that of a curve $C \subset \mathbb{P}^2_K$ containing a line. We may ask in this setting the analogous question: given a family of plane curves specializing to one containing a line—for eample, with K and B as above, the family $\pi : \mathscr{C} \to B$ of plane curves of degree d given by the equation $LF + tG = 0$ for L, F and G general polynomials of degrees 1, $d-1$ and d respectively—where do the flexes of the general fiber of $C \to B$ go? The answer turns put to be in some ways more subtle than that in the case of multiple components; we will not describe it here, for lack of some necessary language, but will mention that (as the reader may verify) the location of the limiting flexes on the line $V(L)$ is *not* the intersection of $V(L)$ with $V(G)$.

To conclude this section, here is an amusing aspect of the geometry of flexes on plane cubics.

Exercise IV-13. Consider a nonsingular plane cubic curve $C \subset \mathbb{P}^2_\mathbb{R}$ over the real numbers. Show that the scheme of flexes will consist, for some pair of integers a and b with $a + 2b = 9$, of a points with residue field $\mathbb{R}$ and b points with residue field $\mathbb{C}$. Deduce in particular that C must have a real flex.

In fact, the number a in this problem is 3. For the pleasure of the reader familiar with the classical theory of elliptic curves, we sketch the argument. Part of it is simple: the exclusion of 5 and 7 follows from the existence of a group law on the set of points of C with residue field $\mathbb{R}$, in terms of which the flexes with residue field $\mathbb{R}$ form a subgroup of the group $\mathbb{Z}/(3) \times \mathbb{Z}/(3)$ of the 9 flexes of $C \times_\mathbb{R} \operatorname{Spec} \mathbb{C}$. To see that $a = 3$, we observe that the $\mathbb{R}$-rational points of C form a compact real one-dimensional Lie group, and hence is isomorphic to $S^1 \times G$ where G is a finite group. For degree reasons, G can have cardinality at most 2.

More generally, if K is any field and $C \subset \mathbb{P}^2_K$ a nonsingular plane cubic, the number of flex points $p \in C$ with residue field K will be 0, 1, 3 or 9. This phenomenon is strictly limited to cubics, however: it follows from

Harris [1979] that for any $d \geq 4$ and any number δ with $0 \leq \delta \leq 3d(d-2)$, $\delta \neq 3d(d-2) - 1$, there exists a field K and a nonsingular plane curve $C \subset \mathbb{P}^2_K$ of degree d whose scheme of flexes contains exactly δ points with residue field K.

In Chapter V we'll discuss another object classically associated to a plane curve $C \subset \mathbb{P}^2_S$: its *dual curve* $C^* \subset (\mathbb{P}^2_S)^*$. We'll encounter many phenomena analogous to those we have just discovered.

IV.2   Blow-ups

Blowing up is a basic tool in classical algebraic geometry. It is used to resolve singularities, to resolve the indeterminacy of rational maps, and to relate birational varieties to one another. Saying that one variety is a blow-up of another along a given subvariety expresses a relationship that is simultaneously close enough to relate the structure of the two intimately, and flexible enough that it is a very common ingredient in the expression of maps between varieties. In this section, we will extend the definition to the category of schemes, defining the notion of the blow-up of an arbitrary (Noetherian) scheme along an arbitrary closed subscheme.

Generalizing the definition of blow-ups in this way actually serves two purposes. First there is the expected benefit: blowing up schemes other than varieties is useful for the same reason blowing up varieties is, that is, for resolving singularities or relating two birational schemes (for example, we will blow up arithmetic schemes in Section IV.2.4).

In addition we will see that, even in the context of maps between varieties, the language of schemes — specifically, being able to talk about blow-ups of a variety X along possibly nonreduced subschemes $Y \subset X$ — represents a highly useful extension of the concept. For example, we will illustrate this in Section IV.2.3 below, where we extend the classical description of nonsingular quadric surfaces as blow-ups of the plane to quadric cones, using this generalized notion of blowing up. Likewise, in Section IV.2.3 we will see a naturally occurring map of varieties that turns out to be a blow-up along a subscheme. These examples are in fact not special: when we broaden the definition of "blow-up" in this way, it turns out that any projective birational morphism of varieties is a blow-up! This is proved in Hartshorne [1977, Theorem II.7.17].

IV.2.1   Definitions and Constructions

For the following, we will assume the reader is familiar with the basic notion of blowing up in the classical context, that is, blowing up varieties along nonsingular subvarieties. (This material is amply covered in, among others, Harris [1995], Hartshorne [1977, Chapter 1], and Shafarevich [1974].) In the simplest circumstances — for example, blowing up a reduced point in the affine plane over an algebraically closed field K — a blow up map may be described exactly as it is classically. We start by reviewing an example, the blow-up of the plane at the origin, to see how the classical construction of the blow-up via gluing may be carried out as well in the category of schemes over a field. Generalizing this to the definition of the blow-up $\mathrm{Bl}_Y(X) \to X$ of an arbitrary scheme X along an arbitrary closed subscheme $Y \subset X$ is simply a matter of expressing this standard construction in a sufficiently natural way. In the following subsection we will give several characterizations of blow-ups in general: a definition, two constructions, and a further description in some special cases, such as the blow-up of a scheme along a regular subscheme (Definition IV-15).

An Example: Blowing up the Plane.

Example IV-14. We start with the blow-up Z of the origin in the affine plane $\mathbb{A}^2_K = \operatorname{Spec} K[x, y]$ over a field K. This can be most concretely described as the union of two open sets, each isomorphic to $\mathbb{A}^2_K$: we let $U' = \operatorname{Spec} K[x', y']$ and $U'' = \operatorname{Spec} K[x'', y'']$, and consider the maps $\varphi' : U' \to \mathbb{A}^2_K$ and $\varphi'' : U'' \to \mathbb{A}^2_K$ dual to the ring homomorphisms

$$(\varphi')^\# : K[x, y] \longrightarrow K[x', y'] \qquad \text{and} \qquad (\varphi'')^\# : K[x, y] \longrightarrow K[x'', y'']$$
$$x \longmapsto x' \qquad\qquad\qquad\qquad x \longmapsto x''y''$$
$$y \longmapsto x'y' \qquad\qquad\qquad\qquad y \longmapsto y''.$$

The map φ' gives an isomorphism between the open subsets

$$U'_x = \operatorname{Spec} K[x', y', x'^{-1}] \quad \text{and} \quad U_x = \operatorname{Spec} K[x, y, x^{-1}],$$

and similarly φ'' gives an isomorphism between the open subsets $U''_y = \operatorname{Spec} K[x', y', \frac{1}{y'}]$ and $U_y = \operatorname{Spec} K[x, y, y^{-1}]$. In particular, they give isomorphisms of the inverse images

$$U'_{xy} = \operatorname{Spec} K[x', y', x'^{-1}, y'^{-1}] \quad \text{and} \quad U''_{xy} = \operatorname{Spec} K[x'', y'', x''^{-1}, y''^{-1}]$$

of the intersection $U_{xy} = U_x \cap U_y = \operatorname{Spec} K[x, y, x^{-1}, y^{-1}]$. We can thus identify the open sets $U'_{xy} \subset U'$ and $U''_{xy} \subset U''$, and so glue together U' and U'' to obtain a scheme

$$Z = U' \cup U'' = \operatorname{Spec} K[x', y'] \bigcup_{U'_{xy} \cong U''_{xy}} \operatorname{Spec} K[x'', y''],$$

where the isomorphism $U'_{xy} \cong U''_{xy}$ is given by the ring homomorphism

$$K[x', y', x'^{-1}, y'^{-1}] \longrightarrow K[x'', y'', x''^{-1}, y''^{-1}]$$
$$x' \longmapsto x''y''$$
$$y' \longmapsto x''^{-1}.$$

We call the union Z, with its structure morphism $\varphi : Z \to \mathbb{A}^2_K$, the *blow-up* of $\mathbb{A}^2_K$ at the origin. The inverse image $E = \varphi^{-1}(0, 0) \subset Z$ of the origin is isomorphic to $\mathbb{P}^1_K$ (this is called the *exceptional divisor* of the blow-up), while φ is an isomorphism everywhere else, that is, $Z \setminus E \cong \mathbb{A}^2_K \setminus \{(0,0)\}$.

One way to think of this construction is to observe that the coordinate rings of the open subsets of the blow-up are enlarged to include the ratios $y' = y/x$ and $x'' = x/y$ respectively. This has a number of consequences. For one thing, the pair of functions x, y on $\mathbb{A}^2_K$ define a map $f : \mathbb{A}^2_K \setminus \{(0,0)\} \to \mathbb{P}^1_K$ on the complement of the origin: in classical language, this is the map $(a, b) \mapsto [a, b]$, or in modern terms it is the map associated to the surjection $\mathcal{O} \oplus \mathcal{O} \to \mathcal{O}$ given by $(f, g) \mapsto xf + yg$. This map cannot be extended to a regular map on all of $\mathbb{A}^2_K$; but if we compose f with the isomorphism $Z \setminus E \cong \mathbb{A}^2_K \setminus \{(0,0)\}$, we see it does extend to a regular map on all of Z. This is because the ideal generated by the (pullbacks of the) functions x and y is locally principal on Z (and generated by a nonzerodivisor), so that where x and y have common zeroes we can simply divide the homogeneous vector $[x, y]$ by their common factor to extend the map. Another effect of the enlarged coordinate rings in the blow-up is to separate the lines through the origin. That is, if L and L' are distinct lines through the origin in $\mathbb{A}^2_K$, the preimages of $L \setminus \{(0,0)\}$ and $L' \setminus \{(0,0)\}$ have doisjoint closures, as shown in the picture (these are just the fibers of the map f).

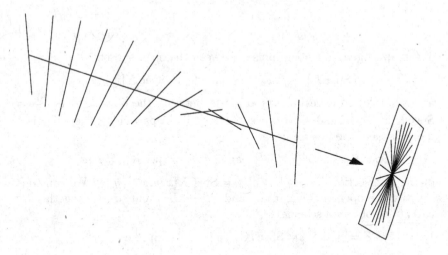

By the same token, if we have a curve $C \subset \mathbb{A}^2_K$ with a node at the origin, the inverse image of the complement of the origin in C is nonsingular in Z, meeting the exceptional divisor at two points.

Definition of Blow-ups in General. We will use these observations as starting points in generalizing the definition of a blow-up to that of an arbitrary scheme along an arbitrary subscheme. The essential fact is that, in the blow-up $\varphi : \mathrm{Bl}_Y(X) \to X$ of a scheme X along the subscheme

$Y \subset X$, the inverse image of Y is locally principal. To formalize this, we start with a definition:

Definition IV-15. Let X be any scheme, $Y \subset X$ a subscheme. We say that Y is an *Cartier subscheme* in X if it is locally the zero locus of a single nonzerodivisor; that is, if for all $p \in X$ there is an affine neighborhood $U = \operatorname{Spec} A$ of p in X such that $Y \cap U = V(f) \subset U$ for some nonzerodivisor $f \in A$. More generally, we say that Y is a *regular subscheme* if it is locally the zero locus of a regular sequence of functions on X.

Definition IV-16. Let X be any scheme, $Y \subset X$ a subscheme. The *blow-up of X along Y*, denoted $\varphi : \operatorname{Bl}_Y(X) \to X$, is the morphism to X characterized by these properties:

(1) The inverse image $\varphi^{-1}(Y)$ of Y is a Cartier subscheme in $\operatorname{Bl}_Y(X)$.

(2) $\varphi : \operatorname{Bl}_Y(X) \to X$ is universal with respect to this property; that is, if $f : W \to X$ is any morphism such that $f^{-1}(Y)$ is a Cartier subscheme in Z, there is a unique morphism $g : W \to \operatorname{Bl}_Y(X)$ such that $f = \varphi \circ g$.

The inverse image $E = \varphi^{-1}(Y)$ of Y in $\operatorname{Bl}_Y X$ is called the *exceptional divisor* of the blow-up, and Y the *center* of the blow-up.

It is clear that these properties uniquely characterize the blow-up $\varphi : \operatorname{Bl}_Y(X) \to X$ of a scheme along a subscheme. It is less clear that the blow-up exists, but we shall soon see that it does.

In the affine case the blow-up can be realized in a very simple way as the closure of the graph of a morphism, and we describe this construction first. We start by generalizing the construction of Example IV-14 to the blow-up at the origin of affine space over an arbitrary ring.

Example IV-17. Let A be any ring and let $\mathbb{A}_A^n = \operatorname{Spec} A[x_1, \dots, x_n]$. Consider the schemes

$$U_i = \operatorname{Spec} T_i \cong \mathbb{A}_A^n,$$

where

$$T_i = A\left[\frac{x_i}{x_i}, \dots, \frac{x_n}{x_i}, x_i\right]$$

is the subalgebra of $T = A[x_1, x_1^{-1}, \dots, x_n, x_n^{-1}]$ generated over A by the functions x_j/x_i and x_i. The rings $(T_i)_{x_j}$ and $(T_j)_{x_i}$ are equal as subrings of T, so we have commuting isomorphisms

$$(U_i)_{x_j} \cong (U_j)_{x_i}.$$

Thus we may form a scheme Z that is the union of the U_i with these open sets identified. Note that the morphisms $U_i \to \mathbb{A}_A^n$ corresponding to the inclusions $A[x_1, \dots, x_n] \hookrightarrow T_i$ agree on the overlap to give a natural structure morphism $\varphi : Z \to \mathbb{A}_A^n$.

This example shows many of the properties of the classical blow-up described in Example IV-14:

(1) Let $U = \mathbb{A}_A^n \setminus V(x_1, \ldots, x_n)$ be the compelement of $V(x_1, \ldots, x_n)$ (the "origin") in $\mathbb{A}_A^n$. We have a morphism

$$\alpha_{(x_1, \ldots, x_n)} : U \to \mathbb{P}_A^{n-1}$$

given by the functions $(x_1, \ldots, x_n)$; or, more formally, by the surjection

$$\mathscr{O}_U^n \longrightarrow \mathscr{O}_U,$$
$$(a_1, \ldots, a_n) \longmapsto \sum a_i x_i.$$

We claim that Z is the closure in $\mathbb{A}_A^n \times_A \mathbb{P}_A^{n-1} = \mathbb{P}_A^{n-1}$ of the graph of α. To see this, we observe that $(U_i)_{x_i} \subset Z$ is the graph of the map

$$\alpha_{(x_i, \ldots, x_n)}\big|_{(U_i)_{x_i}} : (U_i)_{x_i} \to (\mathbb{P}_A^{n-1})_{x_i} = \operatorname{Spec} A\left[\frac{x_1}{x_i}, \ldots, \frac{x_n}{x_i}\right],$$

and that the open sets $(U_i)_{x_i}$ are dense in Z.

(2) The preimage $E = \varphi^{-1}V(x_1, \ldots, x_n) \subset Z$ of $V(x_1, \ldots, x_n) \subset \mathbb{A}_A^n$ under the structure map $\varphi : Z \to \mathbb{A}_A^n$ is isomorphic to $\mathbb{P}_A^{n-1}$; and

$$\varphi : Z \setminus E \xrightarrow{\sim} \mathbb{A}_A^n \setminus V(x_1, \ldots, x_n)$$

is an isomorphism.

(3) Since $(x_1, \ldots, x_n)T_i = (x_i)T_i$, the preimage $E \subset Z$ of the origin $V(x_1, \ldots, x_n) \subset \mathbb{A}_A^n$ is locally defined by a single equation.

Proposition IV-18. *The morphism* $\varphi : Z \to \mathbb{A}_A^n$ *is the blow-up of* $\mathbb{A}_A^n$ *along the subscheme* $V(x_1, \ldots, x_n)$.

Proof. We have already observed that $Z \to \mathbb{A}_A^n$ satisfies condition (1) of Definition IV-16. It remains to show that if $\psi : W \to \mathbb{A}_A^n$ is any morphism such that $\psi^{-1}V(x_1, \ldots, x_n)$ is Cartier, then ψ factors through φ; that is, there exists a map $\alpha : W \to Z$ with $\psi = \varphi \circ \alpha$.

We prove this first when $W = \operatorname{Spec} R$ and R is a local ring. Consider R as an algebra over $A[x_1, \ldots, x_n]$ via the map $\psi^\# : A[x_1, \ldots, x_n] \to R$. Since the ideal $(x_1, \ldots, x_n)R$ is principal, Nakayama's Lemma (Eisenbud [1995, Corollary 4.8]) implies that it is generated by one of the x_i. More concretely, if we write

$$(x_1, \ldots, x_n)R = (\gamma),$$

we can write

$$\gamma = \alpha_1 x_1 + \cdots + \alpha_n x_n$$

for some $\alpha_i \in R$, and likewise $x_i = \beta_i \gamma$. It follows that

$$\gamma = \sum_i \alpha_i x_i = \sum_i \alpha_i \beta_i \gamma,$$

from which we see that at least one of the β_i must be a unit in R; that is, $(x_1, \ldots, x_n)R = (\gamma) = (x_i)$ for some i.

We can now write $x_j = \nu_j x_i$ (where $\nu_j = \beta_j \beta_i^{-1}$) for each j, and we defined the desired map

$$\alpha : W \to U_i \hookrightarrow Z$$

as dual to the homomorphisms of rings

$$A\left[\frac{x_1}{x_i}, \ldots, \frac{x_n}{x_i}, x_i\right] \longrightarrow R,$$

$$\frac{x_i}{x_j} \longmapsto \nu_j.$$

Now suppose that W is an arbitrary scheme, and $\psi : W \to \mathbb{A}_A^n$ a morphism with $\psi^{-1}V(x_1, \ldots, x_n)$ Cartier. For each point $w \in W$, the previous argument yields a map $\alpha : \operatorname{Spec} \mathscr{O}_{W,w} \to Z$ whose image lies in one of the affine open subsets $U_i \cong \mathbb{A}_A^n \subset Z$ covering Z. Such a map can be extended to the Zariski open neighborhood of $w \in W$ on which the images $\alpha^{\#}(x_j/x_i)$ are regular, so we get a covering of W by open sets W_k and morphisms $\alpha_k : W_k \to Z$ such that $\varphi \circ \alpha_k = \psi|_{W_k}$.

We will complete the argument by showing that the maps α_k agree on the overlaps $W_i \cap W_j$, and thus define a morphism α on all of W.

Since the restriction of φ to $Z \setminus E \to \mathbb{A}_A^n \setminus V(x_1, \ldots, x_n)$ is an isomorphism, it will suffice to show that the inverse image $\psi^{-1}(\mathbb{A}_A^n \setminus V(x_1, \ldots, x_n))$ is dense in W. But by hypothesis, $\psi^{-1}V(x_1, \ldots, x_n)$ is a Cartier divisor in W. The following lemma thus completes the argument.

Lemma IV-19. *If $X \subset Y$ is a Cartier subscheme of a scheme, then $Y \setminus X$ is dense in Y (as schemes, not just as topological spaces).*

Proof. We may assume that Y is affine, say $Y = \operatorname{Spec} A$, and that $X = V(f)$ for some nonzerodivisor $f \in A$. To say that there is a proper closed subscheme Y' containing $Y \setminus X$ is to say that the localization map $A \to A_f$ factors through $A/I(Y')$. But since f is a nonzerodivisor, this localization map is a monomorphism. $\square$

Exercise IV-20. (a) Show that the conclusion of Lemma IV-19 fails for

$$X = V(x) \subset Y = \operatorname{Spec} K[x,y]/(xy, y^2).$$

(b) Show more generally that it characterizes Cartier subschemes among all locally principal subschemes of Y.

(c) Show that $\operatorname{Bl}_Y = \varnothing$ if and only if $\operatorname{supp} Y = \operatorname{supp} X$.

The construction of Proposition IV-18 will yield all blow-ups of affine schemes as soon as we understand how blow-ups behave on subschemes, or, more generally, under pullbacks. This follows directly from the definition:

Proposition IV-21. *Let X be any scheme, $Y \subset X$ a subscheme and $\varphi : \operatorname{Bl}_Y(X) \to X$ the blow-up of X along Y. Let $\nu : X' \to X$ be any morphism and set $Y' = \nu^{-1}(Y) \subset X'$. If W is the closure, in the fiber product*

$X' \times_X \mathrm{Bl}_Y X$, of the inverse image $\pi_1^{-1}(X' \setminus Y')$, then $\pi_1 : W \to X'$ is the blow-up of X' along Y'.

This lemma is already interesting in the case $X' = X$, where it asserts that *the inverse image of $X \setminus Y$ in $\mathrm{Bl}_Y X$ is dense.*

Proposition IV-21 is most often applied in case $X' \subset X$ is a closed subscheme. In this case W is simply the closure in $\mathrm{Bl}_Y X$ of the inverse image $\varphi^{-1}(X' \setminus (X' \cap Y))$; it is called the *strict transform*, or *proper transform*, of X' in $\mathrm{Bl}_Y X$. (The full inverse image $\varphi^{-1}(X') \subset \mathrm{Bl}_Y X$ is called the *total transform*.) Thus we may say that, in the blow-up $\mathrm{Bl}_p \mathbb{A}_K^2$, the proper transforms of the lines through the origin $p \in \mathbb{A}_K^2$ are disjoint (note that the proper transforms of the lines map isomorphically to the lines themselves, as they should, since the origin is a Cartier subscheme on each), and that the blow-up of a nodal curve at a node is nonsingular at the points lying over the node.

In case $X' \subset X$ is an open subscheme, Proposition IV-21 says simply that the formation of blow-ups does commute with base change, that is,

$$\varphi^{-1}(X') \cong \mathrm{Bl}_{X' \cap Y} X' \to X'.$$

But more is true: since $\varphi^{-1}(X' \setminus Y)$ is dense, there is a *unique* such isomorphism over X. As a consequence, if $\pi : Z \to X$ is a morphism and suppose we have a cover of X by open sets U such that $\pi^{-1}U \cong \mathrm{Bl}_{U \cap Y} U$ over X, then $Z \cong \mathrm{Bl}_Y X$. In a phrase: blow-ups are determined locally.

Proof of Proposition IV-21. We check first that the inverse image

$$E' = \pi_1^{-1}(Y') \subset W$$

of Y' is a Cartier subscheme of W. It is certainly principal: the inverse image $E = \varphi^{-1}(Y) \subset \mathrm{Bl}_Y X$ is locally principal in $\mathrm{Bl}_Y X$, and $E' \subset W$ is simply its inverse image $\pi_2^{-1}(E)$ under the projection $\pi_2 : W \to \mathrm{Bl}_Y X$. Moreover, since the associated primes of W are exactly the associated primes of X' not containing the ideal of Y', the local defining equation of E in $\mathrm{Bl}_Y X$ cannot pull back to a zero divisor on W.

Next, we have to verify that W has the universal property. Suppose T is any scheme, and $f : T \to X'$ any morphism such that the inverse image $f^{-1}(Y')$ of Y' in T is a Cartier subscheme. In particular, since $f^{-1}(Y') \subset T$ is Cartier, no component or embedded component of T maps to Y'; thus the closure in T of $f^{-1}(X' \setminus Y')$ is all of T.

We have to show that f lifts to a morphism $g : T \to W$ (that is, there exists a morphism $g : T \to W$ such that the composition $\pi_1 \circ g = f$). We do this in three steps. First, let

$$h = \nu \circ f : T \to X$$

be the composition of f with the morphism $\nu : X' \to X$; since the inverse image $h^{-1}(Y) = f^{-1}(Y')$ is Cartier, it follows by the universal property of the blow-up $\mathrm{Bl}_Y X \to X$ that h lifts to a morphism $\tilde{h} : T \to \mathrm{Bl}_Y X$. Next, the pair of maps $f : T \to X'$ and $\tilde{h} : T \to \mathrm{Bl}_Y X$ give a map

$$\tilde{g} : T \to X' \times_X \mathrm{Bl}_Y X$$

whose composition with the projection $\pi_1 : X' \times_X \mathrm{Bl}_Y X \to X'$ is f. Finally, since $\tilde{g}$ maps the inverse image $f^{-1}(X' \setminus Y')$ to W, and the closure in T of $f^{-1}(X' \setminus Y')$ is all of T, it follows that the map $\tilde{g} : T \to X' \times_X \mathrm{Bl}_Y X$ factors through the inclusion of W in $X' \times_X \mathrm{Bl}_Y X$ to give the desired map $g : T \to W$.   □

We are now in a position to blow up any closed subscheme of any affine scheme. If $X = \mathrm{Spec}\, A$ and $f_1, \ldots, f_n \in A$, then $(f_1, \ldots, f_n)$ defines a morphism

$$\alpha_{(f_1, \ldots, f_n)} : U = X \setminus V(f_1, \ldots, f_n) \longrightarrow \mathbb{P}_A^{n-1};$$

more precisely, $(f_1, \ldots, f_n)$ defines a map $\mathscr{O}_X^n \to \mathscr{O}_X$ sending $(a_1, \ldots, a_n)$ to $\sum a_i f_i$, which is an epimorphism exactly on U.

Proposition IV-22. *Let $X = \mathrm{Spec}\, A$ be an affine scheme, and let*

$$Y = V(f_1, \ldots, f_n) \subset X$$

be a closed subscheme. The blow-up of Y in X is the closure in $X \times_A \mathbb{P}_A^{n-1} = \mathbb{P}_A^{n-1}$ of the graph of the morphism

$$\alpha_{(f_1, \ldots, f_n)} : X \setminus Y \to \mathbb{P}_A^{n-1}.$$

Proof. Consider the embedding $X \hookrightarrow \mathbb{A}_A^n = \mathrm{Spec}\, A[x_1, \ldots, x_n]$ given by the ring homomorphism

$$A[x_1, \ldots, x_n] \longrightarrow A,$$
$$x_i \longmapsto f_i.$$

Note that under this embedding we have $X \cap V(x_1, \ldots, x_n) = Y$. By Proposition IV-21, the blow-up of X along Y is the proper transform of X in the blow-up Z of $\mathbb{A}_A^n$ along $V(x_1, \ldots, x_n)$. By Proposition IV-18, on the other hand, the blow-up Z of $\mathbb{A}_A^n$ along $V(x_1, \ldots, x_n)$ is the closure of the graph Γ of the map

$$\alpha_{(x_1, \ldots, x_n)} : \mathbb{A}_A^n \setminus V(x_1, \ldots, x_n) \to \mathbb{P}_A^{n-1}.$$

Since the graph of $\alpha_{(f_1, \ldots, f_n)}$ is simply the intersection of Γ with the preimage of $X \subset \mathbb{A}_A^n$, its closure is the proper transform of $X \subset \mathbb{A}_A^n$ in Z, and the result follows.   □

In this proposition we built in the restriction that the subscheme $Y \subset X$ be defined by finitely many functions f_i, but this is really unnecessary. The reader may check that everything works for infinite sets (though the morphisms go to infinite-dimensional projective spaces).

The Blowup as Proj. We have now proved the existence of the blow-up of an affine scheme along a closed subscheme. We could at this point deduce the existence of blow-ups in general by gluing. However, there is a more elegant construction of blow-ups via global Proj, which accomplishes this in one fell swoop.

Theorem IV-23. *Let X be a scheme and $Y \subset X$ a closed subscheme. Let $\mathcal{I} = \mathcal{I}_{Y,X} \subset \mathcal{O}_X$ be the ideal sheaf of Y in X. If $\mathcal{A}$ is the sheaf of graded $\mathcal{O}_X$-algebras*

$$\mathcal{A} = \bigoplus_{n=0}^{\infty} \mathcal{I}^n = \mathcal{O}_X \oplus \mathcal{I} \oplus \mathcal{I}^2 \oplus \cdots$$

(where the k-th summand is taken to be the k-th graded piece of $\mathcal{A}$), then the scheme $\mathrm{Proj}(\mathcal{A}) \to X$ is the blow-up of X along Y.

Remark. This construction often leads to notational confusion: if $f \in \mathcal{O}_X(U)$ is a regular function vanishing on Y, the symbol "f" could a priori be used to denote either the section of $\mathcal{A}_0 = \mathcal{O}_X$ or the section of $\mathcal{A}_1 = \mathcal{I}$ — two different sections of $\mathcal{A}$. To avoid this, we will often realize $\mathcal{A}$ as a subsheaf of the sheaf

$$\mathcal{O}_X[t] = \bigoplus_{n=0}^{\infty} t^n \mathcal{O}_X,$$

writing

$$\mathcal{A} = \mathcal{O}_X \oplus t\mathcal{I} \oplus t^2\mathcal{I}^2 \oplus \cdots.$$

We will use this notation in the proof below.

Proof. We have to show that the morphism

$$\varphi : B = \mathrm{Proj}(\mathcal{A}) \to X$$

satisfies the two conditions that characterize a blow-up: that the preimage $\varphi^{-1}Y$ of Y in B is Cartier, and that any morphism $f : Z \to X$ with $f^{-1}Y$ Cartier factors uniquely through B. We will write $\mathcal{I}$ for the ideal $\mathcal{I}_Y$ of Y in X.

To show that the preimage of Y in B is Cartier, recall from section I.3.1 that $\varphi^{-1}Y$ is the subscheme of B defined by the ideal sheaf $\mathcal{I}\mathcal{O}_B$. Since the structure sheaf $\mathcal{O}_B$ is the sheaf associated to the sheaf of graded $\mathcal{A}$-modules $\mathcal{A}$, we see that $\mathcal{I}\mathcal{O}_B$ is the sheaf associated to the graded $\mathcal{A}$-module

$$\mathcal{I}\mathcal{A} = \mathcal{I} \cdot \mathcal{O}_B \oplus \mathcal{I} \cdot \mathcal{I} \oplus \mathcal{I} \cdot \mathcal{I}^2 \oplus \cdots$$
$$= \mathcal{I} \oplus \mathcal{I}^2 \oplus \mathcal{I}^3 \oplus \cdots$$

where the term $\mathcal{I} \cdot \mathcal{I}^d = \mathcal{I}^{d+1}$ occurs in degree d. This is the truncation of the graded module

$$\mathcal{A}(1) = \mathcal{O} \oplus \mathcal{I} \oplus \mathcal{I}^2 \oplus \cdots$$

(again the term $\mathscr{I}^{d+1}$ occurs in degree d.) Thus by Exercise III-46, $\mathscr{I}\mathcal{O}_B = \mathcal{O}_B(1)$ is invertible.

It remains to show that if $f : Z \to X$ is a map such that $f^{-1}Y$ is Cartier, then f factors uniquely through B. We will assume for simplicity that $\mathscr{I}$ is coherent. We will realize B as a closed subscheme of $\mathbb{P}(\mathscr{I}) = \operatorname{Proj}\operatorname{Sym}(\mathscr{I})$ and produce the desired map from Z to B by giving a natural map from Z to $\mathbb{P}(\mathscr{I})$ whose image is contained in B.

The maps $\operatorname{Sym}_d(\mathscr{I}) \to \mathscr{I}^d$ give a surjection $\operatorname{Sym}(\mathscr{I}) \to \mathscr{A}$. Its kernel is a sheaf of graded ideals of $\mathscr{A}$ and thus as in section III.2.2 it identifies $B = \operatorname{Proj}\mathscr{A}$ with a closed subscheme of $\mathbb{P}(\mathscr{I})$.

Because $f^{-1}Y$ is Cartier, its ideal $\mathscr{I}\cdot\mathcal{O}_Z$ is invertible. Thus the natural surjection

$$f^*\mathscr{I} = \mathscr{I} \otimes_{\mathcal{O}_X} \mathcal{O}_Z \to \mathscr{I}\cdot\mathcal{O}_Z$$

corresponds as in Theorem III-44 to a map $\alpha : Z \to \mathbb{P}(\mathscr{I})$. Further, by Lemma IV-19, the complement of $f^{-1}Y$ is dense in Z. Since φ is an isomorphism on the complement of $\varphi^{-1}Y$, it follows that $\alpha(Z \setminus f^{-1}Y)$ is contained in B, and thus all of $\alpha(Z)$ is contained in B. The map α is thus the desired map from Z to B.

Both the fact that $f = \varphi\alpha$ and the uniqueness of α follow as well from the density of $Z \setminus f^{-1}Y$ in Z and the last sentence of Exercise III-24.   □

We assumed for simplicity that the ideal sheaf $\mathscr{I}$ was coherent (and not merely quasicoherent); the quasicoherent case could be handled by means of a straightforward generalization of III-44.

Blowing up gives us another way to interpret the projectivized tangent cone to a scheme, which we will use later in this section.

Exercise IV-24. Show that the exceptional divisor in the blow-up $\operatorname{Bl}_p(X)$ of a scheme X at a point $p \in X$ is the projectivized tangent cone $\mathbb{P}TC_p(X)$ to X at p.

Blow-ups along Regular Subschemes. As we mentioned before the statement of Theorem IV-23, the construction of a blow-up may not be as explicit in practice as it appears. The reason is that, even given explicit equations for a scheme X and a subscheme Y, it may not be obvious how to express the *Rees algebra*

$$\mathscr{A} = \bigoplus_{n=0}^{\infty} t^n \mathscr{I}_{Y,X}^n \subset \mathcal{O}_X[t]$$

in terms of explicit generators and relations. (The generators are clear, assuming we know locally generators of the ideal sheaf $\mathscr{I}_{Y,X}$; it's knowing when we have found all the relations that may be tricky.) There is, however, one circumstance in which the Rees algebra has a nice description: when the subscheme $Y \subset X$ is a regular subscheme. We will state the result first in case Y has codimension two.

Proposition IV-25. *Let A be a Noetherian ring and $x, y \in A$; let B be the Rees algebra*

$$B = A[xt, yt] \subset A[t].$$

If $x, y \in A$ is a regular sequence, then

$$B \cong A[X, Y]/(yX - xY)$$

via the map $X \mapsto xt$, $Y \mapsto yt$.

Proof. First we invert x and set $X' = x^{-1}X \in A[x^{-1}][X, Y]$. The element $yX' - Y \in A[x^{-1}][X, Y] = A[x^{-1}][X', Y]$ generates the kernel of the map

$$A[x^{-1}][X', Y] \longrightarrow A[x^{-1}][t],$$
$$X' \longmapsto t,$$
$$Y \longmapsto yt.$$

Since $(yX - xY) = (yX' - Y)$ in the ring $A[x^{-1}][X, Y]$, it suffices to show that x is a nonzerodivisor modulo $yX - xY$ in $A[X, Y]$. Notice that, in the other order, $yX - xY$ is obviously a nonzerodivisor modulo x — it's congruent to yX, the product of two nonzerodivisors! In general, a permutation of a regular sequence is not a regular sequence, but in this setting, as in many others, it is; see Eisenbud [1995, Section 17.1].

In our case we may argue as follows: To show that x is a nonzerodivisor modulo $yX - xY$ we must show that

$$M := \frac{(yX - xY) : (x)}{(yX - xY)} = 0,$$

where $(yX - xY) : (x)$ denotes the ideal $\{f \in A[X, Y] \mid fx \in (yX - xY)\}$. Note that $yX - xY \equiv yX$ modulo x, so $(x, yX - xY)$ is a regular sequence in $A[X, Y]$. Further, $yX - xY$ is clearly a nonzerodivisor (to annihilate it, a polynomial $f(X, Y)$ would have to have leading term in X annihilating x, which is a nonzerodivisor by hypothesis). It follows that the quotient M is isomorphic to the first homology group of the Koszul complex

$$0 \longrightarrow A \xrightarrow{\left(\begin{smallmatrix} -x \\ yX-xY \end{smallmatrix}\right)} A^2 \xrightarrow{(yX-xY \ \ x)} A.$$

By the same argument, this group is isomorphic to

$$\frac{(x) : (yX - xY)}{(x)},$$

which is 0 since $x, yX - xY$ is a regular sequence. (For a more leisurely treatment of this last argument, see Eisenbud [1995, Section 17.1].) $\square$

The heart of the proof above is the statement that if I is generated by a regular sequence of length 2, then the Rees algebra

$$A \oplus I \oplus I^2 \oplus \cdots$$

is isomorphic to the symmetric algebra

$$\text{Sym}_A(I)$$

and this in turn is defined by the determinant of the 2×2 matrix

$$\begin{pmatrix} x & y \\ X & Y \end{pmatrix}.$$

Similar statements are true for larger regular sequences:

Exercise IV-26. If $I = (x_1, \ldots, x_n) \subset A$ is generated by a regular sequence, then

$$A \oplus I \oplus I^2 \oplus \cdots \cong A[X_1, \ldots, X_n]/J$$

where J is generated by the 2×2 minors of the matrix

$$\begin{pmatrix} x_1 & \cdots & x_n \\ X_1 & \cdots & X_n \end{pmatrix}.$$

IV.2.2 Some Classic Blow-Ups

Example IV-27. Let K be a field, and consider the quadric cone

$$Q = \text{Spec } K[x,y,z]/(xy - z^2) \subset \text{Spec } K[x,y,z] = \mathbb{A}^3_K.$$

Let $p = (0,0,0) \in Q$ be the vertex of the cone Q, and let L be a line through p lying on Q, for example $L = V(x,z)$. We would like to describe the blow-ups of Q along both p and L.

We can do this directly, using either Theorem IV-23 or Proposition IV-22. But perhaps the simplest way is to use Proposition IV-21. To begin with, we can verify by either Theorem IV-23 or Proposition IV-22 that the blow-up of $\mathbb{A}^3_K$ at the origin p is the morphism

$$\varphi : \tilde{\mathbb{A}}^3_K = \text{Proj } K[x,y,z][A,B,C]/(xB-yA, xC-zA, yC-zB)$$
$$\longrightarrow \text{Spec } K[x,y,z] = \mathbb{A}^3_K.$$

The exceptional divisor $E = \varphi^{-1}(p) \subset \tilde{\mathbb{A}}^3_K$ is indeed Cartier: for example, we may write the open subset $U_A = \tilde{\mathbb{A}}^3_K \setminus V(A)$ as

$$U_A = \text{Spec } K[x,y,z][b,c]/(xb-y, xc-z) = \text{Spec } K[x,b,c]$$

and in U_A, the exceptional divisor E is the zero locus of (the pullback of) the function x. As in the case of the blow-up of the plane at the origin, the proper transforms $\tilde{L}$ of the lines $L \subset \mathbb{A}_K^3$ through p are all disjoint in $\tilde{\mathbb{A}}_K^3$, and indeed the exceptional divisor E is a copy of $\mathbb{P}_K^2$ whose K-rational points correspond bijectively to the set of these lines via the association $L \mapsto \tilde{L} \cap E$.

Now, when we pull back the defining equation $xy - z^2$ of Q to $\mathbb{A}_K^3$, we find that it factors: it is twice divisible by the defining equation of $\tilde{E}$. For example, in U_A,

$$\varphi^{\#}(xy - z^2) = x^2 b - x^2 c^2 = x^2(b - c^2).$$

We can express this globally as

$$\varphi^{-1}(Q) = V((x, y, z)^2) \cup V(AB - C^2)$$

and by Proposition IV-21 we may conclude that the blow-up $\mathrm{Bl}_p\, Q$ of Q at p is the restriction of φ to the locus $V(AB - C^2) \subset \tilde{\mathbb{A}}_K^3$, that is,

$$\psi : \tilde{Q} = \mathrm{Proj}\, K[x, y, z][A, B, C]/(xB - yA, xC - zA, yC - zB, AB - C^2)$$
$$\longrightarrow \mathrm{Spec}\, K[x, y, z]/(xy - z^2) = Q.$$

We can picture $\tilde{Q}$ as the disjoint union of the (proper transforms of the) lines on Q passing through p:

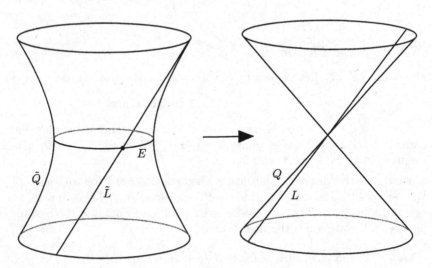

Now, what about the blow-up $\mathrm{Bl}_L\, Q \to Q$ of Q along L? To begin with, note that L is a Cartier subscheme of Q at every point of L except at p, where it is not (p is a singular point of Q, but a nonsingular point of L). It follows that the blow-up $\mathrm{Bl}_L\, Q \to Q$ will be an isomorphism over $Q \setminus \{p\}$, but not an isomorphism. Also, since the inverse image $\psi^{-1}(L) \subset \tilde{Q}$ of L in the blow-up $\tilde{Q} = \mathrm{Bl}_p\, Q \to Q$ of Q at the point p is Cartier, the map $\psi : \tilde{Q} \to Q$ must factor through the blow-up $\mathrm{Bl}_L\, Q \to Q$. It will by now not come as a surprise to the reader to learn that in fact, the two blow-ups are the same! We leave the verification as the following exercise.

Exercise IV-28. Show that the blow-up $\mathrm{Bl}_L\,\mathbb{A}_K^3$ of $\mathbb{A}_K^3$ along the line L may be realized as the map

$$\varphi : \mathrm{Bl}_L\,\mathbb{A}_K^3 = \mathrm{Proj}\,K[x,y,z][A,B]/(xB-zA) \longrightarrow \mathrm{Spec}\,K[x,y,z] = \mathbb{A}_K^3$$

(We may visualize this as the disjoint union of the planes in $\mathbb{A}_K^3$ containing L.) Use this to describe the blow-up $\mathrm{Bl}_L\,Q \to Q$, and show that it is isomorphic to $\mathrm{Bl}_p\,Q \to Q$ as a Q-scheme.

Another surprisingly rich example is the blow-up of a quadric cone of dimension 3.

Example IV-29. Consider now the quadric hypersurface

$$X = V(xw-yz) \subset \mathrm{Spec}\,K[x,y,z,w] = \mathbb{A}_K^4.$$

X is the cone over the nonsingular quadric surface $Q = V(xw-yz) \subset \mathrm{Proj}\,K[x,y,z,w] = \mathbb{P}_K^3$. We want to consider blow-ups of X along three subvarieties: the point $p = (0,0,0,0)$; the plane $\Lambda_1 = V(x=y=0) \subset X$, and the plane $\Lambda_2 = V(x=z=0) \subset X$. What is interesting is that, while all three blow-ups are isomorphisms over $X \setminus \{p\}$, they are all distinct X-schemes; also that the blow-ups $\mathrm{Bl}_{\Lambda_1}\,X$ and $\mathrm{Bl}_{\Lambda_2}\,X$ are isomorphic schemes, but not isomorphic X-schemes.

To begin with, let $\varphi : \tilde{X} \to X$ be the blow-up of X at the point p. This may be described along much the same lines as the blow-up of the quadric surface at a point in the previous example: all the lines on X through the point p are made disjoint; $\tilde{X}$ is nonsingular; and the exceptional divisor is a nonsingular quadric surface naturally identified with $Q \subset \mathbb{P}_K^3$.

The blow-ups X_i of X along the planes Λ_i are described in the following exercise:

Exercise IV-30. Let $\varphi_1 : X_1 = \mathrm{Bl}_{\Lambda_1}\,X \to X$ be the blow-up of X along the plane Λ_1. Show the following assertions.

(a) The scheme X_1 is nonsingular.

(b) The map φ_1 is an isomorphism over $X \setminus \{p\}$.

(c) The fiber $C = \varphi_1^{-1}(p)$ of X_1 over the point p is isomorphic to $\mathbb{P}_K^1$.

(d) The exceptional divisor $E = \varphi_1^{-1}(\Lambda_1)$, which is also the proper transform of Λ_1 in X_1, is isomorphic to the blow-up of $\Lambda_1 \cong \mathbb{A}_K^2$ at the point p.

(e) More generally, the proper transforms $\tilde{\Lambda}_{1,\mu}$ of the planes

$$\Lambda_{1,\mu} = V(x-\mu z, y-\mu w)$$

spanned by the vertex p of X and the lines of one ruling of Q coincide with their total transforms; they are isomorphic to the blow-ups of $\Lambda_{1,\mu}$ at the point p, and intersect pairwise along the curve C.

(f) By contrast, the inverse images $\varphi_1^{-1}(\Lambda_{2,\mu})$ of the planes

$$\Lambda_{2,\mu} = V(x-\mu y,\, z-\mu w)$$

spanned by the vertex p of X and the lines of the other ruling of Q have two irreducible components: the proper transforms $\tilde{\Lambda}_{2,\mu}$ and the curve C. (In particular, they are not Cartier subschemes of X_1.) The proper transforms $\tilde{\Lambda}_{2,\mu}$ map isomorphically to the planes $\Lambda_{2,\mu}$, and are disjoint in X_1; thus we may try to visualize X_1 as the planes $\Lambda_{2,\mu}$ made disjoint.

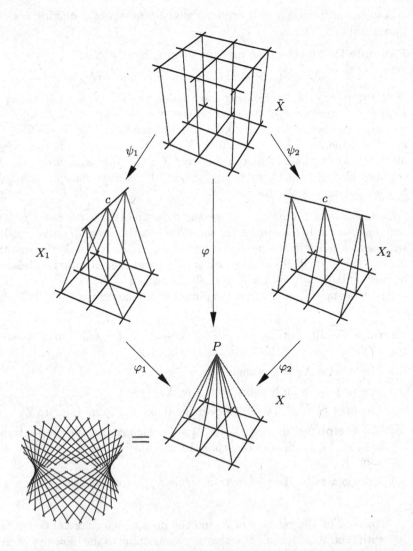

Since the inverse images of the planes Λ_1 and Λ_2 are Cartier subschemes of $\tilde{X}$ (they are of pure codimension one in the nonsingular scheme $\tilde{X}$), the blow-up $\tilde{X} = \mathrm{Bl}_p\, X \to X$ factors through each of the blow-ups $X_i = \mathrm{Bl}_{\Lambda_i}\, X \to X$. In fact:

Exercise IV-31. (a) Show that $\tilde{X} = X_1 \times_X X_2$ as X-schemes.

(b) Show that the induced map $\psi_i : \tilde{X} \to X_i$ is simply the blow-up of X_i along the curve C.

The schemes X_i are certainly not isomorphic to each other as X-schemes, since the inverse image of Λ_2 in X_1 is not Cartier and vice versa, though they are isomorphic as K-schemes (X has an automorphism exchanging the planes Λ_1 and Λ_2). Likewise, neither is isomorphic to $\tilde{X}$ as an X-scheme, since the inverse images of both Λ_1 and Λ_2 are Cartier in $\tilde{X}$.

Exercise IV-32. Show that in fact X_1 and X_2 are not isomorphic to $\tilde{X}$ even as K-schemes. (Hint: one way is to show that X_i contains no two-dimensional subscheme proper over K.)

Exercise IV-33. Here is an interesting way to realize all three of the blow-ups described above. Identify $\mathbb{A}^4_K$ with the affine space associated to the vector space M of 2×2 matrices, or of linear maps $A : V \to W$ between a pair of two-dimensional vector spaces over K:

$$M = \mathrm{Hom}(V, W) = \left\{ \begin{pmatrix} x & y \\ z & w \end{pmatrix} \right\}.$$

Let $\mathbb{P}V^*$ be the projective space of one-dimensional quotients of V^*, that is, one-dimensional subspaces of V, and similarly let $\mathbb{P}W^*$ be the projective space of one-dimensional subspaces of W. Show that X and the blow-ups X_1, X_2 and $\tilde{X}$ are, respectively, the schemes associated to the varieties

$$X = \{A : V \to W \mid \mathrm{rank}\, A \le 1\} \subset \mathbb{A}^4_K,$$
$$X_1 = \{(A, L) \mid L \subset \mathrm{Ker}\, A\} \subset \mathbb{A}^4_K \times \mathbb{P}V^*,$$
$$X_2 = \{(A, L') \mid \mathrm{Im}\, A \subset L'\} \subset \mathbb{A}^4_K \times \mathbb{P}W^*$$
$$\tilde{X} = \{(A, L, M) \mid L \subset \mathrm{Ker}\, A \text{ and } \mathrm{Im}\, A \subset L'\} \subset \mathbb{A}^4_K \times \mathbb{P}V^* \times \mathbb{P}W^*.$$

In fact, the results of Example IV-27 and Example IV-29 apply not only to quadric cones, but to schemes that look locally like them. This is the content of the following exercises, which will require one further definition:

Definition IV-34. Let K be an algebraically closed field of characteristic not equal to 2 and X any scheme over K. We say that a point $p \in X$ is an *ordinary double point* if the formal completion of the local ring $\mathcal{O}_{X,p}$ is

$$\hat{\mathcal{O}}_{X,p} \cong K[\![x_1, \ldots, x_n]\!]/(x_1^2 + x_2^2 + \cdots + x_n^2).$$

For example, an ordinary double point of a curve is what we have been calling a node. More generally, an ordinary double point of an n-dimensional scheme X may be characterized as any point such that the projectivized tangent cone to X at p is a nonsingular quadric hypersurface in $\mathbb{P}T_pX \cong \mathbb{P}_K^n$.

Exercise IV-35. Suppose now that X has dimension 2 and $p \in X$ is an ordinary double point. Let $\tilde{X} = \mathrm{Bl}_p X \to X$ be the blow-up of X at p. Show that the conclusions of Example IV-29 apply as well to X: that $\tilde{X}$ is nonsingular; that the exceptional divisor $E \subset \tilde{X}$ is a conic curve in $\mathbb{P}T_pX \cong \mathbb{P}_K^2$, and that if $C \subset X$ is any curve nonsingular at p then $\mathrm{Bl}_C X \cong \tilde{X}$ as X-schemes.

Exercise IV-36. Keeping the hypotheses of Exercise IV-35, suppose now that X has dimension 3 and $p \in X$ is an ordinary double point. Let $\tilde{X} = \mathrm{Bl}_p X \to X$ be the blow-up of X at p. Show that the conclusions of Example IV-29 apply as well to X: that $\tilde{X}$ is nonsingular; that the exceptional divisor $E \subset \tilde{X}$ is a nonsingular quadric surface $Q \subset \mathbb{P}T_pX \cong \mathbb{P}_K^3$, and that if $S \subset X$ is any surface nonsingular at p then the blow up $\mathrm{Bl}_S X$ has fiber over p isomorphic to $\mathbb{P}_K^1$ (and in particular is not isomorphic to $\tilde{X}$). Show moreover that if S and $S' \subset X$ are two such surfaces, the blow-ups $\mathrm{Bl}_S X$ and $\mathrm{Bl}_{S'} X$ are isomorphic as X-schemes if and only if the projectivized tangent planes $\mathbb{P}T_pS$ and $\mathbb{P}T_pS' \subset \mathbb{P}T_pX$ belong to the same ruling of the quadric Q.

By way of language, for a three-dimensional scheme X with an ordinary double point $p \in X$, the schemes $X' \to X$ obtained (locally around p) as blow-ups of X along surfaces nonsingular at p are called *small resolutions* of X at p. In general, a resolution of singularities $\pi : X' \to X$ — that is, a birational morphsim such that X' is nonsingular — is called *small* if for any subvariety $\Gamma \subset X$ the inverse image $\pi^{-1}(\Gamma)$ has dimension at most

$$\dim(\pi^{-1}(\Gamma)) \leq \frac{\dim(\Gamma) + \dim(X) - 1}{2}.$$

The birational isomorphism between the two small resolutions of a threefold X with an ordinary double point is called a *flop*; see Clemens et al. [1988].

Let X be a scheme and $Y, Z \subset X$ a pair of subschemes. If we blow up X first along one, then along the proper transform of the other, the order in which we do it matters. We can now illustrate this with a simple example, given in the form of a series of exercises.

Exercise IV-37. Let K be a field and $\mathbb{A}_K^3 = \mathrm{Spec}\, K[x, y, z]$. Let L and $M \subset \mathbb{A}_K^3$ be the lines $V(x, y)$ and $V(x, z)$ respectively, and $N = L \cup M = V(x, yz)$ their union. Describe the blow-up $X = \mathrm{Bl}_N \mathbb{A}_K^3 \to \mathbb{A}_K^3$; in particular, show that X has fiber isomorphic to $\mathbb{P}_K^1$ over every point of N, but that it is not nonsingular: it has an ordinary double point p lying over the origin in $\mathbb{A}^3_K$.

Exercise IV-38. Keeping the notations of the preceeding problem, let $Y \to \mathbb{A}^3_K$ be the blow-up of $\mathbb{A}^3_K$ along the line L, $\tilde{M} \subset Y$ the proper transform of M in Y and $X' \to Y$ the blow-up of Y along $\tilde{M}$. Show that the composite map $X' \to Y \to \mathbb{A}^3_K$ factors through the blow-up $X = \mathrm{Bl}_N \mathbb{A}^3_K \to \mathbb{A}^3_K$, and that the induced map $X' \to X$ is one of the small resolutions of the ordinary double point $p \in X$.

Exercise IV-39. Now let $Z \to \mathbb{A}^3_K$ be the blow-up of $\mathbb{A}^3_K$ along the line M, $\tilde{L} \subset Y$ the proper transform of L in Z and $X'' \to Z$ the blow-up of Z along $\tilde{L}$. Show that the composite map $X'' \to Y \to X$ again factors through the blow-up $X \to \mathbb{A}^3_K$, and that the induced map $X' \to X$ is the *opposite* small resolution of the ordinary double point $p \in X$ from $X' \to X$. To see directly that $X' \to X$ and $X'' \to X$ are not isomorphic X-schemes, let N' and N'' be the closures of the inverse image of $L \setminus \{0\}$ in X' and X'', and compare the fibers of N' and N'' over $0 \in \mathbb{A}^3_K$.

IV.2.3 Blow-ups along Nonreduced Schemes

Up to now, we have dealt only with examples of blow-ups $\mathrm{Bl}_Y X \to X$ in which all three objects involved — the original scheme X, the subscheme Y and the blow-up $\mathrm{Bl}_Y X$ — are varieties. In the remaining two parts of this section, we will consider the behavior of blow-ups in the more general setting of schemes, giving examples first of blow-ups along non-reduced subschemes of a scheme X, and then of blow-ups of arithmetic schemes. We will start here by giving some examples of blow-ups of varieties along nonreduced subschemes.

Blowing Up a Double Point. Let $X = \mathbb{A}^2_K = \mathrm{Spec}\, K[x,y]$, and let $\Gamma \subset \mathbb{A}^2_K$ be the subscheme given by the ideal $I = (x^2, y)$. The blow-up $Z = \mathrm{Bl}_\Gamma(\mathbb{A}^2_K)$ will be $\mathrm{Proj}\, A$, where A is the ring

$$A = K[x,y] \oplus I \oplus I^2 \oplus \cdots$$

By Proposition IV-25, we can also write Z as

$$Z = \mathrm{Proj}\, K[x,y][A,B]/(yA - x^2 B)$$

which is covered by the open sets

$$U_A = \mathrm{Spec}\, K[x,y][b]/(y - x^2 b)$$

and

$$U_B = \mathrm{Spec}\, K[x,y][a]/(ya - x^2)$$

where $a = A/B$ and $b = B/A$.

We can see immediately some differences between this scheme and the ordinary blow-up of $\mathbb{A}^2_K$ at the origin. For one thing, though the fiber of each over the origin is isomorphic to $\mathbb{A}^1_K$, the scheme $Z = \mathrm{Bl}_\Gamma(\mathbb{A}^2_K)$ is singular at one point P (the point $a = x = y = 0$ in U_B), while the ordinary blow-up is nonsingular.

We can see more if we express Z in terms of blow-ups with reduced centers. Briefly, the "recipe" for Z in classical language is this (see figure below): first, let Z_1 be the blow-up of $\mathbb{A}^2_K$ at the origin; let $E \subset Z_1$ be the exceptional divisor, that is, the inverse image of the origin. Let P be the point of E lying on the proper transform of the x-axis — that is, the closure of the preimage of the x-axis in $Z_1 \setminus E$. Let Z_2 be the blow-up of Z_1 at P; let $F \subset Z_2$ be the exceptional divisor of this blow up and (by a slight abuse of notation) $E \subset Z_2$ the proper transform of E in Z_2. Then, in classical language, $Z = \mathrm{Bl}_\Gamma(\mathbb{A}^2_K)$ is obtained from Z_2 by blowing down E. In other words:

Proposition IV-40. *The blow-up Z' of $Z = \mathrm{Bl}_\Gamma(\mathbb{A}^2_K)$ at its singular point P is Z_2.*

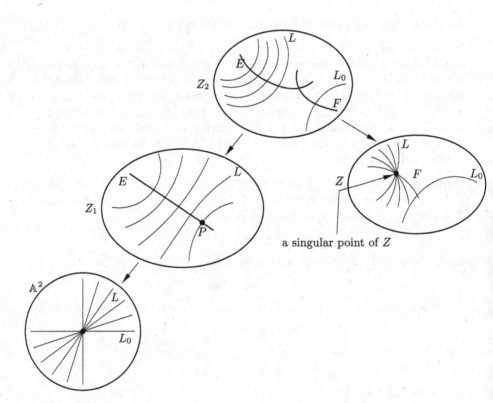

a singular point of Z

'We see from this description that the lines through the origin in the plane are not made disjoint, as they were in the case of the blow-up of $\mathbb{A}_K^2$ at the reduced origin: they are made disjoint in the first blow-up, but then meet each other once more after we blow down $E \subset Z_2$. On the other hand, nonsingular curves through the origin tangent to the x-axis and having different curvatures are separated: after the first blow-up in this sequence they meet transversely at the point P; they are then separated by the second blow-up and are not affected by the blowing down.

Proof of Proposition IV-40. By Exercise IV-35, the blow-up of Z at its singular point is the same as the blow-up of Z at the reduced scheme F associated to the exceptional divisor of $Z = \mathrm{Bl}_\Gamma(\mathbb{A}_K^2) \to \mathbb{A}_K^2$. This scheme F is the total transform in Z of the (reduced) origin in $\mathbb{A}_K^2$, as we see directly from the equations.

On the other hand, we claim that Z_2 may be obtained by first blowing up the reduced origin in $\mathbb{A}_K^2$ to get Z_1, and then blowing up the total transform of Γ in Z_1 — the reverse of the previous process. To see this, observe that by the equations the ideal of Γ in Z_1 is the product of the ideal of $E \subset Z_1$ and the ideal of the point P; since E is Cartier, it follows that $\mathrm{Bl}_{\Gamma'} Z_1 = \mathrm{Bl}_P Z_1$.

With these remarks in place, it now suffices to apply the following lemma:

Lemma IV-41. *Let X be a scheme and Y_1 and $Y_2 \subset X$ closed subschemes. If $f_i : Z_i = \mathrm{Bl}_{Y_i} X \to X$ be the blow-ups of X along Y_1 and Y_2, then*

$$\mathrm{Bl}_{f_1^{-1}(Y_2)} Z_1 \cong \mathrm{Bl}_{f_2^{-1}(Y_1)} Z_2$$

as X-schemes.

Proof. Let $W_1 = \mathrm{Bl}_{f_1^{-1}(Y_2)} Z_1$, and let $g_1 : W_1 \to Z_1$ be the blow-up map; define W_2 and g_2 analogously. Set $h_i = f_i \circ g_i : W_i \to X$. Since $h_1^{-1}(Y_2) = g_1^{-1}(f_1^{-1}(Y_2)) \subset W_1$ is Cartier, the structure map $h_1 : W_1 \to X$ factors through Z_2; that is, there is a map $j_1 : W_1 \to Z_2$ such that $h_1 = f_2 \circ j_1$. Similarly, since $j_1^{-1}(f_2^{-1}((Y_1)) = h_1^{-1}(Y_1) = g_1^{-1}(f_1^{-1}(Y_1)) \subset W_1$ is also Cartier, the map $j_1 : W_1 \to Z_2$ factors through $W_2 = \mathrm{Bl}_{f_2^{-1}(Y_1)} Z_2$, inducing a map $k_1 : W_1 \to W_2$ such that $h_1 = h_2 \circ k_1$. In the other direction, we likewise obtain a map $k_2 : W_2 \to W_1$. Since W_1 has no automorphisms as an X-scheme, $k_2 \circ k_1$ is the identity, and in particular k_1 is an isomorphism. $\square$

Compare this lemma with Exercises IV-37 to IV-39, where we saw that if we replace "total transform" with "proper transform", the order does indeed matter.

Blowing Up Multiple Points. We will consider here a few more examples of blow-ups of the plane along subschemes supported at a point.

Exercise IV-42. For another example, let $\Omega_1 \subset \mathbb{A}_K^2$ be the subscheme defined by the ideal $(y, x^3) \subset K[x, y]$ and Ω_2 the subscheme defined by the ideal $(y^2, x^3) \subset K[x, y]$. Consider the blow-ups $\varphi_i : Z_i = \mathrm{Bl}_{\Omega_i}(\mathbb{A}_K^2) \to \mathbb{A}_K^2$ of the plane at each of these two schemes. Show in that in each case the scheme Z_i is singular, the fiber $\varphi_i^{-1}(P)$ over the origin $P = (x, y) \in \mathbb{A}_K^2$ is isomorphic to $\mathbb{P}_K^1$. Show also that in each case the blow-up map may be factored into a sequence of three blow-ups followed by two contractions, that is, there is a scheme W_i, obtained by blowing up $\mathbb{A}_K^2$ successively at three reduced points, and a map $W_i \to Z_i$ that is constant on the exceptional divisors of the first two blow-ups and is an isomorphism on their complement. What is different about the sequence of points blown up in the two cases?

Not to give a false impression, we should remark that the fibers of blow-ups, even of nonsingular varieties, need not be projective spaces. (Of course, given our assertion that any proper birational morphism is a blow-up, this could hardly be the case.) The subject of the following exercises is a simple example of other behavior.

Exercise IV-43. Let $\mathbb{A}_K^2 = \mathrm{Spec}\, K[x, y]$ be the affine plane over an algebraically closed field K, and let $\Gamma \subset \mathbb{A}_K^2$ be the subscheme given by

$$\Gamma = V(x^3, xy, y^2).$$

Let X be the blow-up $X = \mathrm{Bl}_\Gamma(\mathbb{A}_K^2)$. Show that X is given as

$$X = \mathrm{Proj}\, K[x, y][A, B, C]/I$$

where I is the ideal

$$I = (yA - x^2 B,\ yB - xC,\ AC - xB^2).$$

Hint: blow up $\mathbb{A}_K^3 = \mathrm{Spec}[x, y, z]$ along the subscheme $V(z - xy, x^3, y^2)$, which is a regular subscheme, and consider the proper transform of the plane $V(z)$.

Exercise IV-44. Show that the scheme X of the preceding exercise is nonsingular, with fiber over the origin $(x, y) \in \mathbb{A}_K^2$ a union of two copies of $\mathbb{P}_K^1$ meeting at one point. (In fact, X is the scheme Z_2 of Proposition IV-40.)

It is not the case that we have a one-to-one correspondence between ideals and blow-ups; different ideals may yield the same blow-up. Of course there are many trivial examples of this — for example, any principal ideal yields the trivial blow-up. Only slightly less trivially, let X be any Noetherian scheme, $\Gamma \subset X$ any closed subscheme and $\mathscr{I} \subset \mathscr{O}_X$ its ideal sheaf. Let Γ_n be the subscheme of X defined by the ideal $\mathscr{I}^n$. It follows from the definition via the universal property that the blow-ups $Z_n = \mathrm{Bl}_{\Gamma_n}(X)$ are all isomorphic. Here are some more interesting examples:

Exercise IV-45. Let $\mathbb{A}^2_K = \operatorname{Spec} K[x, y]$ be the affine plane over an algebraically closed field K. Consider the subschemes $\Gamma_n \subset \mathbb{A}^2_K$ given by the ideals

$$I_n = (x^{n+1}, x^{n-1}y, x^{n-2}y^2, \ldots, xy^{n-1}, y^n) = (x, y)^n \cap (x^{n+1}, y).$$

(In other words, I_n is the ideal of polynomials vanishing to order n at the origin, and vanishing to one higher order along the x-axis.) Note that Γ_1 is the scheme Γ of Proposition IV-40.

Show that for $n \geq 2$ the schemes X_n are isomorphic to one another by exhibiting ismorphisms $\varphi_n : X \to X_n$, where $X = \operatorname{Bl}_\Gamma(\mathbb{A}^2_K)$ is the blow-up described in Proposition IV-40.

The j-Function. Here is an example of a blow-up similar to the one we have just described that arises very naturally. It involves the j-function of a plane cubic curve; this is a topic we will not mention officially until the very end of this book, but with which the reader may well be familiar. In any event, we will assume some acquaintance with j in what follows.

We consider the (flat) family $\mathscr{E} \to \mathbb{A}^2_K = \operatorname{Spec} K[a, b]$ of plane cubic curves given by the equation

$$y^2 = x^3 + ax + b.$$

Now, when the curve $C_{a,b}$, given in $\mathbb{A}^2_K = \operatorname{Spec} K[x, y]$ by the equation $y^2 = x^3 + ax + b$, is nonsingular, we associate to it the scalar

$$j(C_{a,b}) = 1728 \frac{4a^3}{4a^3 + 27b^2}.$$

As the reader may know, two such curves $C_{a,b}$ and $C_{a',b'}$ are isomorphic if and only if the values of the j-function are the same. It is thus of some interest to understand how the rational map from $\mathbb{A}^2_K = \operatorname{Spec} K[a, b]$ to $\mathbb{A}^1_K = \operatorname{Spec} K[j]$ behaves — in other words, how the moduli of the curve $C_{a,b}$ behaves when it becomes singular. Most of the time this is clear: if the point (a, b) approaches any point of the curve $4a^3 + 27b^2 = 0$ other than the origin $Q = (a, b) \subset \operatorname{Spec} K[a, b]$, the value of $j(C_{a,b})$ approaches infinity. The question of what happens when $C_{a,b}$ acquires a cusp is more subtle. To put it another way, we have a morphism

$$j : \mathbb{A}^2_K \setminus \{Q\} \longrightarrow \mathbb{P}^1_K$$
$$(a, b) \longmapsto j(C_{a,b})$$

and would like to understand the map in a neighborhood of Q.

The answer is not hard to find: the closure Γ in $\mathbb{A}^2_K \times \mathbb{P}^1_K$ of the graph of the map $j : \mathbb{A}^2_K \setminus \{Q\} \to \mathbb{P}^1_K$ is simply the blow-up

$$\varphi_2 : Z_2 = \operatorname{Bl}_{\Omega_2}(\mathbb{A}^2_K) \to \mathbb{A}^2_K$$

of the plane along the subscheme whose ideal is generated by the numerator and denominator of the expression above for $j(C_{a,b})$. We can also describe it in terms of classical blow-ups as follows:

Exercise IV-46. Factor the projection $\Gamma \to \mathbb{A}_K^2$ into blow-ups and blow-downs at reduced points: explicitly, show that the map j blows up the origin, then the point of intersection of the exceptional divisor with the proper transform of the x-axis, then the intersection of the two exceptional divisors; finally, it blows down the first two exceptional divisors.

From this description we can see many things. For example, consider a *pencil* of cubics specializing to a cusp; that is, restrict the family above to a line through the origin in the plane $\operatorname{Spec} K[a, b]$. Equivalently, consider for some pair α, β the family of curves C_t given by

$$y^2 = x^3 + \alpha t x + \beta t.$$

The limiting value of $j(C_t)$ as t approaches 0 is always $j = 0$ — in terms of the moduli space $\mathcal{M}_1$, the curves approach the curve given by $y^2 = x^3 + 1$ — independently of the slope β/α, as long as $\beta \neq 0$. Conversely, if we want to describe families of plane cubics acquiring a cusp whose j-invariants approach a value other than 0 or ∞, we have to find curves through the origin in the plane $\operatorname{Spec} K[a, b]$ whose proper transform in the triple blow-up W_2 of the plane, described in Exercise IV-46, is separated from the first two exceptional divisors.

In this case the j-function is so explicitly given that we hardly need the geometric analysis. But the qualitative picture is very important: the picture in general when a family a curves of any genus acquires a cusp is the same. For example, if a pencil of plane curves acquires a cusp, the stable limit will always have an elliptic tail of j-invariant either 0 or ∞.

IV.2.4 Blow-ups of Arithmetic Schemes

Since we have defined blow-ups so generally, we can use the construction to relate various arithmetic schemes, as the following examples and exercises illustrate.

We start by blowing up a reduced point in $\mathbb{P}_{\mathbb{Z}}^1$: we let P be the reduced point $P = (3, X) \in \mathbb{P}_{\mathbb{Z}}^1$ and consider the blow-up $Z = \operatorname{Bl}_P(\mathbb{P}_{\mathbb{Z}}^1)$ of $\mathbb{P}_{\mathbb{Z}}^1$ at P. This is straightforward; as before, the only problem is notational. Since the scheme $\mathbb{P}_{\mathbb{Z}}^1 = \operatorname{Proj} \mathbb{Z}[X, Y]$ we are starting with is not affine, we cover it by affine open sets $U_X = \operatorname{Spec} \mathbb{Z}[y] \cong \mathbb{A}_{\mathbb{Z}}^1$ and $U_Y = \operatorname{Spec} \mathbb{Z}[x] \cong \mathbb{A}_{\mathbb{Z}}^1$ where $y = Y/X$ and $x = X/Y$. Since the point to be blown up lies in the complement of U_X, the inverse image of U_X in Z is simply U_X.

Next, we describe the blow-up of U_Y. To avoid confusion, we denote by A and B the two generators 3 and x of the ideal $I = (3, x)$ of $P \in U_Y = \operatorname{Spec} \mathbb{Z}[x]$; we can then write the ring $\mathbb{Z}[x] \oplus I \oplus I^2 \oplus \ldots$ as

$$\bigoplus_{n=0}^{\infty} I^n = (\mathbb{Z}[x])[A, B]/(xA - 3B).$$

We may describe Proj of this ring as the union of the two open subsets W_A and W_B. The first is simpler: setting $b = B/A$, we have

$$W_A = \operatorname{Spec} \mathbb{Z}[x][b]/(x - 3b) = \operatorname{Spec} \mathbb{Z}[b] = \mathbb{A}^1_{\mathbb{Z}},$$

so that the open set $W_A \cong \mathbb{A}^1_{\mathbb{Z}}$, but the map $W_A \to U_Y \cong \mathbb{A}^1_{\mathbb{Z}} \subset \mathbb{P}^1_{\mathbb{Z}}$ is not an isomorphism; rather, it's the map $\operatorname{Spec} \mathbb{Z}[b] \to \operatorname{Spec} \mathbb{Z}[x]$ dual to the ring map sending x to $3b$.

As for the other open set, we have

$$W_B = \operatorname{Spec} \mathbb{Z}[x][a]/(ax - 3),$$

that is to say, W_B is an affine plane conic. For primes $p \neq 3$ the fiber of W_B over $(p) \in \operatorname{Spec} \mathbb{Z}$ is the complement $\operatorname{Spec} \mathbb{Z}/(p)[x, \frac{1}{x}]$ of one point in $\mathbb{A}^1_{\mathbb{Z}/(p)}$ (or equivalently, the complement of two points in $\mathbb{P}^1_{\mathbb{Z}/(p)}$). The fiber of W_B over (3), on the other hand, is the union of two copies of $\mathbb{A}^1_{\mathbb{Z}/(3)}$ meeting at a point.

We have seen that the blow-up Z is a union of three affine opens: two, $U_X = \operatorname{Spec} \mathbb{Z}[y]$ and $W_A = \operatorname{Spec} \mathbb{Z}[b]$, are each isomorphic to $\mathbb{A}^1_{\mathbb{Z}}$, and the third, W_B, is a plane conic in $\mathbb{A}^2_{\mathbb{Z}}$. The identifications among these sets are simple to describe. For example, the open subset $U_{3y} = \operatorname{Spec} \mathbb{Z}[y, \frac{1}{y}, \frac{1}{3}] \subset \operatorname{Spec} \mathbb{Z}[y]$ is identified with the open subset $U_{3b} = \operatorname{Spec} \mathbb{Z}[b, \frac{1}{b}, \frac{1}{3}] \subset \operatorname{Spec} \mathbb{Z}[b]$ via the map dual to the ring isomorphism sending y to $1/3b$; this yields a scheme

$$Z' = U_X \cup W_A = \operatorname{Spec} \mathbb{Z}[y] \underset{\operatorname{Spec} \mathbb{Z}[y, \frac{1}{y}, \frac{1}{3}] = \operatorname{Spec} \mathbb{Z}[b, \frac{1}{b}, \frac{1}{3}]}{\bigcup} \operatorname{Spec} \mathbb{Z}[b]$$

whose fiber over $(p) \in \operatorname{Spec} \mathbb{Z}$ for each prime $p \neq 3$ is a copy of $\mathbb{P}^1_{\mathbb{Z}/(p)}$ (in fact, the inverse image of $\operatorname{Spec} \mathbb{Z} \setminus \{(3)\} = \operatorname{Spec} \mathbb{Z}[\frac{1}{3}]$ in Z' is isomorphic to $\mathbb{P}^1_{\operatorname{Spec} \mathbb{Z}[\frac{1}{3}]}$), and whose fiber over (3) is a disjoint union of two affine lines.

Finally, we glue in the third open set $W_B = \operatorname{Spec} \mathbb{Z}[x][a]/(ax - 3)$, via the identification of the complement of the single point $(3, a, x)$ in W_B with the corresponding open subset of Z' (this is the union of the images in Z' of the open subsets $U_y = \operatorname{Spec} \mathbb{Z}[y, \frac{1}{y}] \subset U_X = \operatorname{Spec} \mathbb{Z}[y]$ and $U_b = \operatorname{Spec} \mathbb{Z}[b, \frac{1}{b}] \subset W_B = \operatorname{Spec} \mathbb{Z}[b]$). This adds one final point: the two components of the fiber of W_B over (3), each isomorphic to $\mathbb{A}^1_{\mathbb{Z}/(3)}$, are each glued onto corresponding components of the fiber Z' over (3) to yield two copies of $\mathbb{P}^1_{\mathbb{Z}/(3)}$ meeting at one point. In sum, the fiber of Z over (p) is $\mathbb{P}^1_{\mathbb{Z}/(p)}$ for $p \neq 3$, and two copies of $\mathbb{P}^1_{\mathbb{Z}/(3)}$ meeting at one point for $p = 3$, as shown on the next page.

There is another way to represent this scheme, which avoids the need for gluing constructions (though we will need the description of the blow-up via gluing to see that it really is the blow-up). This is expressed in the following result:

Proposition IV-47. *The blow-up $Z = \operatorname{Bl}_P(\mathbb{P}^1_{\mathbb{Z}})$ of $\mathbb{P}^1_{\mathbb{Z}}$ at the point $P = (3, X)$ is isomorphic to the plane conic*

$$C = \operatorname{Proj} \mathbb{Z}[S, T, U]/(ST - 3U^2) \subset \mathbb{P}^2_{\mathbb{Z}} = \operatorname{Proj} \mathbb{Z}[S, T, U].$$

$W_A = \operatorname{Spec}\mathbb{Z}[b] = \operatorname{Spec}\mathbb{Z}[x/3]$ $W_B = \operatorname{Spec}\mathbb{Z}[a,x]/(ax-3)$

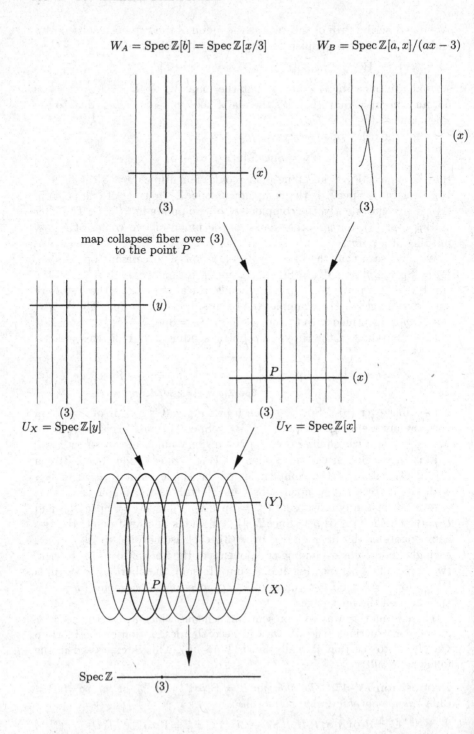

map collapses fiber over (3)
to the point P

$U_X = \operatorname{Spec}\mathbb{Z}[y]$ $U_Y = \operatorname{Spec}\mathbb{Z}[x]$

$\operatorname{Spec}\mathbb{Z}$

Proof. Having already described Z as the union of open sets as above, this is easy: we simply exhibit isomorphisms of these open sets with corresponding open subsets of C and check that they agree on the overlap. First,

$$U_X = \operatorname{Spec}\mathbb{Z}[y] \longrightarrow U_T = \operatorname{Spec}\mathbb{Z}\left[\frac{S}{T}, \frac{U}{T}\right] \Big/ \left(\frac{S}{T} - 3\left(\frac{U}{T}\right)^2\right) = \operatorname{Spec}\mathbb{Z}\left[\frac{U}{T}\right]$$

via the isomorphism sending y to U/T; then

$$W_A = \operatorname{Spec}\mathbb{Z}[b] \longrightarrow U_S = \operatorname{Spec}\mathbb{Z}\left[\frac{T}{S}, \frac{U}{S}\right] \Big/ \left(\frac{T}{S} - 3\left(\frac{U}{S}\right)^2\right) = \operatorname{Spec}\mathbb{Z}\left[\frac{U}{S}\right]$$

via the isomorphism sending b to U/S; and finally

$$W_B = \operatorname{Spec}\mathbb{Z}[a, x]/(ax - 3) \longrightarrow U_U = \operatorname{Spec}\mathbb{Z}\left[\frac{S}{U}, \frac{T}{U}\right] \Big/ \left(\frac{S}{U}\frac{T}{U} - 3\right)$$

via the isomorphism sending a to S/U and x to T/U.   □

Exercise IV-48. Describe in similar terms the blow-up of $\mathbb{P}^1_{\mathbb{Z}}$ at the nonreduced subscheme

$$\Gamma = V(9, X) \subset \mathbb{P}^1_{\mathbb{Z}} = \operatorname{Proj}\mathbb{Z}[X, Y].$$

Use this description to identify the blow-up with the conic in $\mathbb{P}^2_{\mathbb{Z}}$ given by $ST - 9U^2$ in $\mathbb{P}^2_{\mathbb{Z}} = \operatorname{Proj}\mathbb{Z}[S, T, U]$.

In the case of the affine plane over a field, the blow-ups at the subschemes of degree 2 supported at the origin all looked alike, because the automorphism group of $\mathbb{A}^2_K$ acts transitively on nonzero tangent vectors, and hence on subschemes of degree 2. The analogous statement is not true for $\mathbb{A}^1_{\mathbb{Z}}$, however. As we saw in Section II.4.5, there are two types of subschemes of degree 2 supported at such a point, the vertical and the horizontal (or, more accurately, the non-vertical). They may be distinguished by their coordinate rings, which are $\mathbb{Z}/(p)[x]/(x^2)$ and $\mathbb{Z}/(p^2)$, respectively. As the following exercise shows (in conjunction with the preceding exercise), they may also be distinguished by their blow-ups.

Exercise IV-49. Consider the blow-up $Z = \operatorname{Bl}_\Omega(\mathbb{P}^1_{\mathbb{Z}})$ of $\mathbb{P}^1_{\mathbb{Z}}$ at the nonreduced subscheme

$$\Omega = V(3, X^2) \subset \mathbb{P}^1_{\mathbb{Z}} = \operatorname{Proj}\mathbb{Z}[X, Y].$$

Show that the fiber of Z over $(3) \in \operatorname{Spec}\mathbb{Z}$ has two components, one of which is everywhere nonreduced. Use this to show in particular that Z is not isomorphic to any conic in $\mathbb{P}^2_{\mathbb{Z}}$.

Exercise IV-50. Find a curve $C \subset \mathbb{P}^3_{\mathbb{Z}}$ isomorphic to the scheme Z of the preceding exercise.

Hint: First represent Z as a subscheme of $\mathbb{P}^1_{\mathbb{Z}} \times \mathbb{P}^1_{\mathbb{Z}}$, then embed $\mathbb{P}^1_{\mathbb{Z}} \times \mathbb{P}^1_{\mathbb{Z}}$ in $\mathbb{P}^3_{\mathbb{Z}}$ via the Segre embedding. One possible answer is

$$Z = \left\{ [a, b, c, d] : \mathrm{rank}\begin{pmatrix} a & c & 3d \\ b & d & a \end{pmatrix} \leq 1 \right\} \subset \mathbb{P}^3_{\mathbb{Z}};$$

that is, the zero locus in $\mathbb{P}^3_{\mathbb{Z}} = \mathrm{Proj}\,\mathbb{Z}[a, b, c, d]$ of the 2×2 minors of the matrix $\begin{pmatrix} a & c & 3d \\ b & d & a \end{pmatrix}$.

Here are some examples of blow-ups of arithmetic schemes of dimension one, two of which we have already encountered. Recall to begin with that an *order* in a number field K is a subring of the ring of integers in K having quotient field K. In the following three exercises, we'll see that the spectra of orders in a given number field may be related by blowing up.

Exercise IV-51. Let $A = \mathrm{Spec}\,\mathbb{Z}[\sqrt{3}]$ and $B = \mathrm{Spec}\,\mathbb{Z}[11\sqrt{3}]$, as described in Section II.4.2. Show that A is the blow-up of B at the point $(11, 11\sqrt{3})$. (The blow-up along the subscheme (11) is trivial.) Similarly, show that A is the blow-up of the scheme $B' = \mathrm{Spec}\,\mathbb{Z}[2\sqrt{3}]$ at the point $(2, 2\sqrt{3})$.

In the preceding examples, the normalization of the schemes B and B' coincided with the blow up, as is appropriate for schemes we claim are analogues of curves with a simple node and cusp respectively. To see a case where this is not so, we naturally look for a curve with a "tacnode". We will study such a scheme in the following two exercises.

Exercise IV-52. Let A and B be as in the preceding exercise, and let $C = \mathrm{Spec}\,\mathbb{Z}[121\sqrt{3}]$, so that we have morphisms

$$A \longrightarrow B \longrightarrow C.$$

Show that B is the blow-up of C at the (reduced) point $(11, 121\sqrt{3})$. At the same time, exhibit A as the blow-up of C at a nonreduced scheme supported at this point.

Exercise IV-53. To justify the analogies between B and C and curves with a node and tacnode, consider the morphisms $\pi : A \to B$ and $\eta : A \to C$ from A to each. Let $P = (4 + 3\sqrt{3})$ and $Q = (4 - 3\sqrt{3}) \in A$ be the two points lying over the singular points $(11, 11\sqrt{3})$ of B and $(11, 121\sqrt{3})$ of C. Show that the image of the differentials

$$d\pi_P : T_P(A) \longrightarrow T_{(11, 11\sqrt{3})}(B)$$

and

$$d\pi_Q : T_Q(A) \longrightarrow T_{(11, 11\sqrt{3})}(B)$$

do not coincide, but that the images of

$$d\eta_P : T_P(A) \longrightarrow T_{(11, 121\sqrt{3})}(C)$$

and

$$d\eta_Q : T_Q(A) \longrightarrow T_{(11,121\sqrt{3})}(C)$$

do.

The remainder of this section consists of a project for the reader, using several of the techniques we have developed for local analysis to distinguish among arithmetic surfaces.

Example IV-54. Consider the schemes

$$C_1 = \operatorname{Proj}\mathbb{Z}[X,Y,Z]/(XY - 3Z^2), \qquad C_2 = \operatorname{Proj}\mathbb{Z}[X,Y,Z]/(XY - 9Z^2),$$
$$C_3 = \operatorname{Proj}\mathbb{Z}[X,Y,Z]/(XY - 27Z^2), \qquad C_4 = \operatorname{Proj}\mathbb{Z}[X,Y,Z]/(XY - 81Z^2).$$

All four are plane conics, that is, they are the zero loci in $\mathbb{P}_\mathbb{Z}^2$ of homogeneous quadratic polynomials. Moreover, the inverse images of the open subset

$$S = \operatorname{Spec}\mathbb{Z}[\tfrac{1}{3}] = \operatorname{Spec}\mathbb{Z} \setminus \{(3)\} \subset \operatorname{Spec}\mathbb{Z}$$

in all four are isomorphic, via (powers of) the automorphism of the ring $\mathbb{Z}[\tfrac{1}{3}, X, Y, Z]$ given by $(X,Y,Z) \mapsto (3X,Y,Z)$. In particular, each has fiber over (p) a nonsingular conic in $\mathbb{P}_{\mathbb{Z}/(p)}^2$ for $p \neq 3$. Finally, in each case the fiber over (3) is a union of two lines in $\mathbb{P}_{\mathbb{Z}/(3)}^2$.

We claim, however, that *no two of these schemes are isomorphic*; and we will prove this as an illustration of the various techniques developed over the course of this section. The key is the local structure of each scheme around the point $(3, X, Y)$ (which we will, by a slight abuse of notation, call P in each of the four schemes C_i). We start as follows:

Exercise IV-55. Show that C_1 is nonsingular, while C_2, C_3 and C_4 each have P as a unique singular point.

Thus, C_1 cannot be isomorphic to any of the others; and for any two of the others to be isomorphic, a neighborhood of P in each must be isomorphic.

Now, we cannot use the dimension of the tangent space to C_i at P to further distinguish among these: $T_P(C_1)$ is two-dimensional (since C_1 is nonsingular, after all), and $\dim T_P(C_i) = 3$ for each of $i = 2, 3$ and 4. But the tangent cone does provide a useful tool here:

Exercise IV-56. Show that the projective tangent cone to C_2 at P is a nonsingular plane conic, while the tangent cones to C_3 and C_4 at P are each a union of two distinct lines in $\mathbb{P}_{\mathbb{Z}/(3)}^2$.

Thus C_2 cannot be isomorphic to any of the others. Finally, how do we distinguish C_3 and C_4? Blowing up provides the answer:

Exercise IV-57. Let $\tilde{C}_3 = \operatorname{Bl}_P(C_3)$ be the blow-up of C_3 at P, and $\tilde{C}_4$ the blow-up of C_4 at P. Show that $\tilde{C}_3$ is nonsingular, while $\tilde{C}_4$ is not.

For the reader who wishes to pursue this further, the techniques we have are in fact sufficient to prove the following statement:

Exercise IV-58. For any positive integer n, let

$$C_n = \operatorname{Proj} \mathbb{Z}[X, Y, Z]/(XY - 3^n Z^2).$$

Show that for any $n \neq m$, the schemes C_n and C_m are not isomorphic.

Hint: the number of blow-ups required to resolve the singularity of each is $\lfloor \frac{n}{2} \rfloor$; and we can distinguish between n even and n odd by the tangent cone at the singular point before the last blow-up.

In fact, the above analysis shows something more than is claimed: we see that the local rings $\mathcal{O}_{C_n, p}$ (equivalently, the local schemes $\operatorname{Spec} \mathcal{O}_{C_n, p}$) are not isomorphic to one another pairwise.

IV.2.5 Project: Quadric and Cubic Surfaces as Blow-ups

It is a classical fact that a nonsingular quadric surface $Q \subset \mathbb{P}_{\mathbb{C}}^3$ is isomorphic to the surface obtained by blowing up two points in the plane $\mathbb{P}_{\mathbb{C}}^2$ and blowing down the line joining them — in other words, the blow-up of Q at a point is isomorphic to the blow-up of $\mathbb{P}_{\mathbb{C}}^2$ at two points. (This description arises naturally if we consider the graph Γ of the rational map $Q \to \mathbb{P}_{\mathbb{C}}^2$ given by projection from a point on Q.) It is likewise well-known, if less readily seen, that a nonsingular cubic surface $S \subset \mathbb{P}_{\mathbb{C}}^3$ is isomorphic to the blow-up of the plane at six points, no three collinear and not all six on a conic.

In the following series of exercises, we will see how to use our notion of blow-ups along arbitrary subschemes of the plane to extend this description of smooth quadric and cubic surfaces to some singular ones. We start with the case of quadric surfaces. Here we ask: what do we get if, instead of blowing up two points and blowing down the line joining them, we blow up a nonreduced scheme $\Gamma \subset \mathbb{P}_K^2$ of degree 2 and dimension 0, and blow down the unique line containing it? The answer is expressed in the following:

Exercise IV-59. Let K be an algebraically closed field, $Q \subset \mathbb{P}_K^3$ an irreducible quadric, and $p \in Q$ any nonsingular closed point. Show that the blow-up of Q at p is isomorphic to the blow-up of the plane $\mathbb{P}_K^2$ at a subscheme $\Gamma \subset \mathbb{P}_K^2$ of dimension zero and degree 2, with Γ reduced if and only if Q is nonsingular.

The situation over non-algebraically closed fields is illustrated in the following two exercises.

Exercise IV-60. Let $\mathbb{P}_{\mathbb{R}}^3 = \operatorname{Proj} \mathbb{R}[X, Y, Z, W]$ be projective 3-space over the real numbers, and consider the two quadric surfaces $Q_1, Q_2 \subset \mathbb{P}_{\mathbb{R}}^3$ given as the zero loci

$$Q_1 = V(X^2 + Y^2 - Z^2 - W^2) \quad \text{and} \quad Q_2 = V(X^2 + Y^2 + Z^2 - W^2).$$

Show that the blow-up of Q_i at any closed point with residue field $\mathbb{R}$ is isomorphic to the blow-up of the plane $\mathbb{P}^2_{\mathbb{R}}$ at a subscheme $\Gamma_i \subset \mathbb{P}^2_{\mathbb{R}}$ of dimension zero and degree 2, with

$$\Gamma_1 \cong \operatorname{Spec}(\mathbb{R} \times \mathbb{R}) \quad \text{and} \quad \Gamma_2 \cong \operatorname{Spec}(\mathbb{C}).$$

In other words, Γ_1 consists of two points with residue field $\mathbb{R}$ and Γ_2 of one point with residue field $\mathbb{C}$.

We will return to this in Exercise IV-70 below.

Exercise IV-61. More generally, let K be any field, $Q \subset \mathbb{P}^3_K$ a nonsingular quadric, and $p \in Q$ any point with residue field K. Show that the blow-up of Q at p is isomorphic to the blow-up of the plane $\mathbb{P}^2_K$ at a subscheme $\Gamma \subset \mathbb{P}^2_K$ of dimension zero and degree 2. Show moreover that Γ will consist of two points with residue field K if and only if Q contains a line $L \cong \mathbb{P}^1_K \subset Q \subset \mathbb{P}^3_K$, and that in this case $Q \cong \mathbb{P}^1_K \times_K \mathbb{P}^1_K$.

We turn our attention next to cubic surfaces. As in the case of quadrics, we ask: if any nonsingular cubic surface $S \subset \mathbb{P}^3_K$ over an algebraically closed field K is isomorphic to the blow-up of the plane $\mathbb{P}^2_K$ at six points. Indeed, if $\Gamma \subset \mathbb{P}^2_K$ is a collection of six points, no three collinear and not all six on a conic, there will be a four-dimensional vector space of cubics vanishing on Γ. This gives a morphism

$$\mathbb{P}^2_K \setminus \Gamma \to \mathbb{P}^3_K,$$

and by Proposition IV-22 the blow-up $S = \operatorname{Bl}_\Gamma \mathbb{P}^2_K$ of $\mathbb{P}^2$ at Γ is the closure in $\mathbb{P}^2_K \times_K \mathbb{P}^3_K$ of the graph of this morphism. The surface $S \subset \mathbb{P}^2_K \times_K \mathbb{P}^3_K$ projects isomorphically to $\mathbb{P}^3_K$, and its image is a smooth cubic surface; conversely, every smooth cubic $S \subset \mathbb{P}^3_K$ may be obtained in this way.

What happens when the points of Γ come together? A complete answer is naturally more complicated here; we will simply sketch some of the possibilities. A prerequisite for the following exercises is familiarity with the classical theory of smooth cubic surfaces; see for example Griffiths and Harris [1978] or Mumford [1976].

We assume throughout that K is an algebraically closed field.

Exercise IV-62. Let $\Gamma \subset \mathbb{P}^2_K$ be any subscheme of degree 6 consisting of four reduced points and one double point, with Γ not contained in a conic and no subscheme of Γ of degree 3 contained in a line. Show that the blow-up $\operatorname{Bl}_\Gamma \mathbb{P}^2_K$ is isomorphic to a cubic surface with one ordinary double point (defined in Section IV.2.2), and conversely that any cubic surface with one ordinary double point may be realized in this way. (Use the description of the blow-up in Proposition IV-40.) How many lines does such a cubic surface contain?

Exercise IV-63. This time let $\Gamma \subset \mathbb{P}^2_K$ be any subscheme of degree 6 consisting of three reduced points and one curvilinear triple point, again with Γ not contained in a conic and no subscheme of Γ of degree 3 contained in a line. Show that the blow-up $\mathrm{Bl}_\Gamma \, \mathbb{P}^2_K$ is isomorphic to a cubic surface with one double point, but this time the double point is not ordinary. (What is the tangent cone at the double point?) How many lines does such a cubic surface contain?

Exercise IV-64. For an example of a cubic surface with only one line, let $\Gamma \subset \mathbb{P}^2_K$ be any curvilinear subscheme of degree 6 supported at a single point. Suppose that the (unique) subscheme of Γ of degree 3 is contained in a line, but the subscheme of degree 4 is not. Show that the blow-up $\mathrm{Bl}_\Gamma \, \mathbb{P}^2_K$ is isomorphic to a cubic surface that contains a unique line.

IV.3   Fano schemes

IV.3.1   Definitions

In classical geometry, one way to study a projective variety $X \subset \mathbb{P}^n_K$ is via its relation to linear subspaces of $\mathbb{P}^n_K$. Thus, a number of subvarieties of the Grassmannians $\mathbb{G}_K(k, n)$ are associated to such a variety. For example, we can associate to $X \subset \mathbb{P}^n_K$ the loci in $\mathbb{G}_K(k, n)$ of linear spaces that meet X; of tangent spaces to X; of secants to X; or of linear spaces contained in X. All of these subvarieties can now be redefined as subschemes of $\mathbb{G}_S(k, n)$ associated to a subscheme $X \subset \mathbb{P}^n_S$, and as such they are endowed with a richer structure that reflects the geometry of X. Even if we start with a variety $X \subset \mathbb{P}^n_K$ over an algebraically closed field K, the schemes associated to it in this way may be nonreduced.

In this section we will define and study the scheme $F_k(X) \subset \mathbb{G}_S(k, n)$ parametrizing linear spaces of dimension k contained in a scheme $X \subset \mathbb{P}^n_S$; this is called the k-th *Fano scheme* of X. We will try in particular to indicate how and when a nonreduced scheme structure may arise, and how it allows us to extend many classical theorems about Fano varieties. For example, we'll see that, if K is any field and $X \subset \mathbb{P}^3_K$ is any cubic surface not swept out by lines, the Fano scheme of lines on X will have degree exactly 27 over K, though the set of lines contained in X will have cardinality 27 only if X is nonsingular, and even then may not if K is not algebraically closed. More generally, we will see that in many cases the family of Fano schemes associated to a flat family of varieties $\mathscr{X} \subset \mathbb{P}^n_B \to B$ is itself flat over B, and so we will be able to make statements about number and degree in greater generality.

In this chapter we will define Fano schemes by giving their defining ideals, which are the same ideals that were classically used to define the Fano variety; the only difference is that we no longer throw away information by passing to their radicals. However, we will see in Chapter VI that there is a more intrinsic definition of Fano schemes $F_k(X)$ using the functor of points;

and this definition gives us in turn a characterization of various aspects of their geometry (e.g., their tangent spaces) that is more directly related to the geometry of the schemes X. These descriptions are very useful even in case both X and $F_k(X)$ are varieties.

Let S be any scheme, and let $X \subset \mathbb{P}^n_S$ be any subscheme of projective space over S; let $k < n$ be any positive integer. The Fano scheme $F_k(X) \subset \mathbb{G} = \mathbb{G}_S(k,n)$ of X is a scheme parametrizing the linear subspaces of dimension k in $\mathbb{P}^n_S$ lying on X. (As always, the word "parametrize" has a precise meaning, which we will discuss further in Section VI.2.2 below.) We define the $F_k(X)$ first in case the base scheme $S = \operatorname{Spec} R$ is affine. We will describe them in terms of the description given in Section III.2.7 of $\mathbb{G}$ as the union of affine spaces $W_I \cong \mathbb{A}_S^{(k+1)(n-k)}$.

Recall that in this construction we let

$$W = \operatorname{Spec} R[\ldots, x_{i,j}, \ldots] \cong \mathbb{A}_S^{(k+1)(n+1)}$$

(which we think of as the affine space associated to the vector space of $(k+1) \times (n+1)$ matrices), and for each multi-index $I = (i_0, \ldots, i_k) \subset \{0, 1, \ldots, n\}$ let $W_I \cong \mathbb{A}_S^{(k+1)(n-k)} \subset W$ be the closed subscheme given by the ideal $(\ldots, x_{i_\alpha, j_\beta} - \delta_{\alpha, \beta}, \ldots)$ (which we think of as the affine space associated to the subspace of matrixes whose I-th submatrix is the identity). Now, suppose that $G(Z_0, \ldots, Z_n) \in I(X)$ is any homogeneous polynomial in the ideal of X. Applying it to a general linear combination of the rows of a $(k+1) \times (n+1)$ matrix, we obtain a polynomial

$$H_G(u, x) = G\left(\sum u_i x_{0,i}, \sum u_i x_{1,i}, \ldots, \sum u_i x_{k,i}\right)$$

which we may write out as a linear combination of the monomials $u^J = u_0^{j_0} u_1^{j_1} \cdots u_k^{j_k}$ in the variables $u_0, \ldots, u_k$:

$$H_G(u, x) = \sum H_{G,J}(x) \cdot u^J.$$

The coefficient polynomials $H_{G,J}(x)$ are then polynomials in the variables $x_{i,j}$; restricting to the subscheme $W_I \cong \mathbb{A}_S^{(k+1)(n-k)} \subset W$ they are likewise regular functions there. We define the *Fano scheme* $F_k(X)$ to be the subscheme of $\mathbb{G}$ given, in each open subset W_I, by the ideal generated by the polynomials $H_{G,J}(x)$, where G ranges over all elements of the ideal $I(X) \subset R[Z_0, \ldots, Z_n]$ and J indexes monomials of degree d in the variables $u_0, \ldots, u_k$.

Alternatively, for any $(k+1)$-tuple $c = (c_0, \ldots, c_k)$ of elements of R, we may define a polynomial $H_{G,c}(x)$ by

$$H_{G,c}(x) = G\left(\sum c_i x_{0,i}, \sum c_i x_{1,i}, \ldots, \sum c_i x_{k,i}\right)$$

and take the Fano scheme $F_k(X)$ to be the subscheme of $\mathbb{G}$ given in W_I by the ideal generated by the polynomials $H_{G,c}(x)$, where G ranges over $I(X) \subset R[Z_0, \ldots, Z_n]$ and c ranges over R^{n+1}.

To complete this definition we would have to check a number of things: that these subschemes of W_I agree on the overlaps of the W_I, and that the subscheme $F_k(X) \subset \mathbb{G}$ they define does not depend on choice of coordinates (this is easier if we adopt the second way of generating the ideal of $F_k(X) \cap W_I$, but then of course we have to show the two ways yield the same ideal). Finally, we should check that the construction is natural, that is, if $T \to S$ is any morphism and $X_T = X \times_S T \subset \mathbb{P}_T^n$, then the Fano scheme $F_k(X_T) = F_k(X) \times_S T \subset \mathbb{G}_S(k,n) \times_S T = \mathbb{G}_T(k,n)$. This last condition in particular ensures that, given a projective scheme $X \subset \mathbb{P}_S^n$ over an arbitrary (possibly nonaffine) base S, we can define the Fano scheme $F_k(X) \subset \mathbb{G}_S(k,n)$ by restricting to affine open subschemes of S and gluing the results. All of these assertions can either be verified directly from the definitions; but they will follow more readily from the intrinsic characterization of the Grassmannian and of Fano schemes to be given in Section VI.2.2 below.

IV.3.2   Lines on Quadrics

To illustrate the definition of Fano schemes, we will consider a simple case: the lines on the quadric surface $Q = V(X^2 + Y^2 + Z^2 + W^2) \subset \mathbb{P}_K^3$ over an algebraically closed field K. For convenience, we assume the characteristic of K is not 2 (the situation is the same in characteristic 2 as long as we stick to smooth quadrics). Even in this case, we will see some very interesting phenomena; and we will consider some examples over non-algebraically closed fields as well.

Lines on a Smooth Quadric over an Algebraically Closed Field.
As suggested above, we will first write down equations for $F_1(Q)$ in an open subset $W_I \subset \mathbb{G} = \mathbb{G}_K(1,3)$; in this case, symmetry will do the rest. For example, take $W_{X,Y} = W_{1,2}$ the subset of $\mathbb{G}$ corresponding to lines skew to the line $X = Y = 0$; we may identify this with the affine space $\mathbb{A}_K^4 = \operatorname{Spec} K[a,b,c,d]$ associated to the space of matrices of the form

$$(\text{IV.1}) \qquad \begin{pmatrix} 1 & 0 & a & b \\ 0 & 1 & c & d \end{pmatrix}.$$

We then write the restriction H of the polynomial $G(X,Y,Z,W) = (X^2 + Y^2 + Z^2 + W^2)$ to a linear combination $u_0(1,0,a,b) + u_1(0,1,c,d)$ of the rows of this matrix as

$$
\begin{aligned}
H_G(u_0, u_1) &= G(u_0, u_1, u_0 a + u_1 c, u_0 b + u_1 d) \\
&= u_0^2 + u_1^2 + (u_0 a + u_1 c)^2 + (u_0 b + u_1 d)^2 \\
&= (1 + a^2 + b^2)u_0^2 + 2(ac + bd)u_0 u_1 + (1 + c^2 + d^2)u_1^2.
\end{aligned}
$$

The Fano scheme $F_1(Q)$ in $W_{X,Y} \cong \mathbb{A}_K^4$ is defined to be the zero locus of the coefficients of H_G, viewed as a polynomial in u_0 and u_1; that is,

$$F_1(Q) \cap W_{X,Y} = V(1 + a^2 + b^2,\ ac + bd,\ 1 + c^2 + d^2) \subset \operatorname{Spec} K[a,b,c,d].$$

It is not hard to describe the subscheme of $\mathbb{A}_K^4$ defined by these equations. It is reducible, with one (irreducible) component lying in the plane $a = d, b = -c$ and the other in the plane $a = -d, b = c$. Each component is isomorphic via the projection to the plane conic $\operatorname{Spec} K[c,d]/(c^2+d^2+1) \subset \mathbb{A}_K^2 = \operatorname{Spec} K[c,d]$.

We can use this to write down the equations of $F_1(Q)$ in homogeneous coordinates on $\mathbb{G} \subset \mathbb{P}_K^5$. To do this, recall first that the homogeneous coordinates on $\mathbb{P}_K^5$ correspond to the 2×2 minors of a 2×4 matrix; we will label them accordingly $\Pi_{XY}, \Pi_{XZ}, \Pi_{XW}, \Pi_{YZ}, \Pi_{YW}$ and Π_{ZW}. The open subset $W_{X,Y} \subset \mathbb{G}$ is the intersection of $\mathbb{G}$ with the affine open subset $\Pi_{XY} \neq 0$; and the coordinate functions a, b, c and d above on $W_{X,Y} \cong \mathbb{A}_K^4$ are the restrictions of the ratios

$$a = -\Pi_{YZ}/\Pi_{XY}, \quad b = -\Pi_{YW}/\Pi_{XY},$$
$$c = \Pi_{XZ}/\Pi_{XY}, \quad d = \Pi_{XW}/\Pi_{XY}.$$

Also,

$$ad - bc = \Pi_{ZW}/\Pi_{XY},$$

from which we can deduce the defining equation of $\mathbb{G} \subset \mathbb{P}_K^5$:

$$\mathbb{G} = V(\Pi_{ZW}\Pi_{XY} + \Pi_{YZ}\Pi_{XW} - \Pi_{XZ}\Pi_{YW}).$$

Now, from the equations of $F_1(Q) \cap W_{X,Y}$ above, we can see that the Fano scheme $F_1(Q)$ is contained in

$$V(\Pi_{XY}^2 + \Pi_{YZ}^2 + \Pi_{YW}^2, \ \Pi_{YZ}\Pi_{XZ} + \Pi_{YW}\Pi_{XW}, \ \Pi_{XY}^2 + \Pi_{XZ}^2 + \Pi_{XW}^2).$$

Carrying out the same procedure in the other five affine open subsets of $\mathbb{P}^5$ as well yields a complete set of defining equations for $F_1(Q) \subset \mathbb{P}^5$. This is easy because of the symmetry of the equations; we conclude that $F_1(Q)$ has the expression

$$V\big(\Pi_{YZ}^2 - \Pi_{XW}^2, \ (\Pi_{YZ}+\Pi_{XW})(\Pi_{YW}+\Pi_{XZ}), \ (\Pi_{YZ}+\Pi_{XW})(\Pi_{ZW}-\Pi_{XY}),$$
$$\Pi_{YW}^2 - \Pi_{XZ}^2, \ (\Pi_{YW}-\Pi_{XZ})(\Pi_{YZ}-\Pi_{XW}), \ (\Pi_{YW}-\Pi_{XZ})(\Pi_{ZW}-\Pi_{XY}),$$
$$\Pi_{ZW}^2 - \Pi_{XY}^2, \ (\Pi_{ZW}+\Pi_{XY})(\Pi_{YZ}-\Pi_{XW}), \ (\Pi_{ZW}+\Pi_{XY})(\Pi_{YW}+\Pi_{XZ}),$$
$$\Pi_{XY}^2 + \Pi_{YZ}^2 + \Pi_{YW}^2, \ \Pi_{XY}^2 + \Pi_{XZ}^2 + \Pi_{XW}^2\big).$$

It may be easier to understand this if we organize it a little better; the way to do this is suggested by the description above of $F_1(Q) \cap W_{X,Y}$. Let Λ_1 and $\Lambda_2 \cong \mathbb{P}_K^2 \subset \mathbb{P}_K^5$ be the disjoint 2-planes defined by the equations

$$\Lambda_1 = V\big(\Pi_{YZ} + \Pi_{XW}, \ \Pi_{YW} - \Pi_{XZ}, \ \Pi_{ZW} + \Pi_{XY}\big)$$

and

$$\Lambda_2 = V\big(\Pi_{YZ} - \Pi_{XW}, \ \Pi_{YW} + \Pi_{XZ}, \ \Pi_{ZW} - \Pi_{XY}\big).$$

Then we have, simply

$$F_1(Q) = (\Lambda_1 \cup \Lambda_2) \cap \mathbb{G} \subset \mathbb{P}^5$$

as schemes. Each of the planes Λ_i intersects $\mathbb{G}$ in a nonsingular plane conic C_i; so we see that $F_1(Q)$ is simply the union of two conics lying in complementary planes. (In particular, $F_1(Q)$ is simply the closure of the two affine conics in $F_1(Q) \cap W_{X,Y}$ above.) This corresponds to the classical picture of the two rulings of a quadric surface.

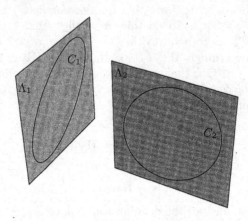

Lines on a Quadric Cone. Next, let us consider what happens to the Fano scheme of lines on a quadric as it varies, and in particular as it degenerates to a singular quadric. Let the base of our family be $B = \operatorname{Spec} K[t] \cong \mathbb{A}^1_K$, and consider first the family of quadrics $\mathscr{Q} \subset \mathbb{P}^3_B$ given by

$$\mathscr{Q} = V(tX^2 + Y^2 + Z^2 + W^2) \subset \operatorname{Proj} K[t][X, Y, Z, W] = \mathbb{P}^3_B.$$

We will denote by $Q_\mu \subset \mathbb{P}^3_K$ the fiber of $\mathscr{Q}$ over the point $(t-\mu) \in B = \mathbb{A}^1_K$. The Fano scheme $F_1(\mathscr{Q})$ is likewise a subscheme of $\mathbb{G}_B(1,3)$, whose fiber over $(t - \mu) \in B = \mathbb{A}^1_K$ is the Fano scheme $F_1(Q_\mu) \subset \mathbb{G}_K(1,3)$ of lines on the quadric $Q_\mu \subset \mathbb{P}^3_K$.

As before, take $W_{X,Y}$ the subset of $\mathbb{G}_B(1,3)$ corresponding to lines skew to the line $X = Y = 0$ and identify this with the affine space $\mathbb{A}^4_B = \operatorname{Spec} K[t][a, b, c, d]$ associated to the space of matrices of the form IV.1.

Write the restriction H of the polynomial $G(X, Y, Z, W) = (tX^2 + Y^2 + Z^2 + W^2)$ to a linear combination of the rows of this matrix as

$$\begin{aligned} H_G(a, x) &= G(u_0, u_1, u_0 a + u_1 c, u_0 b + u_1 d) \\ &= tu_0^2 + u_1^2 + (u_0 a + u_1 c)^2 + (u_0 b + u_1 d)^2 \\ &= (t + a^2 + b^2)u_0^2 + 2(ac + bd)u_0 u_1 + (1 + c^2 + d^2)u_1^2. \end{aligned}$$

The Fano scheme $F_1(\mathscr{Q})$ in $W_{X,Y} \cong \mathbb{A}^4$ is the zero locus of the coefficients, that is,

$$F_1(\mathscr{Q}) \cap W_{X,Y} = V(t + a^2 + b^2, ac + bd, 1 + c^2 + d^2) \subset \operatorname{Spec} K[t][a, b, c, d].$$

For any fixed nonzero scalar $\mu \neq 0 \in K$ the fiber of $F_1(\mathscr{Q}) \cap W_{X,Y}$ over $(t-\mu)$ is a subscheme of $\mathbb{A}^4_K$ isomorphic to the scheme $F_1(Q) \cap W_{X,Y} \subset \mathbb{A}^4_K$

described above (necessarily so, since Q_μ is projectively equivalent to Q by an automorphism of $\mathbb{P}^3_K$ whose action on $\mathbb{G}_K(1,3)$ fixes $W_{X,Y}$). It is reducible, with one component lying in the plane $a = \sqrt{\mu}\, d, b = -\sqrt{\mu}\, c$ and the other in the plane $a = -\sqrt{\mu}\, d, b = \sqrt{\mu}\, c$. Each component is isomorphic via the projection to the plane conic $\operatorname{Spec} K[c,d]/(c^2 + d^2 + 1) \subset \mathbb{A}^2_K = \operatorname{Spec} K[c,d]$.

Now consider the fiber of $F_1(\mathscr{Q}) \cap W_{X,Y}$ over (t), that is, the open subset $F_1(Q_0) \cap W_{X,Y}$ of the Fano scheme $F_1(Q_0)$ of the quadric cone Q_0. This has equation

$$F_1(Q_0) \cap W_{X,Y} = V(a^2 + b^2,\ ac + bd,\ 1 + c^2 + d^2) \subset \mathbb{A}^4_K.$$

It is not hard to see that the support of $F_1(Q_0) \cap W_{X,Y}$ is a single plane conic, lying in the plane $a = b = 0$ and given there by the equation $c^2 + d^2 + 1 = 0$. But $F_1(Q_0)$ is not reduced! Rather, at each point the tangent space is 2-dimensional, spanned by the tangent line to the reduced conic $c^2 + d^2 + 1 = 0$ in the plane $a = b = 0$ and by another vector lying outside this plane:

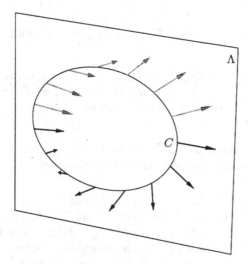

The same picture holds when we consider the entire Fano scheme $F_1(\mathscr{Q}) \subset \mathbb{G}_B(1,3)$ and its fibers $F_1(Q_\mu) \subset \mathbb{G}_K(1,3)$ over $(t - \mu) \in B$. Let $\Lambda_1(\mu)$ and $\Lambda_2(\mu) \cong \mathbb{P}^2_K \subset \mathbb{P}^5_K$ be the disjoint 2-planes defined by the equations

$$\Lambda_1 = V\left(\Pi_{YZ} + \sqrt{\mu}\,\Pi_{XW},\ \Pi_{YW} - \sqrt{\mu}\,\Pi_{XZ},\ \Pi_{ZW} + \sqrt{\mu}\,\Pi_{XY}\right)$$

and

$$\Lambda_2 = V\left(\Pi_{YZ} - \sqrt{\mu}\,\Pi_{XW},\ \Pi_{YW} + \sqrt{\mu}\,\Pi_{XZ},\ \Pi_{ZW} - \sqrt{\mu}\,\Pi_{XY}\right).$$

Then, for $\mu \neq 0$,

$$F_1(Q_\mu) = (\Lambda_1(\mu) \cup \Lambda_2(\mu)) \cap \mathbb{G}_K(1,3) \subset \mathbb{P}^5_K$$

as schemes. As before, each of the planes $\Lambda_i(\mu)$ intersects $\mathbb{G}_K(1,3)$ in a nonsingular plane conic $C_i(\mu)$. But now as μ approaches 0, the two planes $\Lambda_i(\mu)$ have the same limiting position, the plane

$$\Lambda = V(\Pi_{YZ}, \Pi_{YW}, \Pi_{ZW}).$$

As the following exercise will show, the flat limit $\Lambda(0)$ of the schemes $\Lambda(\mu) = \Lambda_1(\mu) \cup \Lambda_2(\mu)$ is a scheme supported on the plane Λ, but having multiplicity 2.

Exercise IV-65. Show that the schemes $\Lambda(\mu) = \Lambda_1(\mu) \cup \Lambda_2(\mu)$ and $\Lambda(0)$ do form a flat family for $\mu \neq 0$, that is, there is a scheme

$$\mathscr{L}^* \subset \mathbb{P}^5_{B^*} = \operatorname{Proj} K[t,t^{-1}][\Pi_{XY}, \Pi_{XZ}, \Pi_{XW}, \Pi_{YZ}, \Pi_{YW}, \Pi_{ZW}]$$

flat over $B^* = B \setminus \{0\} = \operatorname{Spec} K[t,t^{-1}]$, whose fiber over $(t-\mu) \in B^*$ is $\Lambda(\mu)$. Find the equations of $\mathscr{L}$ in $\mathbb{P}^5_{B^*}$; find the equations of the closure $\mathscr{L}$ of $\mathscr{L}^*$ in $\mathbb{P}^n_B$, and thereby of the limit $\Lambda(0)$ of the schemes $\Lambda(\mu)$. Finally, show that the limit of the Fano schemes $F_1(Q_\mu)$ is indeed the Fano scheme $F_1(Q_0)$ of Q_0.

Exercise IV-66. Show that $F_1(Q_0) \subset \mathbb{P}^5_K$ is not contained in a hyperplane.

Exercise IV-67. Let $Q \subset \mathbb{P}^3_K$ be a cone over a nonsingular quadric, $F_1(Q) \subset \mathbb{G}_K(1,3) \subset \mathbb{P}^5_K$ its Fano scheme of lines. Show that $F_1(Q)$ is isomorphic to a double line on a quadric surface, that is, the double line X_1 as described in Section III.3.4.

We'll be able to see the fact that the Fano scheme of a quadric cone is nonreduced more directly in terms of the characterization of $F_1(Q)$ given in Section VI.2.3.

If we let $L = K(t)$ be the function field of the base B of our family, $Q_L \subset \mathbb{P}^3_L$ the fiber of $\mathscr{Q}$ over the generic point $\operatorname{Spec} L \in B$ of our base and $F_1(Q_L) \subset \mathbb{G}_L(1,3)$ the Fano scheme of Q_L, $F_1(Q_L)$ will not be a union of two conics. If we pull it back to the quadratic extension $L' = L(\sqrt{\mu})$ of L—that is, take the fiber product $F_1(Q_L) \times_{\operatorname{Spec} L} \operatorname{Spec} L'$—the scheme we obtain is a union of two conics over $\operatorname{Spec} L$; but $F_1(Q_L)$ itself is irreducible. This is a nice example of a scheme arising in a purely geometric context that is reducible but not absolutely irreducible.

A Quadric Degenerating to Two Planes. Consider now a family of quadrics $\mathscr{Q} \to B$ whose general member is smooth, specializing to a quadric Q_0 consisting of the union of two planes. What is fascinating about this case is not the Fano scheme $F_1(Q_0)$ — after all, since Q_0 is a union of two planes, the locus of lines lying on it is pretty simple — but the geometry of the family. In our main example the Fano scheme $F_1(\mathscr{Q}) \subset \mathbb{G}_B(1,3)$ is not flat over B (and indeed the restriction of the family to the open subset

$B^* = B \setminus \{0\}$ has no flat limit at 0); in other examples a flat limit exists but depends on the particular family and not just on Q_0.

All this is easier to do than to say. To begin with, let $B = \operatorname{Spec} K[s,t] \cong \mathbb{A}_K^2$, and consider the family $\mathscr{Q} \subset \mathbb{P}_B^3 \to B$ given by

$$\mathscr{Q} = V(sX^2 + tY^2 + Z^2 + W^2) \subset \operatorname{Proj} K[s,t][X,Y,Z,W]$$

and let $Q_{\mu,\nu} \subset \mathbb{P}_K^3$ be the fiber of $\mathscr{Q}$ over the point $(s-\mu, t-\nu) \in B$. Let $F_1(\mathscr{Q}) \subset \mathbb{G}_B(1,3)$ be the Fano scheme of $\mathscr{Q}$.

Even without writing down equations, we can see that $F_1(\mathscr{Q})$ is not flat over B: the fibers $F_1(Q_{\mu,\nu})$ of $F_1(\mathscr{Q})$ over B are all one-dimensional, except for the one fiber $F_1(Q_{0,0})$, which is visibly two-dimensional, having support the union of two planes. To see more, we write the equations. As before, we start with the equations of $F_1(\mathscr{Q})$ in the open subset $W_{X,Y}$ of $\mathbb{G}_B(1,3)$; we have

$$F_1(\mathscr{Q}) \cap W_{X,Y} = V(t+a^2+b^2,\ ac+bd,\ s+c^2+d^2) \subset \operatorname{Spec} K[s,t][a,b,c,d].$$

The fiber of $F_1(\mathscr{Q})$ over the origin $(s,t) \in B$ is given in $W_{X,Y}$ by

$$F_1(Q_{0,0}) \cap W_{X,Y} = V(a^2+b^2,\ ac+bd,\ c^2+d^2) \subset \mathbb{A}_K^4.$$

We may also describe this as the union of the two planes Γ_1 and $\Gamma_2 \subset \mathbb{A}_K^4$ given by

$$\Gamma_1 = V(a+\sqrt{-1}b,\ c-\sqrt{-1}d),$$
$$\Gamma_2 = V(a-\sqrt{-1}b,\ c+\sqrt{-1}d).$$

Exercise IV-68. Show that the Fano scheme $F_1(Q_{0,0})$ of a quadric of rank 2 is reduced.

Let us consider now subfamilies of this two-parameter family. To start with, let us fix two nonzero scalars α and $\beta \in K$, and consider the restriction of our family to the line $V(\beta s - \alpha t)$ through the origin in $B = \mathbb{A}_K^2$ with slope β/α; that is, the family $\mathscr{Q}_{\alpha,\beta}$ with base $B' = \operatorname{Spec} K[u]$ given by

$$\mathscr{Q}_{\alpha,\beta} = V(\alpha u X^2 + \beta u Y^2 + Z^2 + W^2) \subset \operatorname{Proj} K[u][X,Y,Z,W]$$

Again, the Fano scheme $F_1(\mathscr{Q}_{\alpha,\beta})$ is not flat over B', for the same reason. What is different here is that the Fano scheme over the complement of the origin in B' *does* have a flat limit. In fact, $F_1(\mathscr{Q}_{\alpha,\beta})$ is reducible, with $F_1(\mathscr{Q}_{0,0})$ one component; and if we remove that component what is left is flat. To see this, we write the equations of $F_1(\mathscr{Q}_{\alpha,\beta})$ in the open subset $W_{X,Y}$ of $\mathbb{G}_B(1,3)$:

$$F_1(\mathscr{Q}_{\alpha,\beta}) \cap W_{X,Y} = V(\alpha u + a^2 + b^2,\ ac+bd,\ \beta u + c^2 + d^2)$$
$$\subset \operatorname{Spec} K[\alpha,\beta][a,b,c,d].$$

Let $\Phi(\gamma) \subset \mathbb{A}_K^4$ be the 2-plane given by

$$\Phi(\gamma) = V(a-\gamma d,\ b+\gamma c)$$

and let $\Psi(\gamma)$ be the disjoint union of $\Phi(\gamma)$ and $\Phi(-\gamma)$. What we see from these equations is that, for every $\mu \neq 0 \in K$, the fiber of $F_1(\mathcal{Q}_{\alpha,\beta})$ over $(u - \mu) \in B'$ is contained in the scheme $\Psi(\sqrt{\alpha/\beta})$, *independently of* μ, and is cut out on $\Psi(\sqrt{\alpha/\beta})$ by the one further equation $\beta u + c^2 + d^2 = 0$. It follows that the flat limit of the Fano schemes $F_1(Q_{\alpha u, \beta u})$ as u approaches 0 is the intersection of $\Psi(\sqrt{\alpha/\beta})$ with the union $V(c^2 + d^2)$ of the two hyperplanes $V(c + \sqrt{-1}\,d)$ and $V(c - \sqrt{-1}\,d)$. This is the union of four lines.

To interpret this geometrically, note that the first of these lines is simply the subscheme of the Grassmannian $\mathbb{G}_K(1, 3)$ of lines lying in the plane $H_1 = V(Z + \sqrt{-1}\,W)$ and passing through the point

$$P_1 = [\sqrt{-1}\sqrt{\beta}, -\sqrt{\alpha}, 0, 0];$$

the second is the subscheme of lines lying in the plane H_1 and passing through the point $P_2 = [\sqrt{-1}\sqrt{\beta}, \sqrt{\alpha}, 0, 0]$; the third the subscheme of lines lying in the plane $H_2 = V(Z - \sqrt{-1}\,W)$ and passing through the point P_1, and the last the subscheme of lines lying in the plane H_2 and passing through the point P_2. Note that the points P_1 and P_2 here may be characterized as the intersection of the double line $M = H_1 \cap H_2$ of the quadric Q_0 with the other quadrics $Q_{\alpha\mu,\beta\mu}$ in the pencil.

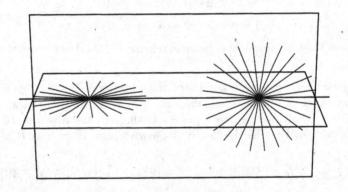

The flat limits of the families $\mathcal{Q}_{\alpha,\beta}$ vary as the ratio β/α varies. In particular, their union is dense in $F_1(Q_{0,0})$. This shows that the Fano scheme $F_1(\mathcal{Q}) \to B$ of the whole family is irreducible, and hence that the restriction of $F_1(\mathcal{Q})$ to the complement of the origin $(s, t) \in B$ does not have a flat limit.

Exercise IV-69. Consider the one-parameter family of quadrics tending to a double plane with equation

$$\mathcal{Q} = V(tX^2 + tY^2 + tZ^2 + W^2) \subset \mathbb{P}_B^3.$$

What is the flat limit of the Fano schemes $F_1(\mathcal{Q}_t)$?

More Examples. To see how the Fano scheme of lines on a quadric over a non-algebraically closed field may behave, we consider the example of quadrics over the real numbers:

Exercise IV-70. Consider the quadrics Q_1, Q_2 and Q_3 in $\mathbb{P}^3_{\mathbb{R}}$ given by the equations

$$Q_1 = V(X^2 + Y^2 - Z^2 - W^2),$$
$$Q_2 = V(X^2 + Y^2 + Z^2 - W^2),$$
$$Q_3 = V(X^2 + Y^2 + Z^2 + W^2).$$

Show that the Fano scheme $F_1(Q_1)$ is the union of two copies of $\mathbb{P}^1_{\mathbb{R}}$, while the Fano scheme of lines on Q_2 is irreducible but not absolutely irreducible. Finally, show that the Fano scheme $F_1(Q_3)$ is the union of two copies of a plane conic not isomorphic to $\mathbb{P}^1_{\mathbb{R}}$.

Here's an example over a function field:

Exercise IV-71. Let $B = \mathbb{P}^9_K$ be the projective space that parametrizes quadric surfaces in $\mathbb{P}^3_K$, and L its function field. Let $\mathscr{Q}_B \subset \mathbb{P}^3_B$ be the universal quadric surface over B. Let $Q_L \subset \mathbb{P}^3_L$ be the fiber of $\mathscr{Q}$ over the generic point $\operatorname{Spec} L$ of B, and $F_L = F_1(Q_L) \subset \mathbb{G}_L(1,3)$ its Fano scheme of lines. Describe F_L. In particular, show that it behaves differently from the examples above, in that is isomorphic to two copies of $\mathbb{P}^1_M$ over a quartic extension M of L, but not over any quadratic extension of L. (In fact, there is a quadratic extension L' of L over which F_L becomes reducible, but the components of $F_L \times_L \operatorname{Spec} L'$ are forms of $\mathbb{P}^1_{L'}$ (in the sense of Section IV.4), but not isomorphic to $\mathbb{P}^1_{L'}$.)

Finally, here is an arithemetic analogue:

Exercise IV-72. Consider the quadrics $\mathscr{Q}_1$, $\mathscr{Q}_2$ and $\mathscr{Q}_3 \subset \mathbb{P}^3_{\mathbb{Z}}$ given by

$$\mathscr{Q}_1 = V(7X^2 + 7Y^2 + Z^2 + W^2)$$
$$\mathscr{Q}_2 = V(7X^2 + 14Y^2 + Z^2 + W^2)$$
$$\mathscr{Q}_3 = V(7X^2 + 49Y^2 + Z^2 + W^2).$$

Describe the Fano scheme $F_1(\mathscr{Q}_i) \subset \mathbb{G}_{\mathbb{Z}}(1,3)$ in each case. In particular, describe the component of $F_1(\mathscr{Q}_i)$ dominating $\operatorname{Spec} \mathbb{Z}$, and its intersection with the fiber $\mathbb{G}_{\mathbb{Z}/(7)}(1,3)$ of $\mathbb{G}_{\mathbb{Z}}(1,3)$ over the point $(7) \in \operatorname{Spec} \mathbb{Z}$.

IV.3.3 Lines on Cubic Surfaces

In the following series of exercises, we will develop some interesting facts about the Fano scheme $F_1(S)$ parametrizing the lines on a cubic surface $S \subset \mathbb{P}^3_K$. To begin with, the following two exercises establish that all nonsingular cubics contain the same number of lines (without deriving the number 27).

Exercise IV-73. Let K be a field, $S \subset \mathbb{P}^3_K$ a nonsingular cubic surface, and $F = F_1(S) \subset \mathbb{G}_K(1,3)$ the Fano scheme of lines on S. Show that F is reduced.

Hint: Take $L \in F_1(S)$ the line given by $X = Y = 0$, and write the cubic polynomial defining S as $X\,Q(Z,W) + Y\,P(Z,W)$ modulo $(X,Y)^2$; show that the condition that the tangent space $T_{[L]}F_1(S)$ be positive-dimensional is that P and Q have a common zero along L. (Compare this to the discussion in Section VI.2.3.)

Exercise IV-74. Let $\mathbb{P}^{19}_K$ be the projective space parametrizing cubic surfaces in $\mathbb{P}^3_K$, and $U \subset \mathbb{P}^{19}_K$ the open subset corresponding to cubics having only finitely many lines (that is, other than cones and scrolls). Let $\mathscr{S}_U \subset \mathbb{P}^3_U$ be the universal cubic in projective 3-space over U, and $\mathscr{F}_U = F_1(\mathscr{S}_U) \subset \mathbb{G}_U(1,3)$ be its Fano scheme of lines. Show that the projection map $\pi : \mathscr{F}_U \to U$ is flat.

Hint: Use the fact that the Fano scheme $F_1(S)$ of lines on a cubic surface $S \subset \mathbb{P}^3_K$ is a local complete intersection, together with Proposition II-32.

In fact, neither of the statements of the two preceding exercises is completely general: that is, for some d and n the Fano variety $F_1(X)$ of lines on a hypersurface $X \subset \mathbb{P}^n_K$ of degree d may be singular and even nonreduced for X nonsingular; and the dimension $\dim F_1(X)$ may jump as X varies, so that the universal Fano scheme $\mathscr{F}$ need not be flat. We will be able to exhibit examples of these behaviors once we have developed a description of tangent spaces to Fano schemes in Chapter VI.

We now consider what happens when our cubic surface becomes singular. There are essentially two cases to consider. In some sense the simpler case is when S has only isolated double points: in this case (as we will see) the Fano scheme $F_1(S)$ is still 0-dimensional of degree 27, and fits into a flat family with the Fano schemes of nonsingular cubics (that is, in terms of Exercise IV-74 above, if we let $U' \subset \mathbb{P}^{19}_K$ be the larger open subset of cubics having at most isolated double points as singularities, and define $\mathscr{S}_{U'} \subset \mathbb{P}^3_{U'}$ and $\mathscr{F}_{U'} \subset \mathbb{G}_{U'}(1,3)$ accordingly, then $\mathscr{F}_{U'} \to U'$ is still flat). On the other hand, when S is a cone or has positive-dimensional singular locus — for example, when S is reducible — $F_1(S)$ will become positive-dimensional; and a further question arises: what may be the flat limit of the Fano schemes $F_1(S_\lambda) \subset \mathbb{G}(1,3)$ as S_λ approaches S?

We start with some examples of cubic surfaces having isolated double points. For this it may be useful to recall the description of such surfaces in Exercise IV-62 above, or in Griffiths and Harris [1978].

Exercise IV-75. Let K be an algebraically closed field, and $S \subset \mathbb{P}^3_K$ a cubic surface having an ordinary double point P (that is, the projectivized tangent cone to S at P is a nonsingular conic). Show that the Fano scheme $F_1(S)$ consists of six nonreduced points (each of multiplicity 2) corresponding to lines on S through P, and 15 reduced points.

Exercise IV-76. Now suppose that S has two ordinary double points P and Q. Show that the scheme $F_1(S)$ consists of one point of multiplicity 4, supported at the point $[\overline{PQ}]$ corresponding to the line $\overline{PQ}$ joining P and Q; eight points of multiplicity 2 corresponding to lines through P or Q other than $\overline{PQ}$; and 7 reduced points. What is the scheme structure of the point $[\overline{PQ}]$?

Exercise IV-77. Now say that S has again just one double point P; but suppose now that P is what is called *an A_2 singularity*: that is, the completion of the local ring $\mathscr{O}_{S,P}$ with respect to the maximal ideal is $\hat{\mathscr{O}}_{S,P} \cong K[\![x,y,z]\!]/(xy - z^3)$ (in particular, the tangent cone is a conic of rank 2, as in Exercise IV-63). Show that the scheme $F_1(S)$ consists of six points of multiplicity 3, corresponding to lines on S through P, and 9 reduced points. Again, what is the scheme structure of the points of multiplicity 3?

In case S is reducible or a cone, the Fano scheme of lines on S is usually pretty obvious; what is of interest is, as we said, the flat limits of the Fano schemes of nearby nonsingular cubics. We consider, in each of the following problems, the same set-up: we take S_0 a reducible cubic or a cone, choose S a general cubic, and let $\{S_\lambda\}_{\lambda \in \mathbb{P}^1}$ be the pencil of cubic surfaces that they span. In each case, we ask what will be the flat limit, as λ tends to 0, of the Fano schemes $F_1(S_\lambda)$.

Exercise IV-78. Take S_0 the union of a nonsingular quadric Q and a plane H, meeting along a nonsingular conic curve C. Let $\{P_1, \ldots, P_6\} = C \cap S$ be the base points of the pencil lying on C. Show that the flat limit, as λ tends to 0, of the Fano schemes $F_1(S_\lambda)$ is reduced of degree 27, consisting of the 12 lines on Q containing one of the points P_i and the 15 lines on H containing 2 of the points P_i.

Exercise IV-79. Now take S_0 the union of three planes H_1, H_2, H_3 in general position. Again, what is the flat limit, as λ tends to 0, of the Fano schemes $F_1(S_\lambda)$?

Exercise IV-80. The same problem, but now take S_0 the cone over a nonsingular plane cubic curve.

Finally, an amusing one on the line(s) on the universal cubic:

Exercise IV-81. Let $B = \mathbb{P}_K^{19}$ be the projective space parametrizing cubic surfaces in $\mathbb{P}_K^3$, L the function field of B, $\mathscr{S}_B \subset \mathbb{P}_B^3$ the universal cubic in projective 3-space over K, and $\mathscr{S}_L$ the fiber of $\mathscr{S}$ over the generic point $\operatorname{Spec} L$ of B. Let $\mathscr{F}_L = F_1(\mathscr{S}_L) \subset \mathbb{G}_L(1,3)$ be its Fano scheme of lines. Show that $\mathscr{F}_L$ consists of one reduced point, whose residue field is a degree 27 extension of L.

Hint: this follows from the fact that the universal Fano variety $\mathscr{F} = \mathscr{F}_1(\mathscr{S}) \subset \mathbb{G}_B(1,3) = \mathbb{P}_K^{19} \times_K \mathbb{G}_K(1,3)$ is irreducible, which in turn follows from the fact that projection on the second factor expresses $\mathscr{F}$ as a $\mathbb{P}_K^{15}$-bundle over $\mathbb{G}_K(1,3)$.

Exercise IV-82. Consider now nonsingular cubic surfaces $S \subset \mathbb{P}_{\mathbb{R}}^3$ over the real numbers. As we have seen, the Fano scheme $F_1(S_{\mathbb{C}}) \subset \mathbb{G}_{\mathbb{C}}(1,3)$ of $S_{\mathbb{C}} = S \times_{\mathrm{Spec}\,\mathbb{R}} \mathrm{Spec}\,\mathbb{C}$ consists of 27 reduced points. It follows that, for some pair of integers a and b with $a + 2b = 27$, $F_1(S) \subset \mathbb{G}_{\mathbb{R}}(1,3)$ will consist of a reduced points with residue field $\mathbb{R}$ and b reduced points with residue field $\mathbb{C}$. Show that a can be 3, 7, 15 or 27, and that no other values are possible. (See Segre [1942].)

IV.4 Forms

Let S be any scheme and X any scheme over S. We say that a scheme Y over S is a *form* of X if for every point $p \in S$ there exists an open neighborhood U of p in S and a flat surjective morphism $T \to U$ of schemes such that $Y \times_S T \cong X \times_S T$ as T-schemes.

To begin with a classic example, consider the conic in the real projective plane $\mathbb{P}_{\mathbb{R}}^2$ given by the equation $X^2 + Y^2 + Z^2 = 0$, that is, the curve $C = \mathrm{Proj}\,\mathbb{R}[X,Y,Z]/(X^2+Y^2+Z^2)$. The curve C has no points defined over $\mathbb{R}$—that is, no points with residue field $\mathbb{R}$—and so cannot be isomorphic to the projective line $\mathbb{P}_{\mathbb{R}}^1$. However, the result of extending the ground field to the algebraic closure $\mathbb{C}$ of $\mathbb{R}$ is $C \times_{\mathrm{Spec}\,\mathbb{R}} \mathrm{Spec}\,\mathbb{C} \cong \mathrm{Proj}\,\mathbb{C}[X,Y,Z]/(X^2 + Y^2 + Z^2) \cong \mathbb{P}_{\mathbb{C}}^1$. Thus C is a form of $\mathbb{P}_{\mathbb{R}}^1$ over $\mathrm{Spec}\,\mathbb{R}$, or, more succinctly, an $\mathbb{R}$-form of the projective line.

As a second example, the reader might check that the field extension $\mathrm{Spec}\,\mathbb{Q}[x]/(x^2 + 1)$ over $\mathrm{Spec}\,\mathbb{Q}$ is a form of the scheme consisting of two distinct points, while $\mathrm{Spec}\,\mathbb{Z}/(2)[x]/(x^2 + 1)$ over $\mathrm{Spec}\,\mathbb{Z}/(2)$ is a form of a double point.

In number theory it is of interest to see how the set of rational points may vary in a family of forms of a given curve. To give a particular case, for any $t \in \mathbb{Q}$ the set of rational solutions (x,y) of what is called *Pell's equation*,

$$ty^2 = x^2 - 1,$$

is the set of $\mathbb{Q}$-rational points on the curve $C_t = \mathrm{Spec}\,\mathbb{Q}[x,y]/(ty^2 - x^2 + 1)$. These curves C_t are forms of $\mathbb{P}^1$ over $\mathrm{Spec}\,\mathbb{Q}$. Likewise, the curves $E_t = \mathrm{Spec}\,\mathbb{Q}[x,y]/(ty^2 - x^3 + 1)$ all have j-invariant 0, and thus are all forms of the curve $E_1 \subset \mathbb{P}_{\mathbb{Q}}^2$ given by $y^2 = x^3 - 1$ (see Section IV.2.3 above and Section VI.2.4 below), but have varying arithmetic properties.

In each of these cases, it is easy to see that the curves given are forms of each other—the curves $E_t \times_{\mathrm{Spec}\,\mathbb{Q}} \mathrm{Spec}\,\mathbb{Q}[\sqrt{t}]$ and $E_1 \times_{\mathrm{Spec}\,\mathbb{Q}} \mathrm{Spec}\,\mathbb{Q}[\sqrt{t}]$ are visibly isomorphic—and less obvious but not hard to see that they are not all isomorphic. (In fact the naive guess—that $E_t \cong E_1$ if and only if

$t \in (\mathbb{Q}^*)^2$, that is, t is the square of a nonzero rational number—is correct, but it is a nontrivial exercise to prove this.)

The set of isomorphism classes of S-forms of projective space $\mathbb{P}^n$, for all n, is in a natural way a group, called the *Brauer group* of S (or, in case $S = \operatorname{Spec} K$ is the spectrum of a field, the Brauer group of K), with the true projective space $\mathbb{P}^n_S$ the identity element. This group may be computed with Galois cohomology; see Serre [1975]. Here is a construction of the Brauer group related to number theory:

Let K be a field, and let A be an n-dimensional *Azumaya algebra* over K—that is, A is an algebra which has dimension n as a vector space, has no nontrivial 2-sided ideals, and has center exactly K. For example, the algebra $M_d(K)$ consisting of all $d \times d$ matrices over K is a d^2-dimensional Azumaya algebra. It follows from the Wedderburn structure theorems that if A is an n-dimensional Azumaya algebra over K then n is a square, say $n = d^2$, and $\overline{K} \otimes A \cong M_d(\overline{K}) \cong \overline{K} \otimes M_d(K)$ (in this sense A is a form of $M_d(K)$.)

Identifying $M_d(K)$ with the endomorphism algebra of a d-dimensional vector space V over K, it is easy to see that the left ideals of $M_d(K)$ each have the form

$$\{a \in M_d(K) \mid \operatorname{Im}(a) \subset W\}$$

for some subspace $W \subset V$. The vector space dimension of the left ideal is then $\dim(V)\dim(W)$. In particular, the left ideals of dimension equal to $d(d-1)$ correspond to the hyperplanes in V, that is, the points of $\mathbb{P}(V)$. The subscheme of the Grassmannian of $d(d-1)$-planes in $M_d(K)$ that are closed under multiplication by $M_d(K)$—that is, are ideals—is isomorphic to $\mathbb{P}^d(K)$ in this way.

At the other extreme, there are (in general) Azumaya algebras A that are division algebras over K—that is, algebras with no left ideals at all. But we can still form the *scheme of left ideals* of A. Of course we must specify it by equations, not as a point set! Let G be the Grassmannian of $d(d-1)$-dimensional subspaces of the vector space A. If $S \subset A_G$ is the universal bundle then the subscheme of G that we want is the largest subscheme X such that the restriction to X of the composite map of vector bundles

$$S \otimes A_G \subset A_G \otimes A_G \xrightarrow{\text{multiplication}} A \longrightarrow A/S$$

on G is zero (this is locally a subscheme defined by the $d(d-1)d^2 \times d$ entries in a matrix representing the composite map.) Extending the ground field, we find that $X_{\overline{K}}$ is the Grassmannian of $d(d-1)$ dimensional ideals of $\overline{K} \otimes A$, that is, projective space! Conversely, it can be shown that every form of projective space over K occurs in this way and that the correspondence between isomorphism classes of forms and Azumaya algebras is one-to-one. For all this see Serre [1975] and Cassels and Fröhlich [1967].

Exercise IV-83. Let A be the quaternion algebra over $\mathbb{R}$; that is, the algebra with basis $1, i, j, k$ and multiplication $i^2 = j^2 = k^2 = -1$, $ij = -ji = k$. Check that its center is $\mathbb{R}$, and that it is a division algebra (and thus an Azumaya algebra over $\mathbb{R}$). Compute the equations of the scheme of 2-dimensional left ideals of A in terms of Plücker coordinates on the Grassmannian of lines in $\mathbb{P}(A) \cong \mathbb{P}_\mathbb{R}^3$. Show directly that this is a form of $\mathbb{P}^1$.

Here is still another way in which a form of $\mathbb{P}^1$ arises, on which the reader can try his hand: Let K be an algebraically closed field, and let $B = \mathbb{P}_K^5$ be the projective space parametrizing conic curves in $\mathbb{P}_K^2$. Let $U \subset B$ be the open subset corresponding to nonsingular plane conics, and let $C_U \subset \mathbb{P}_U^2 = U \times_{\operatorname{Spec} K} \mathbb{P}_K^2$ be restriction to U of the universal conic curve $\mathscr{C} \subset \mathbb{P}_B^2$ over B as described in Section III.2.8. Similarly, let L be the function field of B (or equivalently of U), $\operatorname{Spec} L$ the generic point of B, and $C_L \subset \mathbb{P}_L^2$ the fiber of the universal conic $\mathscr{C}$ over $\operatorname{Spec} L$. We have:

Proposition IV-84. C_U *is a nontrivial form of* $\mathbb{P}_U^1$ *over* U *and* C_L *a nontrivial form of* $\mathbb{P}_L^1$ *over* L.

The point is that, although every smooth conic plane curve in $\mathbb{P}_K^2$ is rational, there is no way of choosing a rational parametrization of each smooth conic consistently over a Zariski open subset of U.

Proof. To establish our claims, it will be enough to show two things: that C_U is a form of $\mathbb{P}^1$ over U, and that $C_L \not\cong \mathbb{P}_L^1$. To see the first, fix a line $M \subset \mathbb{P}_K^2$ and let

$$V = C_U \setminus (C_U \cap (U \times M)) \subset U \times \mathbb{P}_K^2.$$

We claim that the pullback

$$C_V = V \times_U C_U \subset U \times \mathbb{P}_K^2 \times \mathbb{P}_K^2$$

is isomorphic to the product $V \times M \cong \mathbb{P}_V^1$ as V-schemes. The point is, the family $C_V \to V$ has naturally a section—the diagonal—and the presence of a distinguished point on each fiber of $C_V \to V$ allows us parametrize that fiber by projecting from that point onto M. Explicitly, away from the diagonal in C_V we can define a map $\varphi : C_V \to V \times M$ by

$$\varphi : (C, p; q) \longmapsto (C, p; \overline{p,q} \cap M).$$

Exercise IV-85. Show that

(a) φ is a morphism of schemes;

(b) φ extends to a morphism on all of C_V (sending a point (C, p, p) on the diagonal in C_V to the point of intersection with M of the tangent line at p to the fiber of C_V over $(C, p) \in V$);

(c) $\varphi : C_V \to V \times M$ is an isomorphism.

For the second part, observe that if C_L were isomorphic to $\mathbb{P}^1_L$, it would have an L-rational point—that is, a pair of rational functions $F(a, b, c, d, e)$ and $G(a, b, c, d, e) \in K(a, b, c, d, e)$ such that

$$1 + aF^2 + bG^2 + cF + dG + eFG = 0.$$

We can assume (after possibly a change of variables on our original $\mathbb{P}^2_K$) that the denominators of F and G are not in the ideal (c, d, e); so that restricting to the locus $c = d = e = 0$ we get four polynomials $f(a, b)$, $g(a, b)$, $h(a, b)$ and $j(a, b)$ such that

$$1 + a \cdot \left(\frac{f(a, b)}{h(a, b)}\right)^2 + b \cdot \left(\frac{g(a, b)}{j(a, b)}\right)^2 = 0;$$

or, in other words,

$$h(a, b)^2 j(a, b)^2 + a \cdot f(a, b)^2 j(a, b)^2 + b \cdot g(a, b)^2 h(a, b)^2 = 0.$$

Now we simply ask what the degree of each term in this equation is, first as a polynomial in a and then in b, and thus derive a contradiction.   □

Exercise IV-86. Let K be any field. Show that any form X of $\mathbb{P}^1_K$ over Spec K is isomorphic to a plane conic $C \subset \mathbb{P}^2_K$. Conclude in particular that a form X of $\mathbb{P}^1_K$ over Spec K is isomorphic to $\mathbb{P}^1_K$ if any only if it has a point with residue field K.

Exercise IV-87. Using the preceding exercise, show that a form X of $\mathbb{P}^1_K$ over a field K is isomorphic to $\mathbb{P}^1_K$ if any only if it has a zero-dimensional subscheme $\Gamma \subset X$ of odd degree, that is, such that the coordinate ring of Γ has odd dimension as a vector space over K.

Exercise IV-88. Show that there are no nontrivial forms of $\mathbb{A}^1_K$ over Spec K, that is, any form of $\mathbb{A}^1_K$ over Spec K is isomorphic to $\mathbb{A}^1_K$.

Exercise IV-89. Show by example that the conclusions of the three preceding exercises are all false if we do not specify $S = \operatorname{Spec} K$, that is, if we consider forms X of $\mathbb{P}^1_S$ and $\mathbb{A}^1_S$ over a general scheme S. (To find counterexamples, it is enough to take $S = \mathbb{P}^1_K$ a projective line over a field and consider the blow-up of the affine plane $\mathbb{A}^2_K$ at the origin.)

Another generalization of the irrationality of the universal conic — this time asserting that the universal rational normal curve of degree d is rational if and only if d is odd — will be discussed in Exercise VI-38.

For another example of a form of projective space arising in a geometric context, see Exercise IV-71 above.

V

Local Constructions

V.1 Images

In this section we will be concerned with a basic notion: the *image* of a morphism in the category of schemes. As we will see, there are two fundamental properties that we would like the notion of image to have, the *push-pull property* and *invariance under base change*; but the two are incompatible. We will give accordingly two definitions, one straightforward and one less so, each of which is useful in certain situations.

V.1.1 The Image of a Morphism of Schemes

Suppose $\varphi : X \to Y$ is a morphism of schemes. The *set-theoretic image* of φ is defined in the obvious way: it is the subset of Y consisting of those points $y \in Y$ such that there is a point $x \in X$ with $\varphi(x) = y$. The image may or may not be a closed subset: for example, if X is the scheme defined by the ideal $(xy - 1) \subset K[x,y]$ in $\mathbb{A}^2_K = \operatorname{Spec} K[x,y]$, then the image of the projection of X to the affine line $\mathbb{A}^1_K = \operatorname{Spec} K[x]$ is the complement of the origin, an open set. In a certain sense this is because we "forgot" some of the points of the source scheme: if we extend this morphism to a morphism from the closure of X in $\mathbb{A}^1_K \times_K \mathbb{P}^1_K$, the image becomes the whole projective line.

These two examples turn out to be typical. The situation is summed up in the following theorem. To state it, recall that a subset V of a topological space is *constructible* if it is a finite union of locally closed subsets V_i.

Theorem V-1. *If the morphism of Noetherian schemes $\varphi : X \to Y$ is of finite type, the set-theoretic image of φ is constructible in Y. If the morphism φ is projective, the set-theoretic image of φ is closed.*

The first of these statements is due to Chevalley. Note the necessity of the hypothesis that φ is of finite type: the morphism

$$\operatorname{Spec} K[x_1,\dots,x_n]_{(x_1,\dots,x_n)} \longrightarrow \operatorname{Spec} K[x_1,\dots,x_n],$$

coming from the inclusion of the polynomial ring in its localization, does not have constructible image for any n. The second statement, which is quite old, is generally called the Main Theorem of Elimination Theory. The definition of properness, which is a strengthening of the conclusion of the theorem, was essentially made to express this property (see Section III.1). The proof of Theorem V-1 is exactly the same as in the classical case of varieties; see for example Harris [1995] or Hartshorne [1977]. We will not repeat it here.

More novel is the fact that the closure of the image has a natural scheme structure, and this is the fact that we shall explore in this section.

Suppose now we wish to define the image of a morphism $\varphi : X \to Y$ as a scheme. What we would like ideally—though, as we shall see, this is impossible—is to take $\varphi(X)$ to be the smallest subscheme of Y whose inverse image is all of X. In other words, we would characterize $\varphi(X)$ by the *push-pull* property: for every subscheme $Z \subset Y$,

$$Z \supset \varphi(X) \iff \varphi^{-1}(Z) = X.$$

As it happens, though, no such subscheme of Y need exist, as shown by the example of the morphism

$$\varphi : X = \mathbb{A}^2_K \longrightarrow Y = \mathbb{A}^2_K$$

defined by the ring homomorphism $\varphi^\# : K[x,y] \to K[s,t]$ taking x to s and y to st.

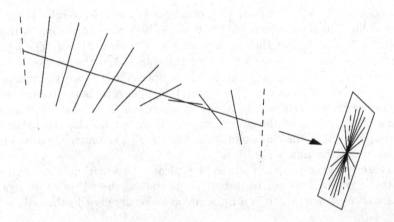

For any $\lambda \in K$, let Y_λ be the complement in Y of the point $(0, \lambda)$. For $\lambda \neq 0$, the inverse image of Y_λ is all of X. Thus, if there were a subscheme $\varphi(X) \subset Y$ satisfying the push-pull property above with $Z = Y$, the support of $\varphi(X)$ would be contained in the set $I = \bigcap_{\lambda \neq 0} Y_\lambda$, which is the union of the point $(0,0)$ and the complement of the x-axis. On the other hand, from $\varphi^{-1}\varphi(X) = X$ it would follow that the support of $\varphi(X)$ contained I, and hence that $\operatorname{supp}\varphi(X) = I$. But I is not a locally closed subset of Y, and so is not the support of any subscheme of Y.

We are, however, close to what we want: there does exist a smallest *closed* subscheme of Y whose inverse image is all of X, and this subscheme does satisfy the push-pull property with respect to closed subschemes $Z \subset Y$.

Definition V-2. If $\varphi : X \to Y$ is a morphism of finite type, then the *scheme-theoretic image*, written $\bar\varphi(X)$, is the closed subscheme of Y whose sheaf of ideals is the sheaf of regular functions on open subsets of Y that pull back to 0 under $\varphi^{\#}$, that is,

$$\bar\varphi(X) = V\big(\operatorname{Ker}(\varphi^{\#} : \mathscr{O}_Y \to \varphi_*\mathscr{O}_X)\big) \subset Y.$$

We say that φ is *dominant* if $\bar\varphi(X) = Y$, or equivalently if the pullback map $\varphi^{\#}$ is a monomorphism.

This condition that a morphism be dominant is *not* just a condition on the underlying map of topological spaces: for example, the inclusion $\varphi : \operatorname{Spec} K \hookrightarrow \operatorname{Spec} K[\epsilon]/(\epsilon^2)$ is a surjection on underlying sets, but the image is a proper closed subscheme, the pullback map $\varphi^{\#}$ is not injective, and the map is not dominant.

Proposition V-3. *If $\varphi : X \to Y$ is a morphism of schemes, the closure of the set-theoretic image is $\bar\varphi(x)_{\mathrm{red}}$.*

Proof. We may reduce at once to the affine case, and prove that if $\varphi^{\#} : B \to A$ is a ring homomorphism then the intersection J of all the primes Q of B that may be written in the form $Q = (\varphi^{\#})^{-1}(P)$ for some prime P of A is the radical of $\operatorname{Ker}(\varphi^{\#})$ (see Section I.2.1). Let $I \subset A$ be the nilradical of A. We have

$$J = \bigcap_{P \subset A \text{ prime}} (\varphi^{\#})^{-1}(P) = (\varphi^{\#})^{-1}(I).$$

Thus if $f \in J$ then $\varphi^{\#}(f) \in I$ is nilpotent, and $f \in \operatorname{rad}(\operatorname{Ker}\varphi^{\#})$. The opposite inequality is immediate. $\qquad\square$

It is sometimes convenient to work in an apparently more general case: If $\varphi : X \to Y$ is any morphism and $X' \subset X$ a closed subscheme, then we define the *scheme-theoretic image* $\bar\varphi(X')$ of X' to be the subscheme defined by the sheaf of ideals I with

$$I(U) = \{f \in \mathscr{O}_Y(U) \mid \varphi^{\#}(f) \in I_{X'}(\varphi^{-1}(U)\}.$$

This "generalization" actually describes a special case, since $\bar{\varphi}(X')$ is the scheme-theoretic image of the composite morphism $X' \to X \to Y$.

If the two schemes X and Y are affine and the map φ is dominant, then the ideal describing the image of a subscheme $X' \subset X$ is obtained in a particularly simple way: we may think of $A(Y)$ as a subring of $A(X)$, and $\bar{\varphi}(X')$ is the subscheme of Y defined by $I_{X'} \cap A(Y)$. We shall be mostly concerned with this case, since it already contains all the new phenomena.

As an example consider the linear projection on the first coordinate

$$\varphi : X = \mathbb{A}^2_K = \operatorname{Spec} K[x,y] \longrightarrow Y = \mathbb{A}^1_K = \operatorname{Spec} K[x]$$

Let X' and X'' be the (abstractly isomorphic) zero-dimensional subschemes given by the ideals $I' = (x, y^2)$ and $I'' = (x^2, y)$. The image of X' is the reduced scheme defined by $(x, y^2) \cap K[x] = (x)$, while the image of X'' is given by $(x^2, y) \cap K[x] = (x^2)$. More generally, the image of the "fat point" $V(x^2, xy, y^2) \subset \mathbb{A}^2_K$ is the double point $V(x^2) \subset \mathbb{A}^1_K$; and the image of the double point $V(x^2, xy, y^2, \alpha x + \beta y) \subset \mathbb{A}^2_K$ is the double point $V(x^2) \subset \mathbb{A}^1_K$ for $\beta \neq 0$, but the reduced point $V(x)$ for $[\alpha, \beta] = [1, 0]$. We see from this example that the scheme structure of the image depends on the relation of the subscheme to the fibers of the morphism φ.

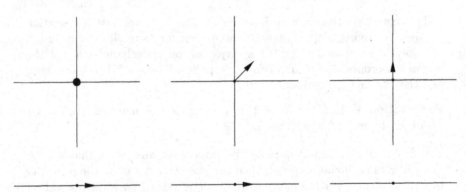

Exercise V-4. Consider now a "family" of such projections of double points: take $B = \operatorname{Spec} K[t] = \mathbb{A}^1_K$ and consider the morphism

$$\varphi : X = \mathbb{A}^2_B = \operatorname{Spec} K[x,y,t] \longrightarrow Y = \mathbb{A}^1_B = \operatorname{Spec} K[x,t].$$

Let $X' \subset X$ be the double line

$$X' = V(x^2, xy, y^2, x + ty) \subset \mathbb{A}^2_B$$

as described in Section II.3.5. Show that the scheme-theoretic image $\bar{\varphi}(X)$ of X' is the double line $V(x^2) \subset Y$, even though the fiber of X' over the origin $(t) \in B$ has image the reduced point $V(x)$ in the fiber of Y over (t). This is in fact the source of some interesting complications, as we will explain in the following section.

Exercise V-5. Show that if $\varphi : X \to Y$ is a morphism and $X' \subset X$ is a reduced subscheme, then $\bar{\varphi}(X')$ is reduced. (Hint: Reduce to the statement that the preimage of a radical ideal under a ring homomomorphism is again radical.)

V.1.2   Universal Formulas

Is there a "formula" for the closure of the image of a map, and if so, what is it? This question, in a somewhat different language, occupied a large number of mathematicians in the past, and the theory is correspondingly rich. In many special cases beautiful and useful formulas were found for the set-theoretic image; the equations they give are usually referred to as *resultants*, a notion we will discuss below. The scheme-theoretic image is significantly more complicated: with a reasonable interpretation of the basic question there can be no formula for the scheme theoretic image! We shall next explain this fact and the opportunities to which it gives rise. We shall keep the discussion informal, but it can be formalized using the notion of family of schemes given in Section II.3.4.

Consider a consequence that a universal formula would have: it would be preserved by base change, or, put more informally, it would specialize on substitution of variables. Some examples will make this clear.

Example V-6. (See figure on the next page.) To begin with, let K be an algebraically closed field, set $B = \operatorname{Spec} K[t] = \mathbb{A}^1_K$, and consider the projection map

$$\varphi : \mathbb{A}^2_B = \operatorname{Spec} K[t, x, y] \to \mathbb{A}^1_B = \operatorname{Spec} K[t, x]$$

defined by the inclusion $K[t, x] \subset K[t, x, y]$. We regard this map informally as a (trivial) family of projection maps $\mathbb{A}^2_K = \operatorname{Spec} K[x, y] \to \mathbb{A}^1_K = \operatorname{Spec} K[x]$ parametrized by $t \in K$. Consider first the closed subscheme $X \subset \mathbb{A}^2_B = \mathbb{A}^3_K$ given as the union of the two disjoint lines $V(y, x)$ and $V(y-1, x+t)$; that is,

$$X = V(y^2 - y, \, yx + yt, \, yx - x, \, x^2 + tx).$$

We will think of X as a family of pairs of points in the (x, y)-plane, parametrized by t, and each with its projection onto the x-axis: for each scalar $a \in K$ we let $X_a \subset \mathbb{A}^2_K$ be the fiber of X over the point $(t - a) \in B$, $Y_a \cong \mathbb{A}^1_K$ the fiber of $\mathbb{A}^1_B$ over $(t - a)$ and $\varphi_a : X_a \to Y_a$ the restriction of φ to X_a.

For each scalar value $a \neq 0 \in K$ the set-theoretic image $\varphi(X_a)$ is the union of the two points $x = 0$ and $x = -a$; for $a = 0$ it is simply the origin $x = 0$. The scheme-theoretic image $\bar{\varphi}(X_a)$ is the subscheme of $\operatorname{Spec} K[x]$ defined by the ideal $I_a := K[x] \cap (y^2 - y, \, yx + ay, \, yx - x, \, x^2 + ax)$. Since

$$K[x, y]/(y^2 - y, \, yx + ay, \, yx - x, \, x^2 + ax) = K[x, y]/(y^2 - y, \, x + ay),$$

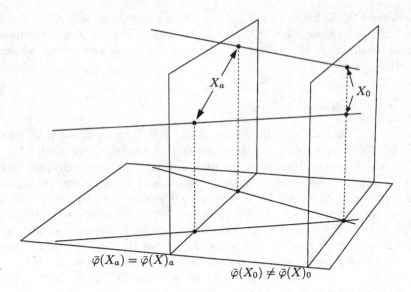

$$\bar{\varphi}(X_a) = \bar{\varphi}(X)_a$$

$$\bar{\varphi}(X_0) \neq \bar{\varphi}(X)_0$$

we see that

$$I_a = \operatorname{Ker}\big(K[x] \to K[x,y]/(y^2 - y, x + ay)\big) = \begin{cases} (x^2 + ax) & \text{if } a \neq 0, \\ (x) & \text{if } a = 0. \end{cases}$$

Thus the scheme-theoretic image $\bar{\varphi}(X_a)$ is the union of two reduced points for $a \neq 0$ and a single reduced point for $a = 0$.

The important thing about this example is that the scheme-theoretic images $\bar{\varphi}(X_a)$ do not fit into *any* family of schemes! That is, there is no polynomial $f(t,x)$ "giving a formula for the scheme-theoretic image" in the sense that for each $a \in K$ the scheme-theoretic image $\bar{\varphi}(X_a)$ is defined by the ideal $f(a,x) = 0$. Indeed, for $a \neq 0$ the scheme $\bar{\varphi}(X_a)$ is defined by the ideal $(x^2 + ax)$, so we would have to have $f(t,x) = g(t,x)(x^2 + tx)$ for some polynomial $g(t,x)$. Since $g(a,x) \neq 0$ for all x when $a \neq 0$, we must have $g(t,x) = g(t)$, a polynomial of one variable vanishing at most when $t = 0$. If now $g(0) = 0$ then $f(0,a)$ would describe the whole line, while if $g(0) \neq 0$ then $f(0,a)$ would describe a double point, and neither of these options is the scheme-theoretic image $\bar{\varphi}(X_a)$.

Perhaps the best we can do in this example is to take the scheme-theoretic image of the whole family, $\bar{\varphi}(X) \subset \operatorname{Spec} K[x,t]$. This image is defined by the ideal $K[t,x] \cap (y^2 - y, \, yx + yt, \, yx - x, \, x^2 + tx)$. To compute this intersection, one shows that the localization map

$$K[t,x,y]/(y^2 - y, \, yx + yt, \, yx - x, \, x^2 + tx)$$

$$\to K[t,t^{-1},x,y]/(y^2 - y, \, yx + yt, \, yx - x, \, x^2 + tx)$$

$$= K[t,t^{-1},x]/(x^2 + tx)$$

is a monomorphism, and it follows at once that the intersection is $(x^2 + tx)$.

We thus see that the scheme-theoretic image $\bar{\varphi}(X_0)$ of the fiber of X over the origin $(t) \in B$ is properly contained in the fiber $\bar{\varphi}(X)_0$ of the scheme-theoretic image $\bar{\varphi}(X)$ over the origin. In particular, the fiber of any closed subscheme of $\mathbb{A}_B^1$ containing $\bar{\varphi}(X_a)$ for $a \neq 0$ will properly contain $\bar{\varphi}(X_0)$, so that the scheme-theoretic images $\bar{\varphi}(X_a)$ cannot form a family in any sense. The equation $x^2 + tx$ of the scheme-theoretic image $\bar{\varphi}(X)$ gives the "correct" defining ideal for $\bar{\varphi}(X_a)$ when we specialize t to any $a \neq 0$, while for $a = 0$ it gives an ideal defining a scheme a little larger than $\bar{\varphi}(X_0)$. This choice of "approximation" for a defining equation of scheme-theoretic images is a resultant, in a sense that we shall describe.

Example V-7. (See figure below.) To see an example of the same phenomenon involving nonreduced schemes, let K, $B = \operatorname{Spec} K[t] = \mathbb{A}_K^1$, and

$$\varphi : \mathbb{A}_B^2 = \operatorname{Spec} K[t, x, y] \to \mathbb{A}_B^1 = \operatorname{Spec} K[t, x]$$

be as before, and consider the closed subscheme $X \subset \mathbb{A}_B^2 = \mathbb{A}_K^3$ given by

$$X = V(x^2, xy, y^2) \cap V(ty + x) = V(ty + x, y^2).$$

Viewed as a subscheme of $\mathbb{A}_B^2 = \mathbb{A}_K^3$, it is the intersection of the first order infinitesimal neighborhood of the t-axis with a helical surface winding around the axis; it is a double line supported on the t-axis. As before, we will think of X as a family of double points in the plane, each with its projection onto the y-axis: for each scalar $a \in K$ we let $X_a \subset A_K^2$ be the fiber of X over the point $(t - a) \in B$, $Y_a \cong \mathbb{A}_K^1$ the fiber of $\mathbb{A}_B^1$ over $(t - a)$ and $\varphi_a : X_a \to Y_a$ the restriction of φ to X_a.

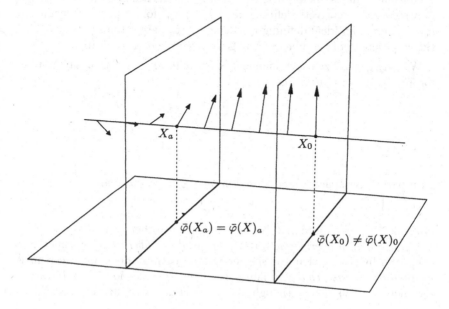

For each scalar value $t = a \in K$ the set-theoretic image $\varphi(X_a)$ is the point $x = 0$. The scheme-theoretic image $\bar{\varphi}(X_a)$ is defined by $I_a := K[x] \cap (ay + x, y^2)$. Since

$$K[x,y]/(ay + x, y^2) = \begin{cases} K[x]/(x^2) & \text{if } a \neq 0, \\ K[x,y]/(x,y^2) & \text{if } a = 0, \end{cases}$$

we see that

$$I_a = \text{Ker}\big(K[x] \rightarrow K[x,y]/(ay+x, y^2)\big) = \begin{cases} (x^2) & \text{if } a \neq 0, \\ (x) & \text{if } a = 0. \end{cases}$$

Thus the scheme-theoretic image $\bar{\varphi}(X_a)$ is a double point for $a \neq 0$ and a simple point for $a = 0$.

Just as in the previous example, we see that the scheme-theoretic images $\bar{\varphi}(X_a)$ cannot be the fibers of any closed subscheme of $\mathbb{A}^1_B$ over B. In particular, the scheme-theoretic image of the whole family, $\bar{\varphi}(X) \subset \text{Spec}\, K[x,t]$ is defined by the ideal $K[t,x] \cap (ty+x, y^2)$, which is readily seen to be simply (x^2) (as before, this follows from the fact that the localization map

$$K[t,x,y]/(ty+x, y^2) \rightarrow K[t,t^{-1},x,y]/(ty+x, y^2) = K[t,t^{-1},x]/(x^2)$$

is a monomorphism). We thus see that the scheme-theoretic image $\bar{\varphi}(X_0)$ of the fiber of X over the origin $(t) \in B$ is properly contained in the fiber $\bar{\varphi}(X)_0$ of the scheme-theoretic image $\bar{\varphi}(X)$ over the origin, so that the fiber of any closed subscheme of $\mathbb{A}^1_B$ containing $\bar{\varphi}(X_a)$ for $a \neq 0$ will properly contain $\bar{\varphi}(X_0)$. Once again it follows that the scheme-theoretic images $\bar{\varphi}(X_a)$ cannot form a family of schemes.

Note also that as before the equation x^2 of the scheme-theoretic image $\bar{\varphi}(X)$ gives the "correct" defining ideal for $\bar{\varphi}(X_a)$ for every $a \neq 0$, while for $a = 0$ it gives an ideal defining a scheme a little larger than $\bar{\varphi}(X_0)$; again, the equation x^2 of the fiber $\bar{\varphi}(X)_a$ is an example of a resultant.

We may generalize the phenomena that we have just seen by saying that for any family of morphisms

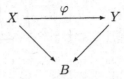

with parameter space B and closed point $b \in B$, we have an inclusion

$$\bar{\varphi}(X_b) \subset \bar{\varphi}(X)_b,$$

but this inclusion need not be an equality: in other words, *the fiber of the image may properly contain the image of the fiber*. This is sometimes expressed by saying that the scheme-theoretic image does not necessarily *commute with base change*. (The base change in question is pullback via the inclusion $\{b\} \hookrightarrow B$, though the same issues arise for any morphism

$B' \to B$.) Note by contrast that the set-theoretic image does commute with base change; this is just set-theory. A still more general form of this statement is expressed by the following result:

Proposition V-8. *If*

$$
\begin{array}{ccc}
X' = X \times_Y Y' & \xrightarrow{\ \psi'\ } & X \\
{\scriptstyle \varphi'} \downarrow & & \downarrow {\scriptstyle \varphi} \\
Y' & \xrightarrow[\ \psi\ ]{} & Y
\end{array}
$$

is a pull-back diagram of morphisms of schemes, then $\bar{\varphi}'(X') \subset \psi^{-1}\bar{\varphi}(X)$. If the morphism φ is finite, these two schemes have the same underlying set, the closure $\overline{\varphi'(X')}$ of the set theoretic image. In particular, when φ is finite the set-theoretic image is closed. If ψ is flat, then $\bar{\varphi}'(X') = \psi^{-1}\bar{\varphi}(X)$.

Note that, in the previous case, $Y' = Y_b$ was the fiber of a morphism $Y \to B$, but we need not assume that in general.

Proof. We reduce at once to the affine case, and consider this diagram of coordinate rings:

$$
\begin{array}{ccc}
A' = A \otimes_B B' & \xleftarrow{\ \psi'^{\#}\ } & A \\
{\scriptstyle \varphi'^{\#}} \uparrow & & \uparrow {\scriptstyle \varphi^{\#}} \\
B' & \xleftarrow[\ \psi^{\#}\ ]{} & B
\end{array}
$$

In this setting the first assertion becomes the inequality

$$
\mathrm{Ker}(\varphi'^{\#}) \supset \psi^{\#}(\mathrm{Ker}(\varphi^{\#}))B',
$$

which is immediate from the commutativity of the diagram.

The second statement of the proposition becomes in this affine case the assertion that, if A is a finite B-module, then the radicals of the ideals $\mathrm{Ker}(\varphi'^{\#})$ and $\psi^{\#}(\mathrm{Ker}(\varphi^{\#}))B'$ are equal. Given the inequality above, and the fact that the radical of an ideal is the intersection of the primes containing it, we must show that if P' is a prime ideal of B' containing $\psi^{\#}(\mathrm{Ker}(\varphi^{\#}))B'$ then $P' \supset \mathrm{Ker}(\varphi'^{\#})$.

The preimage $P = (\psi^{\#})^{-1}P'$ of P' in B contains $\mathrm{Ker}(\varphi^{\#})$, so we have $A \otimes_B B_P \neq 0$. Since A is a finitely generated B-module, Nakayama's Lemma Eisenbud [1995, Section 4.1] shows that $A \otimes_B B_P/P_P \neq 0$. Since the unit element of B maps to the unit element of A and B_P/P_P is a field, we see that the induced map $B_P/P_P \to A \otimes_B B_P/P_P$ is a monomorphism. Now consider the diagram

$$(A \otimes_B B') \otimes B'_{P'}/P'_{P'} = A \otimes_B B_P/P_P \otimes_{B_P/P_P} B'_{P'}/P'_{P'} \longleftarrow A \otimes_B B_P/P_P$$

$$B'_{P'}/P'_{P'} \longleftarrow B_P/P_P$$

obtained by localizing and factoring out P' and P. Sincè B_P/P_P is a field and the right-hand vertical arrow is a monomorphism, the left-hand vertical arrow is also a monomorphism (every module is flat over a field!). It follows that the kernel of $\varphi'^{\#}$ is contained in P' as required.

The closedness of the set-theoretic image of φ when φ is finite follows if we take Y' to be a point of Y in the closure of the set-theoretic image (and thus in the scheme- theoretic image) and pull back via the inclusion morphism of this point.

For the last statement, suppose that ψ is flat, that is, B' is a flat B-module. Tensoring B' with the exact sequence

$$0 \longrightarrow \mathrm{Ker}(\varphi^{\#}) \longrightarrow B \longrightarrow \mathrm{Im}(\varphi^{\#}) \longrightarrow 0$$

and the inclusion $\mathrm{Im}(\varphi^{\#}) \hookrightarrow A$ gives back an exact sequence and an inclusion; and we see in particular that $(\mathrm{Ker}(\varphi^{\#}))B'$, which is the image of

$$B' \otimes_B \mathrm{Ker}(\varphi) \to B' \otimes_B B = B'$$

is the kernel of φ', as required.   □

The second ("equality") statement of the proposition fails in general, for example in the fiber product diagram of affine schemes corresponding to the diagram of rings

$$0 \longleftarrow K(t)$$

$$K \xleftarrow{\ 0 \leftarrow\!\mid t\ } K[t]$$

However, the equality statement does hold whenever the map φ is projective — the proof, which would take us too far afield, may be reduced again to Nakayama's Lemma using the deep fact that, when φ is projective, $\varphi_*(\mathcal{O}_X)$ is a finite $\mathcal{O}_Y$-module; see Hartshorne [1977], Corollary II.5.20.

Despite the nonexistence of a universal formula for the scheme-theoretic image, there are, as we have mentioned, many formulas giving equations that define it set-theoretically, and each of them gives a scheme containing the scheme-theoretic image, as one sees from the previous proposition. One way to produce such a formula is to choose a universal model, a family of morphisms parametrized by a scheme B, say

$$X_0 \xrightarrow{\ \varphi_0\ } Y_0 \xrightarrow{\ \pi\ } B$$

such that the morphisms we are interested in occur as pullbacks, say

$$X = (\pi\varphi_0)^{-1}(b) \xrightarrow{\varphi} Y = \pi^{-1}(b),$$

with morphism φ the restriction of φ_0. We then take the actual scheme-theoretic image of φ_0 and restrict it to Y. We call the defining equations of the scheme-theoretic image of φ_0 *resultants*. By Proposition V-8, these equations define a scheme in Y containing the scheme-theoretic image, and in some cases, such as when φ_0 is finite, they define a scheme that has the closure of the set-theoretic image as underlying set.

It turns out that resultants are often conveniently described as determinants. Before coming to this, we explain a general context in which these determinants arise.

V.1.3   Fitting Ideals and Fitting Images

There is, at least in a restricted context, an alternative notion of the image of a morphism $\varphi : X \to Y$ of schemes, which we will call the *Fitting image* and denote by $\varphi_{\mathrm{Fitt}}(X)$. This has the virtue that it does commute with base change, but the defect that (since, as we will see below, the Fitting image $\varphi_{\mathrm{Fitt}}(X)$ may properly contain the scheme-theoretic image $\bar{\varphi}(X)$) it does not have the push-pull property.

Fitting Ideals. To set this up, we need the sheaf-theoretic version of Fitting's lemma. Let X be a scheme, let $\mathscr{F}$ be a coherent sheaf on X, and let

$$\mathscr{E}_1 \xrightarrow{\Psi} \mathscr{E}_0 \longrightarrow \mathscr{F} \longrightarrow 0$$

be an exact sequence with $\mathscr{E}_0 \cong \mathscr{O}_X^n$ and $\mathscr{E}_1 \cong \mathscr{O}_X^m$ free sheaves. (Here we allow m to be infinite.) For any integer l, we define *the ideal of $l \times l$ minors* of Ψ to be the sheaf $I_l(\Psi) \subset \mathscr{O}_X$ of ideals generated by the $l \times l$ minors of a matrix representative of Ψ. This is independent of the choice of isomorphisms $\mathscr{E}_0 \cong \mathscr{O}_X^n$ and $\mathscr{E}_1 \cong \mathscr{O}_X^m$, and can also be defined when $\mathscr{E}_1$ and $\mathscr{E}_0$ are merely quasicoherent, since it can also be described as the image of the natural map

$$\bigwedge^l \mathscr{E}_1 \otimes \bigwedge^l \mathscr{E}_0^* \longrightarrow \mathscr{O}_X$$

induced by Ψ. The key fact about the ideals $I_l(\Psi)$ is Fitting's Lemma. See Eisenbud [1995, page 497] for a proof.

Lemma V-9 (Fitting's Lemma). *If*

$$\mathscr{E} \xrightarrow{\Psi} \mathscr{O}_X^n \longrightarrow \mathscr{F} \longrightarrow 0$$

and

$$\mathscr{E}' \xrightarrow{\Psi'} \mathscr{O}_X^{n'} \longrightarrow \mathscr{F} \longrightarrow 0$$

are exact sequences, then for each $k \in \mathbb{Z}$

$$I_{n-k}(\psi) = I_{n'-k}(\psi').$$

In view of the lemma, we may make the following definition:

Definition V-10. The k-th *Fitting ideal* $\mathrm{Fitt}_k \, \mathscr{F}$ of the sheaf $\mathscr{F}$ is the ideal given locally as $I_{n-k}(\Psi)$ for any presentation of $\mathscr{F}$ as above. The zero locus $V(\mathrm{Fitt}_k \, \mathscr{F}) \subset X$ will be called the k-th *Fitting scheme* of $\mathscr{F}$.

It is instructive to compare the geometry of the zeroth Fitting ideal with the more naive notion of the *support* of a sheaf $\mathscr{F}$. Briefly, for any coherent sheaf $\mathscr{F}$ on a scheme X we define the *annihilator* $\mathrm{ann}\,\mathscr{F} \subset \mathscr{O}_X$ by taking $(\mathrm{ann}\,\mathscr{F})(U) \subset \mathscr{O}_X(U)$ to be the annihilator of $\mathscr{F}(U)$ as an $\mathscr{O}_X(U)$-module for each open set $U \subset X$. We define the *support* of $\mathscr{F}$, denoted $\mathrm{supp}(\mathscr{F})$, to be the zero locus $V(\mathrm{ann}\,\mathscr{F}) \subset X$ of the annihilator as a subscheme of X.

(Note that "support" is used in two different senses: the support of a scheme is its underlying set or topological space, while the support of a sheaf is a scheme. In those cases where we wish to refer to the underlying set of the support of a sheaf $\mathscr{F}$ — that is, $\mathrm{supp}(\mathrm{supp}\,\mathscr{F})$ — we will call it simply the "set-theoretic support".)

For any coherent sheaf $\mathscr{F}$ on a scheme X we have

$$(\mathrm{Fitt}_0 \, \mathscr{F}) \cdot \mathscr{F} = 0,$$

so that $\mathrm{Fitt}_0 \, \mathscr{F} \subset \mathrm{ann}\,\mathscr{F}$. In the other direction, if $\mathscr{F}$ admits a presentation

$$\mathscr{O}_X^n \longrightarrow \mathscr{F} \longrightarrow 0$$

then $\mathrm{Fitt}_0 \, \mathscr{F} \supset (\mathrm{ann}\,\mathscr{F})^n$. See for example Eisenbud [1995, Proposition 20.7] for both these inequalities. It follows from the first of these properties that

$$V(\mathrm{Fitt}_0 \, \mathscr{F}) \supset \mathrm{supp}\,\mathscr{F},$$

that is, the zeroth Fitting scheme contains the support of $\mathscr{F}$; and from the second that the underlying sets of the subschemes $V(\mathrm{Fitt}_0 \, \mathscr{F})$ and $\mathrm{supp}\,\mathscr{F}$ are equal, that is,

$$\left| V(\mathrm{Fitt}_0 \, \mathscr{F}) \right| = \left| \mathrm{supp}\,\mathscr{F} \right|.$$

The difference lies in the scheme structure: the zeroth Fitting ideal may be properly contained in the annihilator of $\mathscr{F}$, and the zeroth Fitting scheme correspondingly may properly contain the support of $\mathscr{F}$.

For example, consider the scheme $X = \mathbb{A}_K^1 = \mathrm{Spec}\,K[x]$, with subschemes $Y = V(x)$ and $Z = V(x^2)$. The sheaves

$$\mathscr{F} = \mathscr{O}_Z \quad \text{and} \quad \mathscr{G} = \mathscr{O}_Y \oplus \mathscr{O}_Y$$

each have length 2; but the support of $\mathscr{F}$ is Z, while the support of $\mathscr{G}$ is the smaller scheme Y. By contrast, we see that the zeroth Fitting schemes are equal:

$$V(\mathrm{Fitt}_0 \, \mathscr{F}) = V(\mathrm{Fitt}_0 \, \mathscr{G}) = Z.$$

Exercise V-11. (a) If X is any scheme, $Y \subset X$ a subscheme, the zeroth Fitting ideal $\mathrm{Fitt}_0\, \mathscr{O}_Y$ of the structure sheaf of Y is simply the ideal sheaf $\mathscr{I}_Y$ of Y, and the 0th Fitting scheme correspondingly is Y itself.

(b) Show that the zeroth Fitting ideal of a direct sum is the product of the zeroth Fitting ideals of the summands, that is, for any pair of coherent sheaves $\mathscr{F}$ and $\mathscr{G}$ on X,

$$\mathrm{Fitt}_0(\mathscr{F} \oplus \mathscr{G}) = \mathrm{Fitt}_0\, \mathscr{F} \cdot \mathrm{Fitt}_0\, \mathscr{G}.$$

Show by example that this is not true if $\mathscr{F} \oplus \mathscr{G}$ is replaced by an arbitrary extension of $\mathscr{F}$ by $\mathscr{G}$, that is, a sheaf $\mathscr{H}$ such that there exists an exact sequence

$$0 \longrightarrow \mathscr{F} \longrightarrow \mathscr{H} \longrightarrow \mathscr{G} \longrightarrow 0.$$

(c) Deduce from the first two parts of this exercise that if X is a regular one-dimensional scheme (for example, $X = \mathbb{A}^1_K$ or $X = \mathrm{Spec}\,\mathbb{Z}$) and $\mathscr{F}$ is any sheaf whose set-theoretic support is a closed point $p \in X$, then

$$\mathrm{Fitt}_0\, \mathscr{F} = \mathfrak{m}_p^l,$$

where l is the length of $\mathscr{F}$.

Fitting Images. Suppose now that $\varphi : X \to Y$ is a finite morphism of schemes. The direct image $\varphi_* \mathscr{O}_X$ is then a coherent sheaf on Y, and by our definition the scheme-theoretic image $\bar{\varphi}(X)$ is the support $\mathrm{supp}(\varphi_*(\mathscr{O}_X))$ of this sheaf. Taking the zeroth Fitting scheme instead of the support gives us the promised alternative definition of image that commutes with base change:

Definition V-12. For any finite morphism $\varphi : X \to Y$ of schemes, the *Fitting image* of φ is the zeroth Fitting scheme of the direct image $\varphi_*(\mathscr{O}_X)$, that is,

$$\varphi_{\mathrm{Fitt}}(X) = V(\mathrm{Fitt}_0\, \varphi_*(\mathscr{O}_X)).$$

For any closed subscheme $Z \subset X$, we will likewise define the Fitting image $\varphi_{\mathrm{Fitt}}(Z)$ to be the Fitting image of φ restricted to Z, that is, the zeroth Fitting scheme of the direct image $\varphi_*(\mathscr{O}_Z)$

Exercise V-13. Consider Examples V-6 and V-7. In each case, show that the Fitting image of the fiber X_0 of X over the origin is a double point, not a reduced point, so that we have

$$\varphi_{\mathrm{Fitt}}(X_0) = \varphi_{\mathrm{Fitt}}(X)_0.$$

Exercise V-14. Consider the projection and inclusion morphisms

$$X = \mathrm{Spec}\, K[\epsilon]/\epsilon^2 \xrightarrow{\varphi} \mathrm{Spec}\, K \xrightarrow{\iota} \mathrm{Spec}\, K[t].$$

Show that

$$\iota_{\mathrm{Fitt}}(\varphi_{\mathrm{Fitt}}(X)) \subsetneqq (\iota \circ \varphi)_{\mathrm{Fitt}}(X).$$

V.2   Resultants

V.2.1   Definition of the Resultant

The oldest — and still one of the most important — applications of the ideas we have introduced in the last two subsections is the *resultant of two polynomials in one variable*. To define this, say we are given two polynomials

$$f(x) = a_0 x^m + \cdots + a_m,$$
$$g(x) = b_0 x^n + \cdots + b_n,$$

with coefficients a_i, b_i in a field K. The goal was to describe the condition on the coefficients a_i, b_i for the two polynomials to have a common factor; that is, a common root in the algebraic closure of K. In applications it is also natural to look at families of polynomials. That is, we may take the coefficients a_i, b_i to be regular functions on a base scheme B, so that we think of f and g as "families of polynomials in one variable parametrized by B", and we wish to describe the locus in B over which f and g have a common factor. More precisely, we want to describe the image in B — in whichever sense! — of the scheme $V(f, g) \subset \mathbb{A}_B^1$.

There are, roughly speaking, four ways to interpret this problem: we could ask formulas, in terms of the a_i's and b_i's, for functions generating the ideal of

(1) the *reduced image* of $V(f, g)$ — that is, the reduced scheme associated to the scheme-theoretic image;

(2) the scheme-theoretic image of $V(f, g)$;

(3) the Fitting image of $V(f, g)$; or

(4) the pullback of the image from a suitable universal family.

As shown by examples such as V-6 and V-7 above, no such formula can exist for the images in the first two senses. We shall begin by describing the classical approach, and ultimately show that (for the correct choice of "universal family") it coincides with the third and fourth options, and commutes with base change.

To carry out the classical approach, we begin by choosing a universal family. We will work not over a field but over an arbitrary ring S. Let A be the polynomial ring

$$A = S[a_0, \ldots, a_m, b_0 \ldots, b_n].$$

With f and g defined as above, we set

$$X := V(f, g) \subset \mathbb{A}_A^1$$

and let

$$\varphi : X \subset \mathbb{A}_A^1 = \operatorname{Spec} A \times_S \mathbb{A}_S^1 \longrightarrow \operatorname{Spec} A =: Y.$$

Since X is defined by two polynomials, we might expect it to have codimension 2 in $\mathbb{A}_A^1$. In this case, assuming that the map is "generically finite" (that is, finite over some open set) its image would have codimension 1 in Spec A. In the classical situation, where S is field, A has unique factorization, so the closure of the image would be described by one equation, called the *resultant* of f and g.

It turns out that the conclusion of this suggestive argument is correct. We will see in due course that the scheme-theoretic image $\bar{\varphi}(X)$ of φ is reduced, and coincides with the Fitting image; so that in the case of this universal family there is no ambiguity about what is meant by the image of X.

Perhaps the most direct way to write down this equation is the following. Consider the ring $\tilde{A} \subset A[a_0^{-1}, b_0^{-1}]$ defined as

$$\tilde{A} = A\left[\frac{a_1}{a_0}, \ldots, \frac{a_m}{a_0}, \frac{b_1}{b_0}, \ldots, \frac{b_n}{b_0}\right].$$

Let

$$B = \tilde{A}[\alpha_1, \ldots, \alpha_m, \beta_1, \ldots, \beta_n]/I,$$

where I is the ideal generated by the $m + n$ elements

$$(-1)^i \sigma_i(\alpha) - \frac{a_i}{a_0} \quad \text{and} \quad (-1)^i \sigma_i(\beta) - \frac{b_i}{b_0},$$

σ_i being the i-th elementary symmetric function. We may describe B intuitively as the ring obtained from A by adjoining the roots of the polynomials f and g. Note that $\tilde{A} \subset B$ because the elementary symmetric functions are algebraically independent.

The expression

$$\mathscr{R} = \prod_{i,j}(\alpha_i - \beta_j)$$

is a symmetric function in the α_i and separately in the β_j. Since $(-1)^i a_i/a_0$ is the i-th elementary symmetric function in the α_i, and similarly for $(-1)^j b_j/b_0$ and the β_j, the function $\mathscr{R}$ can be written as a polynomial in the ratios a_i/a_0 and b_j/b_0, and thus $\mathscr{R}$ is an element of $\tilde{A}$. Each α_i occurs n times in $\mathscr{R}$ and once in an elementary symmetric function, and similarly for the β's, so $\mathscr{R}$ will be bihomogeneous of degrees n, m in the a_i/a_0 and b_j/b_0 respectively. Thus

$$R_{m,n}(f, g) := a_0^n b_0^m \mathscr{R} = a_0^n b_0^m \prod_{i,j}(\alpha_i - \beta_j),$$

is in A; it is a bihomogeneous polynomial, of degree n in the a_i and of degree m in the b_j, called the *resultant* of f and g.

In general, if f_0 and g_0 are polynomial of degrees m and n over an S-algebra S_0, we write $R_{m,n}(f_0, g_0) \in S_0$ for the result of substituting the coefficients of f_0 and g_0 into $R_{m,n}(f,g)$ — that is, the image of $R_{m,n}(f,g)$ under the homomomorphism

$$A = S[a_0, \ldots, a_m, b_0, \ldots, b_n] \to S_0$$

sending the a_i and b_j to the coefficients of f and g. We call $R_{m,n}(f_0, g_0)$ the *resultant* of f_0 and g_0. When we view S_0 as an algebra over different rings S, the resultant does not depend on S; in particular, we would have obtained the same result for any S_0 by taking $S = \mathbb{Z}$. Thus, for each m and n, we can speak of "the resultant" as an element $R_{m,n}$ of the ring $\mathbb{Z}[a_0, \ldots, a_m, b_0, \ldots, b_n]$, and "the resultant of two polynomials" $f, g \in S[x]$ as the image of $R_{m,n}$ under the corresponding homomomorphism

$$\mathbb{Z}[a_0, \ldots, a_m, b_0, \ldots, b_n] \to S.$$

If the leading coefficients of f_0 and g_0 do not vanish — that is, if these polynomials really have the stated degrees — then $R_{m,n}(f_0, g_0) = 0$ if and only if f_0 and g_0 have a common factor, and the classical goal is fulfilled. We shall see that this even works if at most one of the leading coefficients vanish; but it turns out that $R_{m,n}(f,g)$ is contained in the ideal (a_0, b_0), and thus vanishes identically on pairs of polynomials where both these coefficients are 0. In fact, the map φ is not finite in this case, since over the origin in $\mathbb{A}_S^{m+n+2}$ the fiber is the affine line, and the set-theoretic image of X is not closed.

V.2.2 Sylvester's Determinant

We will next relate the zero locus of the resultant of two polynomials to the Fitting image of their common zero locus. We start by computing the equation of the Fitting image $\varphi_{\mathrm{Fitt}}(X)$ where defined. To do this, let S be any ring and $f, g \in S[x]$ two polynomials, of degrees m and n respectively. In order to define the Fitting image of $X = V(f, g) \subset \mathbb{A}_S^1$ under the projection map $\pi : \mathbb{A}_S^1 \to \operatorname{Spec} S$, we must assume that this map is finite, or equivalently that the ideal $(f, g) \subset S[x]$ contains a monic polynomial; see Eisenbud [1995, Proposition 4.1]. For simplicity, we assume that f itself is monic.

To compute the ideal $\mathrm{Fitt}_0 \, \pi_*(\mathscr{O}_X)$ defining the Fitting image of X, we first realize $\pi_*(\mathscr{O}_X)$ as the S-module $S[x]/(f, g)$. This module is the cokernel of the map

$$S[x] \oplus S[x] \xrightarrow{(f,g)} S[x]$$

and the source and target may both be regarded as (infinitely generated) free S-modules. But in order to compute the Fitting ideal we need a finite presentation. Since we have assumed that f is monic of degree m, the S-submodule

$$S[x]_{<m} = S \oplus Sx \oplus Sx^2 \oplus \cdots \oplus Sx^{m-1} \subset S[x]$$

maps onto $S[x]/(f,g)$; and since $S[x]_{<m}$ is isomorphic as S-module to the S-algebra $S[x]/(f)$, we can also realize the module $S[x]/(f,g)$ as the cokernel of the map induced by multiplication by g:

$$G : S^m \cong S[x]_{<m} \cong S[x]/(f) \xrightarrow{\times g} S[x]/(f) \cong S[x]_{<m} \cong S^m.$$

Thus the Fitting ideal $\mathrm{Fitt}_0\, \pi_*(\mathscr{O}_X)$ is generated by the determinant of an $m \times m$ matrix representative of the map $G : S^m \to S^m$.

Now, it is not hard to write down such a matrix representative of G explicitly, and thus to give an explicit formula for the Fitting image. But if we use the freedom we have to compute the Fitting ideal from *any* free presentation, we can get a picture that preserves more of the symmetry between f and g.

To carry this out, consider the free module

$$S[x]_{<m+n} = S \oplus Sx \oplus \cdots \oplus Sx^{m+n-1} \cong S^{m+n}.$$

It clearly maps onto $S[x]/(f,g)$, and the kernel of this map is the set of polynomials $h \in S[x]$ of degree at most $m+n-1$ that lie in the ideal $(f,g) \subset S[x]$. We claim that any such $h \in S[x]_{<m+n} \cap (f,g)$ can be expressed as a linear combination

$$h = a \cdot f + b \cdot g$$

for some polynomials a of degree at most $n-1$ and b of degree at most $m-1$. To see this, note to start with that we must have

$$h = a' \cdot f + b' \cdot g$$

for some $a', b' \in S[x]$. Now, since f is monic of degree m, we can divide b' by f and write

$$b' = qf + b$$

where b is a polynomial of degree at most $m-1$. Adding the expression

$$0 = (qg) \cdot f - (qf) \cdot g$$

to the expression above, we get another expression for h:

$$h = (a' + qg) \cdot f + b \cdot g.$$

Set $a = a' + qg$. Since the degrees of h and bg are both at most $m+n-1$, the degree of af must be at most $m+n-1$ as well; and since f is monic of degree m it follows that a must have degree at most $n-1$, as we claimed.

It follows that the module $S[x]/(f,g)$ is the cokernel of the map

$$S[x]_{<n} \oplus S[x]_{<m} \longrightarrow S[x]_{<m+n}$$
$$(a,b) \longmapsto af + bg.$$

It is easy to write down a matrix representative of this map. With respect to the obvious bases for source and target it is the *Sylvester matrix*

$$\mathrm{Syl}_{(m,n)}(f,g) = \begin{pmatrix} a_0 & a_1 & \cdots & \cdots & a_{m-1} & a_m & 0 & \cdots & 0 \\ 0 & a_0 & a_1 & \cdots & & a_{m-1} & a_m & \cdots & 0 \\ \vdots & \ddots & \ddots & \ddots & & & \ddots & \ddots & \\ 0 & \cdots & 0 & a_0 & a_1 & \cdots & \cdots & a_{m-1} & a_m \\ b_0 & b_1 & \cdots & b_{n-1} & b_n & 0 & \cdots & \cdots & 0 \\ 0 & b_0 & b_1 & \cdots & b_{n-1} & b_n & \cdots & \cdots & 0 \\ \vdots & \ddots & \ddots & \ddots & & & \ddots & \ddots & \\ \vdots & & \ddots & \ddots & \ddots & & & \ddots & \ddots \\ 0 & \cdots & & 0 & b_0 & b_1 & \cdots & b_{n-1} & b_n \end{pmatrix},$$

where there are n rows of a's and m rows of b's.

The next result shows that the zero locus of the resultant does in fact coincide with the Fitting image for the universal family of pairs of polynomials above.

Theorem V-15. *Let*

$$A = \mathbb{Z}[a_0, \ldots, a_m, b_0, \ldots, b_n],$$

and let

$$f = a_0 x^m + \cdots + a_m$$

and

$$g = b_0 x^n + \cdots + b_n \in A[x]$$

be the generic polynomials in one variable of degrees m and n. Let $\varphi :$ $\mathbb{A}_A^1 \to \operatorname{Spec} A$ be the projection map. The scheme-theoretic image of $X = V(f,g) \subset \mathbb{A}_A^1$ under φ has defining ideal generated by $R_{m,n}(f,g)$, which is equal to the Sylvester determinant $\det(\mathrm{Syl}_{m,n}(f,g))$.

Proof. To simplify notation, set $R' = \det(\mathrm{Syl}_{m,n}(f,g)) \in A$, and set $R = R_{m,n}(f,g)$. In algebraic language, we must show that R is equal to R' and generates the ideal $A \cap (f,g)A[x]$.

First, we show that $R' \in (f,g)A[x]$. If we do column operations on $\mathrm{Syl}_{m,n}(f,g)$, adding x^{m+n-t} times the t-th column to the last for $t = 1, \ldots, m+n-1$ then since the number of columns is $m+n$, we get a new matrix with the same determinant R'. But the last column of the new matrix is

$$\begin{pmatrix} x^{n-1}f \\ \vdots \\ f \\ x^{m-1}g \\ \vdots \\ g \end{pmatrix},$$

so the determinant is in $(f,g)A[x]$ as required.

Next suppose that $P \in A$ lies in $(f,g)A[x]$; we shall show that P is divisible by R in A. Note that f and g are bihomogeneous: they are homogeneous separately in the a_i and the b_j. It follows that the bihomogeneous components of any polynomial in (f,g) are again in (f,g), so we may assume that P is bihomogeneous, say of bidegree (d,e). (Note that R' is itself bihomogeneous of bidegree (n,m).)

To analyze the situation we embed A in a larger polynomial ring. Let $B_0 = \mathbb{Z}[\alpha_1,\ldots,\alpha_m,\beta_1,\ldots,\beta_n]$ and let $B = B_0[a_0,b_0]$. We map A to B by sending the a_i to the coefficients of the polynomial

$$f' = a_0 \prod_{j=1}^{m}(x - \alpha_j),$$

and sending the b_i to the coefficients of $g' = b_0 \prod_{j=1}^{n}(x - \beta_j)$. The coefficients of f' are $\pm a_0 \sigma_i(\alpha)$, where σ_i is the i-th elementary symmetric function, and similarly for g'. Since a_0, b_0 and the elementary symmetric functions of the α_i and β_j are algebraically independent, the same is true for

$$a_0, a_0\sigma_1(\alpha),\ldots,a_0\sigma_m(\alpha), b_0, b_0\sigma_1(\beta),\ldots,b_0\sigma_m(\beta),$$

so the map $A \to B$ is indeed an embedding. Recall that R was defined as the element $a_0^n b_0^m \prod_{i,j}(\alpha_i - \beta_j) \in B$, which happens to lie in the subring A.

We now return to the polynomial $P \in A \cap (f,g)A[x]$. Since P is bihomogeneous in the $a_i = \pm a_0 \sigma_i(\alpha)$ and the $b_j = \pm a_0 \sigma_j(\beta)$ of bidegree (d,e) we may write $P = a_0^d b_0^e h$ for some $h \in B_0$. Since the elementary symmetric function $\sigma_i(\alpha)$ is a linear polynomial in each α_i, and similarly for the β_j, we see that h has degree $\le d$ in each α_i and $\le e$ in each β_j.

For given indices i,j, let L be the quotient field of the domain $B/(\alpha_i - \beta_j)$. Let $\bar{P}$ be the image of P in L. Since f and g have a common root in L, the constant polynomial $\bar{P} \in (f,g)L[x]$ has this root too, so $\bar{P} = 0$; thus P is divisible by $\alpha_i - \beta_j$ in B, and it follows that h is divisible by $\alpha_i - \beta_j$ in B_0. Thus h is divisible by $\prod_{i,j}(\alpha_i - \beta_j)$. In particular, we see that $d \ge m$ and $e \ge n$, and thus $P = a_0^d b_0^e h$ is divisible by R in B; we write $P = RQ$ in B.

If h is any polynomial function of the α's and β's that is separately symmetric in the α's and in the β's, of degree $\le u$ in each α_i and degree $\le v$ in each β_j, then because the elementary symmetric functions generate the ring of all symmetric functions, h is actually a polynomial function in the coefficients a_i/a_0 and b_j/b_0, of degree $\le u$ in the first variables and degree $\le v$ in the second. The polynomial $a_0^d b_0^e h$ is thus in A for any d,e larger than the degree of h in α and the degree of h in β, respectively.

Because of the form of P and R, we may write $Q = a_0^{d-m} b_0^{e-n} q$ for some polynomial $q \in B_0$ of degree $\le d - m$ in each α_i and $\le e - n$ in each β_j. Further, q is symmetric in the α_i and in the β_j separately, so Q may be written as a polynomial in the $\sigma_i(\alpha)$ and $\sigma_j(\beta)$ of degree $\le d - m$ and $\le e - n$ in the two sets of variables. By the remark above, $Q \in A$, as required.

If we take $P = R'$, then we see that R divides R' in A; but since both have the same degree, they are equal up to a sign. (Evaluating both sides at the pair of polynomials $f(x) = x^m$ and $g(x) = x^n$ shows that they are in fact equal.) In particular, R is in $A \cap (f, g)A[x]$, so by what we have shown R generates this ideal, and we are done.   □

We cannot apply Proposition V-8 directly to the situation of Theorem V-15 because, as we have already noted, the map $V(f, g) \to \operatorname{Spec} A$ is not finite. However, if we first restrict to an open subset of A over which it is finite, then all is well:

Corollary V-16. *Let* $B = \mathbb{Z}[a_0, \ldots, a_m, b_0, \ldots, b_n][a_0^{-1}]$, *and let*

$$f = a_0 x^m + \cdots + a_m \quad \text{and} \quad g = b_0 x^n + \cdots + b_n \in B[x]$$

be the generic polynomials in one variable of degrees m and n such that f has a unit leading coefficient. The projection map $\varphi : V(f, g) \to \operatorname{Spec} B$ is finite. The scheme-theoretic image of $V(f, g) \subset \mathbb{A}_B^1$ under φ has defining ideal generated by $R_{m,n}(f, g)$. Thus if f_0, g_0 are any two polynomials of degrees m, n over an algebraically closed field L, one of which has unit leading coefficient, then f_0 and g_0 have a common root in L if and only if $R_{m,n}(f_0, g_0) = 0$.

Proof. With A as in Theorem V-15, the ring B is a flat A-algebra, so by Proposition V-8 the first statement of the corollary follows from the corresponding statement in Theorem V-15. To see the finiteness, note that $B[x]/(f)$ is already finite over B, since it is generated as a module by 1, x, $\ldots$, x^{m-1}. Thus finiteness holds for the factor ring $B[x]/(f, g)$ whose spectrum is $V(f, g)$.

Because $V(f, g)$ is finite over $\operatorname{Spec} B$ we may apply the set-theoretic part of Proposition V-8, and the last statement of the corollary follows in the situation where f has unit leading coefficient; the case where g has unit leading coefficient would follow by a similar argument.   □

Corollary V-17. *Let S be a ring and suppose that $f, g \in S[x]$ are polynomials in one variable over S. If $f(x) = a_0 \prod_{i=1}^m (x - \alpha_i)$ factors completely over S, then*

$$R_{m,n}(f, g) = a_0^n \prod_{i=1}^m g(\alpha_i).$$

Proof. The resultant is the specialization (from the ring over which the generic polynomials f, g are defined and factor) of the expression

$$a_0^n b_0^m \prod_{i,j} (\alpha_i - \beta_j) = a_0^n \prod_i g(\alpha_i).$$ □

Exercise V-18. The fact that the resultant is 0 if both f and g have vanishing leading coefficient may seem at first an unfortunate anomaly, but this aspect disappears if we "compactify" the affine line as the projective line over A (strictly speaking this is the relative compactification over $\operatorname{Spec} A$). If we watch what happens to the roots of the polynomial f as the leading coefficient of f goes to 0, we see that at least one of them moves to ∞; that is, if we homogenize and regard f as defining a subset of m points the projective line, then the limiting position of this zero set contains the point ∞. Thus it is reasonable that if the leading coefficients of both f and g are 0, then (as polynomials of degree m and n) they share ∞ as a common root. Prove that if $F = a_0 x^m + a_1 x^{m-1} y + \cdots + a_m y^m$ and $G = b_0 x^n + \cdots + b_n y^n$ are the generic homogeneous forms of degrees m, n over $A = \mathbb{Z}[a_0, \ldots, a_m, b_0, \ldots, b_n]$, then the image scheme of $V(F, G) \subset \mathbb{P}_A^1$ under the natural projection to $\operatorname{Spec} A$ has defining ideal generated by the resultant $R_{m,n}(F, G) := R_{m,n}(F(x, 1), G(x, 1))$. Show that if F_0, G_0 are nonzero homogeneous forms of degrees m, n over any field L, then F_0 and G_0 have a common zero in $\mathbb{P}_L^1$ if and only if $R_{m,n}(F_0, G_0) = 0$.

Exercise V-19. Show that in Theorem V-15 the ring of integers can be replaced by any commutative ring K as follows: let A be the polynomial ring over the integers as in the theorem, and let $B = A \otimes_{\mathbb{Z}} K$.

(a) Show that it is enough to prove that the sequence

$$0 \to (R_{m,n}(f, g)) \to A \to A/(R_{m,n}(f, g)) \to 0$$

remains exact when we tensor over $\mathbb{Z}$ with K.

(b) Show that for each prime p the abelian group $A/(R_{m,n}(f, g))$ has no p-torsion. (You might use the fact that $R_{m,n}(f, g)$, regarded as a polynomial in a_m, b_n over a smaller polynomial ring, has "leading term" $a_m^n b_n^m$ in a suitable sense.) Use the result of (b) to prove (a).

Exercise V-20. The resultant computation above seems quite special, but it can be used to compute the (set-theoretic) image of an arbitrary finite map $\varphi : X \to Y$. Pass to an affine cover and suppose that $Y = \operatorname{Spec} A'$ for some ring A', while $X = \operatorname{Spec} A'[x_1, \ldots, x_d]/I$. Reduce by induction to the case $d = 1$. Let $f_1, \ldots, f_e$ be generators of I; suppose the maximum of their degrees in x_1 is n. Show that I contains a monic polynomial f of degree m, say. Let $t_1, \ldots, t_e$ be new indeterminates, and let $g = \sum_i t_i f_i \in A'[t_1, \ldots, t_e]$. Let $R = R_{m,n}(f, g)$ be the resultant, and let J be the ideal generated by the coefficients in A' of the monomials in t occurring in R. Show that J defines the image of φ set-theoretically, and is contained in the ideal defining the scheme-theoretic image. (In practice people do not usually compute the image this way, but rather using the technique of Gröbner bases, which, unlike resultants, actually computes the whole ideal of the scheme-theoretic image; see Cox et al. [1997, Section 2.8] or Eisenbud [1995, Chapter 15] for this method.)

V.3   Singular Schemes and Discriminants

V.3.1   Definitions

In this section we will take up the smooth and singluar points of a morphism of schemes. To motivate our definition, consider a map $f : M \to N$ of differentiable manifolds, and assume that $\dim M \geq \dim N$. The simplest behavior of such a map occurs at points $x \in M$ where the differential Df_x is surjective: restricted to a suitable neighborhood U of such a point x, the map f looks like the projection onto one factor of a product, and we say that x is a smooth point of f.

In the category of schemes, the Zariski open sets are too large to permit product structures, and (as we shall see) there are also complications arising from the fact that points may have different residue fileds. It is still possible to define smoothness of a morphism of schemes in terms of local product structure — see for example Altman and Kleiman [1970] — but in the present context it will make more sense to adopt a characterization of smooth and singular points that generalizes the differential characterization for manifolds.

To carry this out we introduce the module of Kähler differentials of a homomomorphism of rings, and its global version, the relative cotangent sheaf of a morphism of schemes. If $\psi : A \to B$ is a map of rings, we define $\Omega_{B/A}$, the *module of A-linear Kähler differentials*, to be the free B-module generated by symbols db for all $b \in B$, modulo the relations

$$d(b_1 b_2) = b_1 \, db_2 + b_2 \, db_1 \quad \text{for all } b_1, b_2 \in B$$

and

$$d\psi_a = 0 \quad \text{for all } a \in A.$$

These relations ensure that the map

$$B \longrightarrow \Omega_{B/A},$$
$$b \longmapsto db,$$

is an A-linear derivation; in fact, it is the universal A-linear derivation in a suitable sense. (See Eisenbud [1995] for details.) It is easy to deduce from this, for example, that if $B = A[x_1, \ldots, x_n]/(f_1, \ldots, f_n)$ then $\Omega_{B/A}$ in the cokernel of the Jacobian matrix

$$\left(\frac{\partial f_i}{\partial x_j} \right).$$

To globalize, let $\varphi : X \to Y$ be a morphism of schemes. We define the *relative cotangent sheaf* of $\varphi : X \to Y$, written $\Omega_{X/Y}$, to be the sheaf whose value on an open affine subset U of X mapping to an open affine subset V of Y is the module of $\mathscr{O}_Y(V)$-linear Kähler differentials of $\mathscr{O}_X(U)$. The collection of open sets in X just specified forms a base $\mathscr{B}$ for the open sets of X. The axioms for a $\mathscr{B}$-sheaf (see Section I.1.3) are easily checked — they amount to the statment that the construction of modules of differentials is compatible with localization — so the data just given really does define a sheaf.

The *relative dimension* of a morphism $\varphi : X \to Y$ of schemes at a point $x \in X$ is defined to be the difference $\dim(X, x) - \dim(Y, \varphi(x))$.

Definition V-21. Let $\varphi : X \to Y$ be a morphism of Noetherian schemes, and suppose that φ is flat, of finite type, and has constant relative codimension d. We define the *singular scheme* $\operatorname{sing} \varphi \subset X$ of φ to be the subscheme of X defined by the d-th Fitting ideal of the relative cotangent sheaf of φ; that is,

$$\operatorname{sing} \varphi = V(\operatorname{Fitt}_d \Omega_{X/Y}) \subset X.$$

In case the morphism $\varphi|_{\operatorname{sing} \varphi} : \operatorname{sing} \varphi \to Y$ is finite, we define the *discriminant scheme* $\Delta(\varphi) \subset Y$ to be the Fitting image of $\operatorname{sing} \varphi$ in Y, that is,

$$\Delta(\varphi) = V\left(\operatorname{Fitt}_0(\varphi_* \mathcal{O}_{\operatorname{sing} \varphi})\right) \subset Y.$$

To understand this definition, we first return to the most classical case, that of a map $\varphi : X \to Y$ of irreducible, nonsingular varieties over an algebraically closed field. In this case the condition of constant relative dimension is automatic, while flatness translates into the condition that the dimension of the fibers $\varphi^{-1}(y)$ is constant, and equal to the relative dimension.

The notion of a singular scheme being a local one, we may pass to the affine case and assume that $Y = \operatorname{Spec} A$ and write

$$X = \operatorname{Spec} A[x_1, \ldots, x_n]/(f_1, \ldots, f_m).$$

As we noted, in this case the module of A-linear Kähler differentials $\Omega_{X/Y}$ is the cokernel of the $n \times m$ Jacobian matrix

$$\begin{pmatrix} \partial f_1/\partial x_1 & \cdots & \partial f_m/\partial x_1 \\ \vdots & & \vdots \\ \partial f_1/\partial x_n & \cdots & \partial f_m/\partial x_n \end{pmatrix}.$$

Thus the support of the scheme $\operatorname{sing} \varphi$ consists exactly of those points $x \in X$ where the rank of the Jacobian matrix is less than $n - d$ — in other words, the locus where the differential $D\varphi$ of the map fails to be surjective, just as in the classical definition.

However, the scheme-theoretic setting presents many new phenomena. For example, consider a finite extension of fields $K \hookrightarrow K'$, and let

$$\varphi : X = \operatorname{Spec} K' \longmapsto Y = \operatorname{Spec} K$$

be the corresponding morphism of one-point schemes. Here the relative dimension is zero, so the map φ is singular if and only if $\Omega_{X/Y} \neq 0$; by a classical result in field theory, this is the case if and only if the extension $K \hookrightarrow K'$ is not separable.

V.3.2   Discriminants

We now want to consider in more detail the case of a morphism $\varphi : X \to Y$ where the singular scheme $\operatorname{sing} \varphi$ is finite over Y, and to describe the discriminant scheme $\Delta(\varphi) \subset Y$ associated to such a map. We will focus primarily (but not exclusively) on the case where X is a closed subscheme of $\mathbb{A}_Y^1$, given as the zero locus of a single polynomial $f(x)$ whose coefficients are regular functions on Y; this will lead us to the definition of the *discriminant* of a polynomial in one *discriminant!of a polynomial*variable. In this setting, by analogy with the discussion of resultants in the preceding section, we may view the problem of defining the discriminant scheme as that of giving formulas for the equations defining the set of points $y = [\mathfrak{p}] \in Y = \operatorname{Spec} A$ over which the polynomial f has repeated factors, that is, such that the reduction $\bar{f} \in A/\mathfrak{p}[x]$ of $f \in A[x]$ mod $\mathfrak{p}$ has multiple roots in the algebraic closure of $\kappa(y)$. As in the discussion of resultants, we will have both the general definition above using Fitting ideals, and (in the restricted case $X = \operatorname{Spec} A[x]/(f)$) a more classical notion of discriminant defined in effect as the pullback of the (reduced) branch scheme of a suitable "universal family" of polynomials; and we will ultimately show that they coincide where the latter is defined.

To set up the classical construction, we need first of all to define our universal branched cover. Let

$$A = \mathbb{Z}[a_0, \ldots, a_m],$$

and let

$$f = a_0 x^m + \cdots + a_m \in A[x]$$

be the generic polynomial in one variable of degree m. Extending our polynomial ring, we define

$$B_0 = \mathbb{Z}[\alpha_1, \ldots, \alpha_m] \qquad \text{and} \qquad B = B_0[a_0]$$

and map A to B by sending a_i to the i-th coefficient of the polynomial $f = a_0 \prod_i (x - \alpha_i)$, which is $\pm a_0 \sigma_i(\alpha)$. As in the proof of Theorem V-15, these coefficients are algebraically independent, and we regard A as a subring of B.

In B we may form the polynomial

$$D_1 = \prod_{i<j} (\alpha_i - \alpha_j).$$

This polynomial vanishes when two of the roots of f are equal, but it is not a solution of our problem because it is not a polynomial in A. For one thing, it is not invariant under permutations of the roots. But it is easy to see that if π is a permutation of the roots, D_1 applied to the permuted roots is equal to $\operatorname{sgn}(\pi)D_1$, where $\operatorname{sgn}(\pi) = \pm 1$ is the sign of the permutation. Thus D_1^2 is invariant under under permutations of the roots, and is expressible as a polynomial in the symmetric functions $\sigma_i(\alpha) = a_i/a_0$. Each α_i occurs to degree $\leq m - 1$ in D_1, and thus to degree $\leq 2m - 2$ in D_1^2, from which we see that

$$D_m(f) = a_0^{2m-2} D_1^2 = a_0^{2m-2} \prod_{i<j} (\alpha_i - \alpha_j)^2$$

defines an element of A. The polynomial $D_m(f)$ is called the *discriminant* of f.

Proposition V-22. *With notation as above, we have*

$$D_m(f) = \frac{(-1)^{m(m-1)/2}}{a_0} R_{m,m-1}(f, f').$$

If $f_0 \in L[x]$ is a monic polnomial of degree m in one variable over a field L, then $D_m(f_0) = 0$ if and only if f_0 has a multiple root in the algebraic closure of L.

Proof. Applying Corollary V-17 to the ring B defined above we see that

$$R_{m,m-1}(f, f') = a_0^{m-1} \prod_{i=1}^{m} f'(\alpha_i).$$

But

$$f'(x) = \sum_{j=1}^{m} f(x)/(x - \alpha_j),$$

so

$$f'(\alpha_i) = a_0 \prod_{j \neq i} (\alpha_i - \alpha_j)$$

and

$$
\begin{aligned}
R_{m,m-1}(f, f') &= a_0^{m-1} \prod_{i=1}^{m} f'(\alpha_i) = a_0^{2m-1} \prod_{j \neq i} (\alpha_i - \alpha_j) \\
&= (-1)^{m(m-1)/2} a_0^{2m-1} \prod_{i<j} (\alpha_i - \alpha_j)^2 \\
&= (-1)^{m(m-1)/2} a_0 D_m(f),
\end{aligned}
$$

whence the formula for $D_m(f)$. The second statement of the corollary follows from the fact that the resultant of a monic polynomial and another polynomial over a field vanishes if and only if the two have a common root, and the usual computation that shows, over a field, that a polynomial $f(x)$ and its derivative have a common root if and only if f has a multiple root. □

As a corollary of this proposition, we have the promised identity between the classical and modern definitions, in the restricted case $X = \operatorname{Spec} A[x]/(f)$.

Corollary V-23. *Let A be any ring, $f \in A[x]$ a monic polynomial of degree m with coefficients in A, and $\varphi : X = \operatorname{Spec} A[x]/(f) \to Y = \operatorname{Spec} A$ the corresponding morphism. The discriminant scheme of φ is the zero locus of the discriminant of f, that is,*

$$\Delta(\varphi) = V(D_m(f)) \subset Y.$$

For example, if $B = \mathbb{Z}[\alpha]$ is an order in a number field, and

$$f(x) = x^n + a_{n-1}x^{n-1} + \cdots + a_0 \in \mathbb{Z}[x]$$

is the irreducible monic polynomial satisfied by α, the discriminant of f is the defining equation of the discriminant scheme of the morphism

$$\operatorname{Spec} B \to \operatorname{Spec} \mathbb{Z}.$$

Exercise V-24. Show that the singular scheme of this morphism is the *different* of f, as defined for example in Lang [1994].

Exercise V-25. Consider the orders $A = \mathbb{Z}[\sqrt{3}]$, $B = \mathbb{Z}[11\sqrt{3}]$, $C = \mathbb{Z}[2\sqrt{3}]$, and $D = \mathbb{Z}[121\sqrt{3}]$, discussed in Section II.4.2 and Exercise IV-51. What are the discriminant schemes of the morphisms $\operatorname{Spec} A \to \mathbb{Z}$, $\operatorname{Spec} B \to \mathbb{Z}$, $\operatorname{Spec} C \to \mathbb{Z}$, and $\operatorname{Spec} D \to \mathbb{Z}$? How does this relate to the pictures of these schemes drawn previously?

V.3.3 Examples

To illustrate our definition of the discriminant scheme of a morphism, we will calculate it in a number of specific examples. For all of the following morphisms $\varphi : X \to Y$, the target space will be the affine line $Y = \operatorname{Spec} K[t]$ over a field K of characteristic zero. For all but the last example, the morphism φ will be finite; and for all but the last two, X will be a subscheme of the affine line $\mathbb{A}^1_Y = \operatorname{Spec} K[t][x]$, finite and flat over Y. Specifically, we take

$$\varphi : X = \operatorname{Spec} K[t, x]/(f) \to Y = \operatorname{Spec} K[t],$$

where

$$f(x) = x^k + a_{k-1}(t)x^{k-1} + \ldots + a_1(t)x + a_0(t)$$

is a monic polynomial in x with coefficients in $K[t]$. In all the cases we consider the discriminant scheme, being a subscheme of $\mathbb{A}^1_K$ supported at the origin, is determined by its degree.

Example V-26. We start with the example $f(x) = x^k - t^m$, that is, the map

$$\varphi_{k,m} : X = X_{k,m} = \operatorname{Spec} K[t, x]/(x^k - t^m) \to Y = \operatorname{Spec} K[t].$$

We will denote the degree of the discriminant scheme of this map by $\delta_{k,m}$ and calculate it in three ways.

Note before we start that if

$$\mu_\alpha : Y = \operatorname{Spec} K[t] \longrightarrow Y = \operatorname{Spec} K[t]$$

is the map given by $t \mapsto t^\alpha$, we have a fiber product diagram

$$
\begin{array}{ccc}
X_{k,\alpha m} & \longrightarrow & X_{k,m} \\
{\scriptstyle \varphi_{k,\alpha m}} \downarrow & & \downarrow {\scriptstyle \varphi_{k,m}} \\
Y & \xrightarrow{\ \mu_\alpha\ } & Y
\end{array}
$$

so that, by the invariance of the discriminant scheme under pullback,

$$\delta_{k,\alpha m} = \alpha \cdot \delta_{k,m}$$

for any α and m.

For our first calculation of $\delta_{k,m}$, we use the definition of the discriminant of the polynomial $f(x)$ as the resultant of f and f' and apply Sylvester's determinant. Thus, for example, in case $k = 2$ the discriminant is

$$
\delta_{2,m} = \begin{vmatrix} 1 & 0 & t^m \\ 2 & 0 & 0 \\ 0 & 2 & 0 \end{vmatrix} = 4t^m,
$$

so that (the characteristic of K being 0) $\delta_{2,m} = m$. More generally, for arbitrary k we have $a_{k-1} = \cdots = a_1 = 0$ and $a_0 = -t^m$. Therefore

$$
\delta_{k,m} = \begin{vmatrix}
1 & 0 & \cdots & 0 & 0 & 0 & -t^m & 0 & \cdots & 0 & 0 \\
0 & 1 & \cdots & 0 & 0 & 0 & 0 & -t^m & \cdots & 0 & 0 \\
\vdots & \vdots & \ddots & \vdots & \vdots & \vdots & \vdots & \vdots & \ddots & \vdots & \vdots \\
0 & 0 & \cdots & 1 & 0 & 0 & 0 & 0 & \cdots & -t^m & 0 \\
0 & 0 & \cdots & 0 & 1 & 0 & 0 & 0 & \cdots & 0 & -t^m \\
k & 0 & \cdots & 0 & 0 & 0 & 0 & 0 & \cdots & 0 & 0 \\
0 & k & \cdots & 0 & 0 & 0 & 0 & 0 & \cdots & 0 & 0 \\
\vdots & \vdots & \ddots & \vdots & \vdots & \vdots & \vdots & \vdots & \ddots & \vdots & \vdots \\
0 & 0 & \cdots & k & 0 & 0 & 0 & 0 & \cdots & 0 & 0 \\
0 & 0 & \cdots & 0 & k & 0 & 0 & 0 & \cdots & 0 & 0 \\
0 & 0 & \cdots & 0 & 0 & k & 0 & 0 & \cdots & 0 & 0
\end{vmatrix}
$$

$$= (-1)^{k-1} k^k \cdot t^{m(k-1)},$$

and hence

$$\delta_{k,m} = m(k-1).$$

A second way to calculate $\delta_{k,m}$ is to use the expression of the discriminant as the product of the pairwise differences of the roots of a polynomial. To start, suppose that k divides m; say $m = kl$, and suppose also that K

contains a primitive l-th root of unity ζ. Then we can factor

$$f(x) = x^k - t^{kl} = \prod_{i=0}^{k-1} (x - \zeta^i t^l).$$

The discriminant of f is then the product

$$\Delta(f) = \prod_{0 \le i < j \le k-1} (\zeta^j t^l - \zeta^i t^l)^2 = \prod_{0 \le i < j \le k-1} (\zeta^j - \zeta^i) t^{2l},$$

which vanishes to order $\delta_{k,m} = k(k-1)l = m(k-1)$ at the origin. Thus $\delta_{k,m} = m(k-1)$ whenever $k|m$; and by the formula $\delta_{k,\alpha m} = \alpha \cdot \delta_{k,m}$ established above it follows that $\delta_{k,m} = m(k-1)$ for all k and m.

Exercise V-27. The last argument uses the presence in our field K of all the l-th roots of unity. Use a base change argument (bearing in mind the definition of degree!) to extend the result to any field K of characteristic not dividing l.

Finally, we can describe the discriminant scheme of $\varphi = \varphi_{k,m}$ directly from the definition. To begin with, the sheaf of relative differentials of the map φ is

$$\Omega_{X/Y} = \mathcal{O}_X\{dx\}/(kx^{k-1}dx),$$

where $\mathcal{O}_X\{dx\}$ denotes the free $\mathcal{O}_X$-module with generator dx. The resolution of $\Omega_{X/Y}$ is thus

$$0 \longrightarrow \mathcal{O}_X \xrightarrow{\nu} \mathcal{O}_X \longrightarrow \Omega_{X/Y} \longrightarrow 0$$

where the map ν is multiplication by x^{k-1}. The zeroth Fitting ideal of $\Omega_{X/Y}$ is accordingly generated by the 1×1 minor of the matrix (x^{k-1}), so that the singular locus of the map $\varphi_{k,m}$ is

$$\operatorname{sing} \varphi = V(\operatorname{Fitt}_0 \Omega_{X/Y}) = V(x^{k-1}) \subset X$$

and $\mathcal{O}_{\operatorname{sing} \varphi} \cong \Omega_{X/Y}$ as sheaves of $\mathcal{O}_X$-modules.

To resolve the pushforward of the structure sheaf of $\operatorname{sing} \varphi$, we can thus push forward the exact sequence above. The pushforward $\varphi_* \mathcal{O}_X$ is the locally free $\mathcal{O}_Y$-module generated by the elements $\alpha_0 = 1$, $\alpha_1 = x$, ..., $\alpha_{k-1} = x^{k-1}$. Multiplication by x^{k-1} takes the first generator $\alpha_0 = 1$ to the last one $\alpha_{k-1} = x^{k-1}$, the second generator $\alpha_1 = x$ to $x^k = t^m \alpha_0$, the third generator $\alpha_2 = x^2$ to $x^{k+1} = t^m \alpha_1$, and so on. The resolution of the pushforward $\varphi_* \mathcal{O}_{\operatorname{sing} \varphi}$ is thus

$$0 \longrightarrow \mathcal{O}_Y^{\oplus k} \xrightarrow{\varphi_* \nu} \mathcal{O}_Y^{\oplus k} \longrightarrow \varphi_* \mathcal{O}_{\operatorname{sing} \varphi} \longrightarrow 0$$

where the map $\varphi_* \nu$ is given by the matrix

$$
\varphi_* \nu = \begin{pmatrix}
0 & t^m & 0 & \cdots & 0 & 0 \\
0 & 0 & t^m & \cdots & 0 & 0 \\
0 & 0 & 0 & \cdots & 0 & 0 \\
\vdots & \vdots & \vdots & \ddots & \vdots & \vdots \\
0 & 0 & 0 & \cdots & 0 & t^m \\
1 & 0 & 0 & \cdots & 0 & 0
\end{pmatrix}.
$$

The discriminant scheme of $\varphi = \varphi_{k,m}$ is the zero locus of the determinant $(-1)^{k-1} t^{m(k-1)}$ of this matrix, and so we see once again that $\delta_{k,m} = m \times (k-1)$.

Note that in this setting the discriminant scheme measures not only the number of sheets in a branched cover that come together (that is, the nonreducedness of the special fiber, as measured by the "$k-1$" factor in the expression $\delta_{k,m} = m(k-1)$), but also how fast they are coming together (the "m" factor). Thus, while in all the examples of the form $X = \operatorname{Spec} K[t,x]/(x^2 - t^m) \to Y = \operatorname{Spec} K[t]$ the fiber over the origin in Y is the same double point, the discriminant has different degrees depending on the speed with which the two points $x = \pm \sqrt{t^m}$ approach each other as t approaches 0.

Here is an arithmetic analogue of this:

Exercise V-28. Reinterpret the results of Exercise V-25 in light of the calculation in Example V-26 of the discriminant of a general projection of a node, cusp and tacnode.

We will also see applications of Example V-26 in the discussion of *dual curves* (Section V.4.2).

Although the examples we have seen so far may seem special, we can use any of the three approaches to describe the discriminant of any finite flat morphism $\varphi : X \to Y$ where Y is nonsingular and one-dimensional and X is locally embeddable in $\mathbb{A}^1_Y$, that is, for any point $p \in X$ the local ring $\mathcal{O}_{X,p}$ is of the form $\mathcal{O}_{Y,\varphi(p)}[x]/(f)$. For example, following the second approach we may make a base change $Y' \to Y$ so that the polynomial f factors completely into linear factors over $\mathcal{O}_Y$ (that is, the pullback $X' = X \times_Y Y'$ is a union of k irreducible components X_i, each mapping isomorphically to Y). The discriminant of the pullback morphism $\varphi' : X' \to Y'$ will then be given by the product of the pairwise differences of the factors of f, so that its degree will be the sum of the degrees of the pairwise intersections $X_i \cap X_j$; and the degree of the discriminant of φ at each point $y \in Y$ will be simply the sum of the degrees of the discriminant scheme of φ' at the points y' of Y' lying over y, each divided by the order of ramification of $Y' \to Y$ at that point.

Our next example, accordingly, will be the simplest example of a morphism $X \to Y$ such that X is not locally embeddable in $\mathbb{A}^1_Y$.

Example V-29. Take $Y = \mathbb{A}^1_K = \operatorname{Spec} K[t]$ as before, and take X the union of three copies of $\mathbb{A}^1_K$ meeting at one point with three-dimensional Zariski tangent space; the map φ will map each component of X isomorphically to Y. We may realize X as the union of the three coordinate axes in $\mathbb{A}^3_K = \operatorname{Spec} K[x,y,z]$, and take $\varphi : X \to Y$ the restriction to X of the projection map $\mathbb{A}^3_K \to \mathbb{A}^1_K$ given by $t \mapsto x+y+z$: that is, we set

$$X = \operatorname{Spec} K[t][x,y,z]/(xy, xz, yz, t-x-y-z) \longrightarrow Y = \operatorname{Spec} K[t].$$

We can simplify the expression for X at the expense of some symmetry, writing

$$X = \operatorname{Spec} K[t][x,y]/(xy, x(t-x-y), y(t-x-y)).$$

The sheaf $\Omega_{X/Y}$ is then given as

$$\Omega_{X/Y} = \mathcal{O}_X\{dx, dy\}/(ydx + xdy, (t-2x)dx - xdy, -ydx + (t-2y)dy)$$

Since there are two generators and three relations, the resolution of $\Omega_{X/Y}$ takes the form

$$\mathcal{O}_X^{\oplus 3} \xrightarrow{\ \nu\ } \mathcal{O}_X^{\oplus 2} \longrightarrow \Omega_{X/Y} \longrightarrow 0,$$

where the map ν is given by the matrix

$$\nu = \begin{pmatrix} y & t-2x & -y \\ x & -x & t-2y \end{pmatrix}.$$

In terms of the original description $X = \operatorname{Spec} K[x,y,z]/(xy,xz,yz)$ of X, the 2×2 minors of this matrix are

$$-xy - x(t-2x) = x^2,$$
$$y(t-2y) + xy = -y^2, \text{ and}$$
$$(t-2x)(t-2y) - xy = -x^2 - y^2 + z^2,$$

so that the zeroth Fitting ideal $\operatorname{Fitt}_0 \Omega_{X/Y}$ is simply the square (x^2, y^2, z^2) of the maximal ideal of the origin in X. The singular locus $\operatorname{sing} \varphi$ of the map φ is thus the first-order neighborhood of the origin in $\mathbb{A}^3_K$; in particular, it has degree 4. For any sheaf $\mathcal{F}$ on Y supported at the origin in $\mathbb{A}^1_K$, the zeroth Fitting ideal will be simply t^m, where m is the vector space dimension of $\Gamma(\mathcal{F})$. The degree of the discriminant is thus equal to the degree 4 of the singular locus.

It is also easy to describe the direct image $\varphi_*\mathcal{O}_{\operatorname{sing}(\varphi)}$ directly: since the fiber of the projection $\operatorname{sing} \varphi \to Y$ over the origin has degree 3, we must have

$$\varphi_*(\mathcal{O}_{\operatorname{sing}\varphi}) = \mathcal{O}_Y/(t^2) \oplus (\mathcal{O}_Y/(t))^{\oplus 2},$$

and we can calculate the zeroth Fitting ideal $\operatorname{Fitt}_0(\varphi_*\mathcal{O}_{\operatorname{sing}\varphi})$ accordingly.

If X had consisted of three *coplanar* lines meeting at a point — say, if we took $\varphi : X = \operatorname{Spec} K[t,x]/(x^3 - t^3) \to Y = \operatorname{Spec} K[t]$ — then by the calculation made previously we would have $\deg(\Delta(\varphi)) = 6$, while in the spatial case we have $\deg(\Delta(\varphi)) = 4$. This difference is what underlies the example in Section II.3.4 whose general fiber is isomorphic to the three coordinate lines in $\mathbb{A}^3_K$ and whose special fiber is supported on three coplanar lines: the jump in the discriminant, in a sense, forces the appearance of the embedded point in the flat limit. We saw the same phenomenon emerge when we considered Hilbert polynomials in Sections III.3.1 and III.3.2.

Our last examples will be in the form of exercises. In them we consider the simplest cases of a morphism $\varphi : X \to Y$ of relative dimension one, such that the singular locus $\operatorname{sing} \varphi$ is finite over Y.

Exercise V-30. Let K be a field of characteristic 0, let $Y = \operatorname{Spec} K[t] = \mathbb{A}^1_K$, and let $\varphi : X \to Y$ be the family of curves

$$\varphi : X = \operatorname{Spec} K[t][x,y]/(xy - t^m) \subset \mathbb{A}^2_Y \longrightarrow Y.$$

Show that the discriminant scheme $\Delta(\varphi) = V(t^m) \subset Y$.

In the preceding exercise, the singular curves of the families considered have the simplest possible curve singularity, called a *node*. To generalize it, we need first of all to make a definition:

Definition V-31. Let $C \to \operatorname{Spec} K$ be a curve over an algebraically closed field K and $p \in C$ a closed point of C. We will say that p is a *node, cusp* or *tacnode* of C if the formal completion $\hat{\mathcal{O}}$ of the local ring $\mathcal{O}_{C,p}$ with respect to its maximal ideal is isomorphic to $K[\![x,y]\!]/(y^2 - x^2)$, $K[\![x,y]\!]/(y^2 - x^3)$ or $K[\![x,y]\!]/(y^2 - x^4)$ respectively.

These singularities may also be characterized geometrically: a node, for example, is a point of C at which two smooth branches cross transversely; a tacnode is one where two smooth branches are simply tangent, that is, intersecting in a scheme of degree 2. In the case of a plane curve $C \subset \mathbb{P}^2_K$, we will say that a node or tacnode p of C is *ordinary* if neither branch individually has intersection multiplicity 3 or more with its projective tangent line. (The reader may verify that if $p \in C$ is a cusp, this will always be the case.)

Exercise V-32. Let K and Y be as in Exercise V-30, and find the discriminant scheme $\Delta(\varphi) \subset Y$ for the following families $\varphi : X \to Y$ of curves:

(a) (a family acquiring a cusp)

$$\varphi : X = \operatorname{Spec} K[t][x,y]/(y^2 - x^3 - t^m) \subset \mathbb{A}^2_Y \longrightarrow Y.$$

(b) (a family acquiring a tacnode)

$$\varphi : X = \operatorname{Spec} K[t][x,y]/(y^2 - x^4 - t^m) \subset \mathbb{A}^2_Y \longrightarrow Y.$$

(c) (a family acquiring an ordinary k-fold point)

$$\varphi : X = \operatorname{Spec} K[t][x,y]/(x^k + y^k - t^m) \subset \mathbb{A}^2_Y \longrightarrow Y.$$

V.4   Dual Curves

V.4.1   Definitions

We consider now another object associated to a plane curve $C \subset \mathbb{P}_S^2$: its *dual curve* C^*, a subscheme of the dual projective plane $(\mathbb{P}_S^2)^*$ as defined in Section III.2.3. In classical algebraic geometry, the definition is simple enough: for K an algebraically closed field and $C \subset \mathbb{P}_K^2$ nonsingular of degree $d \geq 2$, the dual curve is defined to be the set of projective tangent lines $\mathbb{T}_p C$ to C, regarded as points of the dual projective plane $(\mathbb{P}_K^2)^*$. More generally, for a curve C without multiple components and containing no lines the dual curve is defined to be the closure of the locus of tangent lines $\mathbb{T}_p C$ to C at nonsingular points p of C. (In the following, we will refer to this locus as the "classical dual".)

As in the case of flexes (Section IV.1), what we will do here is to propose a natural set of defining equations for C^*, which will yield a definition of the dual $C^* \subset (\mathbb{P}_S^2)^*$ of a plane curve $C \subset \mathbb{P}_S^2$ over an arbitrary scheme S. Also as in the case of flexes, C^* will be a closed subscheme of $(\mathbb{P}_S^2)^*$, flat over S over the open subset of S where it does not contain the fiber of $(\mathbb{P}_S^2)^*$, and invariant under base change. Our definition will agree with the classical definition for nonsingular plane curves over fields of characteristic zero, but not for singular ones: if p is an isolated singular point of a plane curve $C \subset \mathbb{P}_K^2$, the line in $(\mathbb{P}_K^2)^*$ dual to p will be a component of C^* in our definition, though it is not part of the classical dual. Again, this is necessary if we want the duals of a family of plane curves to form a closed family; if we want to recover information about the classical dual, we simply discard the extra components. (The need for characteristic zero will be explained below.)

To make our definition, we have first to introduce one auxiliary object. Recall from Section III.2.8 that in the product $\mathbb{P}_S^2 \times_S (\mathbb{P}_S^2)^*$ we have the *universal line* Σ, whose fiber over each point $l \in (\mathbb{P}_S^2)^*$ is the corresponding line $l \subset \mathbb{P}_S^2$: in terms of homogeneous coordinates X, Y, Z on $\mathbb{P}_S^2$ and A, B, C on $(\mathbb{P}_S^2)^*$,

$$\Sigma = V(AX + BY + CZ) \subset \mathbb{P}_S^2 \times_S (\mathbb{P}_S^2)^*.$$

Now let $C \subset \mathbb{P}_S^2$ be any plane curve, and assume for the moment that C contains no lines (that is, there is no point $s \in S$ and line $l \subset \mathbb{P}_{\kappa(s)}^2$ in the fiber of $\mathbb{P}_S^2$ over s contained in C; this is stronger than supposing that the equation of C does not have a linear factor). We define the *universal line section* Γ_C of C to be the intersection

$$\Gamma_C = \pi_1^{-1}(C) \cap \Sigma \subset \mathbb{P}_S^2 \times_{S'} (\mathbb{P}_S^2)^*.$$

By our hypothesis that C contains no lines, the map $\pi_2 : \Gamma_C \to (\mathbb{P}_S^2)^*$ will be finite (and hence flat) of degree d; and we define the *dual curve* $C^* \subset (\mathbb{P}_S^2)^*$ of C to be the discriminant scheme of this map, as in Definition V-21.

Finally, if C does contain lines, we simply define C^* to be the closure in $(\mathbb{P}^2_S)^*$ of the discriminant scheme of the restriction of π_2 to the open subset of $(\mathbb{P}^2_S)^*$ where it is finite.

A few introductory remarks are in order. To begin with, C^* is by definition a closed subscheme of $(\mathbb{P}^2_S)^*$, and its formation commutes with base change: if $S' \to S$ is any morphism and $C' = S' \times_S C \subset \mathbb{P}^2_{S'}$, then we will have also

$$\Gamma_{C'} = S' \times_S \Gamma_C \subset \mathbb{P}^2_{S'} \times_S (\mathbb{P}^2_{S'})^*,$$

and hence $C'^* = S' \times_S C^*$. In particular, for any point $s \in S$, the fiber C^*_s of C^* over s will be the dual of the fiber $C_s \subset \mathbb{P}^2_{\kappa(s)}$ of C over s.

The support of C^* is easy to describe: by the definition of the discriminant scheme of a finite morphism, it is the set of lines $l \in (\mathbb{P}^2_S)^*$ such that the intersection of the corresponding line $l \subset \mathbb{P}^2_S$ with C is singular over $\kappa(l)$. (This means either $l \cap C$ is nonreduced, or — in case $\kappa(l)$ has characteristic $p > 0$ and is not algebraically closed — has a point whose residue field in an inseparable extension of $\kappa(l)$.) In particular, as we indicated, if $p \in C \subset \mathbb{P}^2_K$ is any singular point, the line in $(\mathbb{P}^2_K)^*$ dual to p — that is, the locus of lines in $\mathbb{P}^2_K$ passing through p — will be contained in C^*.

Finally, we remark that the "dual curve" need not be a curve! If $C \subset \mathbb{P}^2_S$ is nonreduced — that is, it has a multiple component — then the dual C^* will be all of $(\mathbb{P}^2_S)^*$. Even if $C \subset \mathbb{P}^2_S$ is nonsingular and reduced it may have nonreduced fibers C_s over some points of S. For example, the fiber of the curve $C = V(X^2 + Y^2 + Z^2) \subset \mathbb{P}^2_{\mathbb{Z}}$ over $(2) \in \operatorname{Spec}\mathbb{Z}$ is a double line. In such cases the dual C^* will contain the corresponding fibers of $(\mathbb{P}^2_S)^*$, and so will not be a "plane curve" as we have defined it.

Exercise V-33. Verify that the dual C^* of the plane curve

$$C = V(X^2 + Y^2 + Z^2) \subset \mathbb{P}^2_{\mathbb{Z}}$$

mentioned above is, by our definition, the subscheme

$$C^* = V(4A^2 + 4B^2 + 4C^2) \subset (\mathbb{P}^2_{\mathbb{Z}})^*,$$

so is not a plane curve in $(\mathbb{P}^2_{\mathbb{Z}})^*$.

In the example of Exercise V-33, our definition seems at first willfully perverse: why shouldn't we take the dual to be the zero locus of $A^2 + B^2 + C^2$ instead? Indeed, as long as the base S is nonsingular of dimension one (as in the exercise) any fibers of $(\mathbb{P}^2_S)^* \to S$ contained in C^* will be components of C^*; we can discard them to arrive at a scheme $\tilde{C}^*$ flat over S. But we will see below how to construct examples of curves $C \subset \mathbb{P}^2_S$ with nonreduced fibers C_s over isolated points $s \in S$ where this simply cannot be avoided: the fiber $(\mathbb{P}^2_{\kappa(s)})^*$ of $(\mathbb{P}^2_S)^*$ over s will be contained in the closure of $C^* \cap (\mathbb{P}^2_{S\setminus\{s\}})^*$. If we want the definition to behave well with respect to base change, this means the dual of C_s has to be all of $(\mathbb{P}^2_{\kappa(s)})^*$.

Exercise V-34. Let K be a field, and $C \subset \mathbb{P}^2_K$ a curve, smooth over K. Show that the dual C^* is reduced, unless every tangent line to C is multiply tangent — that is, has intersection multiplicity 3 or more with C, or is tangent to C at more than one point.

In fact, if the characteristic of K is zero it cannot happen that every tangent line is multiply tangent (see for example Harris [1995, Proposition 15.3]), so the dual of any nonsingular plane curve will be reduced. But there are examples of plane curves over fields of finite characteristic such that every point is a flex, and other examples where every tangent line is tangent at several points. In such a case, our definition yields a nonreduced dual curve C^*.

Exercise V-35. Let K be any field of characteristic zero, and $C \subset \mathbb{P}^2_K$ a curve of degree d having no multiple components. Show that the dual curve $C^* \subset (\mathbb{P}^2_K)^*$ is a plane curve of degree exactly $d(d-1)$.

(Hint: Use the formula given in Proposition V-22 for the discriminant to describe the intersection of C^* with a general line in $(\mathbb{P}^2_K)^*$.)

V.4.2 Duals of Singular Curves

In case $C \subset \mathbb{P}^2_K$ has isolated singular points, as we said, our definition diverges from the classical: any line $l = p^* \subset \mathbb{P}^2_K$ passing through a singular point of C will correspond to a point in the support of the dual curve C^*. What is the multiplicity of this component? The following exercise gives the answer in some cases, and derives as a consequence one of the classical Plücker formulas (see Coolidge [1931]).

Exercise V-36. Using Example V-26 on discriminants, show that if $p \in C$ is a node, cusp, tacnode or ordinary triple point, then the line in $p^* \subset (\mathbb{P}^2_K)^*$ dual to p appears with multiplicity 2, 3, 4 or 6 respectively in C^*. Deduce the Plücker formula for the degree d' of the classical dual of an irreducible plane curve $C \subset \mathbb{P}^2_K$ over a field of characteristic zero of degree d, having as singularities δ nodes, κ cusps, λ tacnodes and τ ordinary triple points:

$$d' = d(d-1) - 2\delta - 3\kappa - 4\lambda - 6\tau.$$

(Note that the same count is valid if C is reducible, as long as no component of C is a line.) Find a curve singularity $p \in C$ for which p^* appears with multiplicity 5 in the dual curve C^*.

V.4.3 Curves with Multiple Components

As in the case of flexes (Section IV.1), a very different sort of question emerges when we consider curves with multiple components. Here, as we have said, the definition of the dual curve C^* yields not a curve in $(\mathbb{P}^2_K)^*$, but rather the whole dual plane $(\mathbb{P}^2_K)^*$. But we can consider a family of generically nonsingular curves specializing to a multiple curve and ask: in such a family, where do the tangent lines go?

We will illustrate this problem with the same sort of curve as we used in Section IV.1; but we will start with a simpler example: instead of a family of quartics specializing to a double conic we will consider a family of conics degenerating to a double line. To set it up, let K be an algebraically closed field of characteristic zero, and set $S = \operatorname{Spec} K[t]$. Let $L(X, Y, Z)$ and $Q(X, Y, Z) \in K[X, Y, Z]$ be homogeneous polynomials of degrees 1 and 2 respectively, and assume that their zero loci in $\mathbb{P}^2_K$ intersect transversely, that is, that $V(Q, L) \subset \mathbb{P}^2_K$ is reduced. Consider the curve $\mathscr{C} \subset \mathbb{P}^2_S$ given by

$$\pi : \mathscr{C} = \operatorname{Proj}(K[t])[X, Y, Z]/(L(X, Y, Z)^2 + tQ(X, Y, Z))$$
$$\subset \operatorname{Proj}(K[t])[X, Y, Z] = \mathbb{P}^2_S \to S.$$

Let $\mathscr{C}^* \subset (\mathbb{P}^2_S)^*$ be the dual of the curve $\mathscr{C} \subset \mathbb{P}^2_S$. As with the scheme of flexes of the curve $\mathscr{C}$, the scheme $\mathscr{C}^*$ will have two components: one, the fiber of $(\mathbb{P}^2_S)^*$ over the origin $(t) \in S$; and the other the closure $\mathscr{C}'$ of the inverse image in $\mathscr{C}^*$ of the punctured line $T = \operatorname{Spec} K[t, t^{-1}] = S \setminus \{(t)\} \subset S$ (equivalently, the closure of the dual of the curve $\mathscr{C} \cap \mathbb{P}^2_T \subset \mathbb{P}^2_T$). The scheme $\mathscr{C}'$ will be flat over all of S, with fiber over a point $(t - \mu) \in S$ other than the origin the dual $(C_\mu)^*$ of the curve $C_\mu = V(F + \mu G) \subset \mathbb{P}^2_K$; it will therefore have as fiber over the origin a scheme $C'_0 \subset \mathbb{P}^2_K$ of dimension 1 and degree 2, which we will call the "limiting position" of the duals of the nearby nonsingular curves C_μ as μ approaches zero. Once more, we can translate the naive question, "where do the tangent lines to a conic go when the conic degenerates into a double line?" into the precise problem: determine the support of the flat limit C'_0. The answer is expressed in the following

Proposition V-37. *The fiber C'_0 of $\mathscr{C}'$ over the origin $(t) \in S$ consists of the union of the two lines dual to the two points $t = L = Q = 0$ of intersection of the line $L = 0$ and the conic $Q = 0$ in the plane $t = 0$.*

Proof. We will do this by explicit calculation. To begin with, since the characteristic of K is not 2, we can choose affine coordinates (x, y) on the plane so that the line $V(L)$ is the x-axis $y = 0$, and the conic $V(Q)$ (if it is nonsingular) the zero locus of $y^2 - x^2 + 1$. The equation of the curve $\mathscr{C} \subset \mathbb{P}^2_S$ is then

$$\mathscr{C} = V(y^2 + t(y^2 - x^2 + 1))$$

and after replacing the coordinate t on $S = \mathbb{A}^1_K$ with the coordinate $u = t/(1 - t)$ in a neighborhood of the origin and multiplying through by $1 - t$, we may rewrite this as

$$\mathscr{C} = V(y^2 - t(x^2 - 1))$$

(note that if $V(Q)$ were singular, we could have taken this as the equation of $\mathscr{C}$ originally). Now, we can express any line in $\mathbb{P}_K^2$ not passing through the point $[0, 1, 0]$ (that is, the point at infinity on the y-axis) as the zero locus of an equation

$$y - ax - b;$$

a and b are then affine coordinates on the corresponding subset $\mathbb{A}_K^2 \subset (\mathbb{P}_K^2)^*$. The equations of the universal line section $\Gamma_{\mathscr{C}}$ of $\mathscr{C}$ in the open subset $\mathbb{A}_K^2 \times \mathbb{A}_K^2 \subset \mathbb{P}_K^2 \times (\mathbb{P}_K^2)^*$ are then

$$\Gamma_{\mathscr{C}} = V(y - ax - b, y^2 - t(x^2 - 1)) \subset \operatorname{Spec} K[a, b, x, y]$$
$$= V((ax + b)^2 - t(x^2 - 1)) \subset \operatorname{Spec} K[a, b, x].$$

Now, we may expand out the equation of $\Gamma_{\mathscr{C}}$ as

$$(ax + b)^2 - t(x^2 - 1) = (a^2 - t)x^2 + 2abx + (b^2 + t),$$

from which we see that the equation of the dual curve $\mathscr{C}^*$ is

$$(2ab)^2 - 4(a^2 - t)(b^2 + t) = -4ta^2 + 4tb^2 + 4t^2 = -4t(a^2 - b^2 - t).$$

This has, as expected, two components: the entire fiber of $\mathbb{P}_S^2$ over the origin, and the curve $\mathscr{C}'$ given as the zero locus

$$\mathscr{C}' = V(a^2 - b^2 - t) \subset \mathbb{P}_S^2.$$

The intersection of this second component with the fiber of $\mathbb{P}_S^2$ over the origin is then

$$C_0' = V(a^2 - b^2) = V(a + b) \cup V(a - b) \subset \mathbb{P}_K^2.$$

In other words, it is the union of the two lines in $(\mathbb{P}_K^2)^*$ dual to the points $(1, 0)$ and $(-1, 0)$ of intersection of the line (y) and the conic $x^2 - 1$ in the plane.   □

In fact, we can see this result from the real picture: if we draw the family of conics $y^2 = t(x^2 - 1)$ specializing to the double line $y = 0$, we see readily that every line through either of the points $(\pm 1, 0)$ is a limit of tangent lines to the curves $y^2 = t(x^2 - 1)$ for small values of t; we may also write this family of tangent lines directly. For example, the line $y + x + 1 = 0$ through the point $(-1, 0)$ is the limit of the line

$$y + x + \sqrt{1 - t} = 0,$$

which is tangent to the curve $y^2 = t(x^2 - 1)$. It is also clear from the picture that any line meeting the x-axis in a point other than $(\pm 1, 0)$ (other than the x-axis itself) will be transverse to the curve $y^2 = t(x^2 - 1)$ for small t.

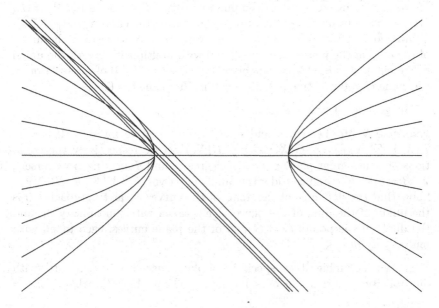

The same sort of phenomena occur with arithmetic schemes. In fact, we have already seen an example: the conic $\mathscr{C} = V(X^2 + Y^2 + Z^2) \subset \mathbb{P}^2_{\mathbb{Z}}$, which we may view as a family of conics over $S = \operatorname{Spec}\mathbb{Z}$, specializing from a nonsingular conic over the generic point of $\operatorname{Spec}\mathbb{Z}$ to a double line over the point $(2) \in \operatorname{Spec}\mathbb{Z}$. Here the function 2 on $\operatorname{Spec}\mathbb{Z}$ plays the role of the variable t; we can take $L(X, Y, Z) = X + Y + Z$ and $Q(X, Y, Z) = -XY - YZ - XZ$ to arrive at the curve $\mathscr{C} = V(L^2 + 2Q)$. The only difference here is that fibers $L_{(2)}$ and $Q_{(2)}$ over (2) in the conic Q and the line L do not intersect in two points, but rather meet in a single point with residue field $\mathbb{F}_4$. The result is expressed in the

Exercise V-38. Show that the limit of the duals of the fibers of $\mathscr{C}$ over $\operatorname{Spec}\mathbb{Z} \setminus \{(2)\}$ is the line in $(\mathbb{P}^2_{\mathbb{F}_4})^*$ corresponding to the one point of intersection of $L_{(2)}$ and $Q_{(2)}$ in $\mathbb{P}^2_{\mathbb{F}_4}$.

If you are curious about the duals of the quartic curves in the family specializing to a double conic, here is the situation:

Exercise V-39. Let K be an algebraically closed field of characteristic zero, and set $S = \operatorname{Spec}K[t]$. Let $Q(X, Y, Z)$ and $G(X, Y, Z) \in K[X, Y, Z]$ be a homogeneous quadric polynomial and a homogeneous quartic polynomial respectively such that the curves $V(Q)$ and $V(G)$ intersect transversely, and $\mathscr{C} \subset \mathbb{P}^2_S$ the curve given by

$$\pi : \mathscr{C} = \operatorname{Proj}K[t][X, Y, Z]/(Q(X, Y, Z)^2 + tG(X, Y, Z))$$
$$\subset \operatorname{Proj}K[t][X, Y, Z] = \mathbb{P}^2_S \to S.$$

Let $\mathscr{C}^* \subset (\mathbb{P}_S^2)^*$ be the dual of the curve $\mathscr{C} \subset \mathbb{P}_S^2$, and $\mathscr{C}'$ the closure of the inverse image in $\mathscr{C}^*$ of the punctured line $T = \operatorname{Spec} K[t, t^{-1}] = S \setminus \{(t)\} \subset S$ (equivalently, the closure of the dual of the curve $\mathscr{C} \cap \mathbb{P}_T^2 \subset \mathbb{P}_T^2$). Show that the fiber C_0' of $\mathscr{C}'$ over the origin $(t) \in S$ consists of the union of the dual of the plane conic $t = Q = 0$ with multiplicity 2 with the union of the eight lines dual to the eight points $t = Q = G = 0$ of intersection of the conic $Q = 0$ and the quartic $G = 0$ in the plane $t = 0$.

The general situation is this:

Exercise V-40. Let m, e and $d = me$ be positive integers. Let $F = F(X, Y, Z)$ and $G = G(X, Y, Z) \in K[X, Y, Z]$ be respectively the equations of nonsingular plane curves of degrees e and d meeting transversely, let $B = \operatorname{Spec} K[t]$ and consider the family of curves $\mathscr{C} = V(F^m + tG) \subset \mathbb{P}_B^2$. Show that the flat limit of the duals of the curves C_t as t approaches 0 is the union of the dual of the curve $V(F)$, taken with multiplicity m, and the duals of the points $F = G = 0$ of the plane curves, each taken with multiplicity $m - 1$.

There is no ambiguity in specifying a plane curve Γ as its support "with multiplicity m": this can only mean the scheme $V(F^m)$, where $\Gamma_{\mathrm{red}} = V(F)$.

V.5   Double Point Loci

Let $\varphi : X \to Y$ be a morphism of varieties. The *double point locus* of φ was classically defined to be the closure in $X \times X$ of the locus of pairs of distinct points $p, q \in X$ with common image $\varphi(p) = \varphi(q) \in Y$. In this section we will give a scheme-theoretic definition that, as we will see, captures more of the geometry of the map.

For the following, we will let $\varphi : X \to Y$ be a separated morphism of schemes. As we saw in Chapter III, the separated hypothesis means that the diagonal Δ_X is a closed subscheme of the fiber product $X \times_Y X$, and is satisfied for all affine and projective schemes.

Definition V-41. The *double point scheme* D_φ of a separated morphism $\varphi : X \to Y$ is the scheme

$$D_\varphi = V(\operatorname{ann} \mathscr{I}_{\Delta_X}) \subset X \times_Y X$$

associated to the ideal $\operatorname{ann} \mathscr{I}_{\Delta_X} \subset \mathscr{O}_{X \times_Y X}$ of functions f with $f\mathscr{I}_{\Delta_X} = 0$.

To understand this definition, recall that if $\varphi : X \to Y$ is a morphism of S-schemes (in the examples below S will be $\operatorname{Spec} K$ or $\operatorname{Spec} \mathbb{Z}$) and $Y \to S$ is separated, then the fiber product $X \times_Y X$ is a closed subscheme of $X \times_S X$: it is the (scheme-theoretic) inverse image

$$X \times_Y X = (\varphi \times \varphi)^{-1}(\Delta_Y)$$

where $\Delta_Y \subset Y \times_S Y$ is the diagonal. (Note that we have a fiber square

$$
\begin{array}{ccc}
X \times_Y X & \longrightarrow & Y \\
\downarrow & & \downarrow{\scriptstyle \Delta} \\
X \times_S X & \longrightarrow & Y \times_S Y;
\end{array}
$$

that is, $X \times_Y X$ is the fibered product of $X \times_S X$ and Y over $Y \times_S Y$.) Now, away from the diagonal, the double point scheme D_φ is simply the fiber product $X \times_Y X$; in particular, the closed points of its support away from the diagonal are simply the pairs of distinct points $p, q \in X$ such that $\varphi(p) = \varphi(q)$. In this sense it generalizes the classical double point locus, though it may have nontrivial scheme structure even in case φ is a morphism of varieties. In addition, as our examples will show, it may have components supported on the diagonal $\Delta_X \subset X \times_S X$ — that is, its support may properly contain the double point locus — and may even be nonempty in cases where the classical double point locus is empty.

Example V-42. Consider the maps from $X = \mathbb{A}_K^1 = \operatorname{Spec} K[t]$ to $Y = \mathbb{A}_K^2 = \operatorname{Spec} K[x, y]$ given by polynomials of degree 3, mapping X onto plane cubic curves with a node or a cusp. Specifically, consider for each value of the parameter λ the map $\varphi_\lambda : X \to Y$ given by the ring homomorphism

$$
\begin{aligned}
\varphi_\lambda^\# : K[x, y] &\longrightarrow K[t] \\
x &\longmapsto t^2 - \lambda \\
y &\longmapsto t^3 - \lambda t.
\end{aligned}
$$

The image of φ_λ is the plane curve C_λ with equation $y^2 = x^2(x + \lambda)$.

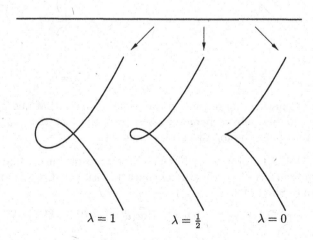

$\lambda = 1$ \qquad $\lambda = \frac{1}{2}$ \qquad $\lambda = 0$

The fiber product $X \times_Y X$ is the spectrum $\operatorname{Spec} R$ of the algebra

$$\begin{aligned} R &= K[t_1] \otimes_{K[x,y]} K[t_2] \\ &= K[t_1, t_2] / \big((t_1^2 - \lambda) - (t_2^2 - \lambda), (t_1^3 - \lambda t_1) - (t_2^3 - \lambda t_2) \big) \\ &= K[t_1, t_2] / \big((t_1 - t_2)(t_1 + t_2, t_1^2 - \lambda) \big). \end{aligned}$$

By our definition, then, the double point scheme D_φ of $\varphi = \varphi_\lambda$ is given by

$$\begin{aligned} D_\varphi &= V\big(\operatorname{ann}((t_1 - t_2)/((t_1 - t_2)(t_1 + t_2, t_1^2 - \lambda))) \big) \\ &= V((t_1 + t_2, t_1^2 - \lambda)) \subset \operatorname{Spec} K[t_1, t_2] = X \times_K X. \end{aligned}$$

Assuming that the characteristic of K is not 2, for $\lambda \neq 0$, D_φ consists of the two reduced points $(\sqrt{\lambda}, -\sqrt{\lambda})$ and $(-\sqrt{\lambda}, \sqrt{\lambda})$. For $\lambda = 0$, on the other hand, D_φ is a double point supported at the origin. (Note that D_{φ_0} is the flat limit of the schemes D_{φ_λ} for $\lambda \neq 0$ as λ approaches 0.)

To see the difference between the classical and modern approaches, observe that for $\lambda \neq 0$ the scheme $X \times_Y X$, as pictured in the figure below, is reduced; hence the double point scheme is reduced as well and coincides with the classical double point locus. For $\lambda = 0$, however, they differ: the scheme $X \times_Y X$ consists of the diagonal in $X \times_K X$, plus an embedded point supported at the origin. In the classical language, we see only the reduced scheme associated to $X \times_Y X$, and so miss the embedded point; thus the double point locus of φ_0 is empty. Scheme theory, by contrast, does see the nonreduced structure of $X \times_Y X$ and as a result the double point scheme of φ_0 is nonempty, reflecting the fact that φ_0 fails to be an immersion at $t = 0$.

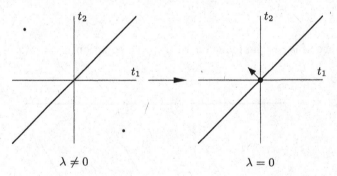

$$\lambda \neq 0 \qquad\qquad\qquad\qquad \lambda = 0$$

To make somewhat more precise the remark above that the nonemptiness of the double point scheme of φ_0 reflects the fact that φ_0 is not an immersion, we have the following two exercises.

Exercise V-43. Let $\varphi : X \to Y$ be a finite map of nonsingular varieties over an algebraically closed field K of characteristic zero, and D_φ its double point scheme. Show that

$$\operatorname{supp}(D_\varphi \cap \Delta_X) = \{(p,p) : \operatorname{Ker}(d\varphi_p) \neq 0\} \subset X \times_K X.$$

Exercise V-44. Let $\varphi : X \to Y$ again be a finite map of irreducible varieties and D_φ its double point scheme. Using Harris [1995, Theorem 14.9], we see that in fact φ is an embedding if and only if $D_\varphi = \varnothing$, that is, if and only if

$$(\varphi \times \varphi)^{-1}(\Delta_Y) = \Delta_X$$

as schemes (which is to say, $X \to X \times_Y X$ is an isomorphism).

Here is an exercise showing that the same geometric ideas apply as well in the context of arithmetic schemes: the schemes in question are spectra of rings of integers in number fields (they were first introduced in Section II.4.2), but as we will see they behave very much like the curves described in the example above.

Exercise V-45. Consider the maps

$$\varphi : \operatorname{Spec} \mathbb{Z}[11\sqrt{3}] \longrightarrow \mathbb{A}^1_{\mathbb{Z}}$$

and

$$\mu : \operatorname{Spec} \mathbb{Z}[2\sqrt{3}] \longrightarrow \mathbb{A}^1_{\mathbb{Z}}$$

associated to the maps $\mathbb{Z}[t] \to \mathbb{Z}[11\sqrt{3}]$ and $\mathbb{Z}[t] \to \mathbb{Z}[2\sqrt{3}]$. Describe the double point scheme of each map, and in particular reconfirm the assertion of Exercise II-36 that the singularity of the former is a node, while that of the latter is a cusp.

Finally, we point out also that nonreduced structures on the double point locus arise naturally even away from the diagonal in $X \times X$, for example for maps φ that are immersions. For example, consider the case where Y is a curve and $\varphi : X \to Y$ its normalization. Suppose that two points $P, Q \in X$ map to the same point R of the image curve $Y = \varphi(X)$.

Exercise V-46. Show that if R is a node, the double point scheme will be reduced at the point $(P, Q) \in X \times X$. By contrast, show that if R is a tacnode (that is, Y has two smooth branches at R, simply tangent to one another) D_φ will be nonreduced at the point $(P, Q) \in X \times X$. Can you describe the scheme structure?

VI

Schemes and Functors

At the end of the first chapter of this book we discussed a way of embedding the category of schemes into the larger category of contravariant functors F from the category of schemes to the category of sets.

This embedding is useful in at least three ways:

(1) The effect of some basic constructions, such as products, is much easier to describe on functors of points than on schemes.

(2) In trying to construct a certain scheme, it is often easy to construct the functor that would be the functor of points of that scheme, if the scheme existed; the construction problem is then reduced to the problem of proving that the functor is representable and the use of Yoneda's Lemma (VI-1). The process is exactly analogous to the use of distributions in analysis: there, when trying to prove the existence of a nice function solving a given differential equation, one first proves the existence of a solution that is a distribution, and then is left with the (possibly more tractable) *regularity* problem of proving that the distribution is represented by integration against a function.

(3) Many aspects of the geometry of schemes can be extended to the category of functors, so that it is sometimes useful to forget about representing a functor and work in that category (or some suitable subcategory) directly.

In this chapter we illustrate these points, first with some basic constructions, and then with some examples coming from the desire to parametrize families of schemes. We shall see in Section VI.2.4 that some of these lead to functors that are not actually schemes, though they are rather close.

VI.1   The Functor of Points

We start where we left off at the end of Chapter I. Recall that the *functor of points* of a scheme X is the functor

$$h_X : (\text{schemes})^\circ \to (\text{sets})$$

where $(\text{schemes})^\circ$ and (sets) represent the category of schemes with the arrows reversed and the category of sets; h_X takes each scheme Y to the set

$$h_X(Y) = \text{Mor}(Y, X)$$

and each morphism $f : Y \to Z$ to the map of sets

$$h_X(Z) \to h_X(Y)$$

defined by sending an element $g \in h_X(Z) = \text{Mor}(Z, X)$ to the composition $g \circ f \in \text{Mor}(Y, X)$. We say that a functor $F : (\text{schemes})^\circ \to (\text{sets})$ is *representable* if it is of the form h_X for some scheme X. By Yoneda's Lemma below, X is unique if it exists; in this case we say that X *represents F*. The set $h_X(Y)$ is called the set of *Y-valued points* of X (if $Y = \text{Spec}\, T$ is affine, we will often write $h_X(T)$ instead of $h_X(\text{Spec}\, T)$ and call it the set of T-valued points of X).

Recall also that this construction defines a functor

$$h : (\text{schemes}) \to \text{Fun}((\text{schemes})^\circ, (\text{sets}))$$

(where morphisms in the category of functors are natural transformations), sending

$$X \mapsto h_X$$

and associating to a morphism $f : X \to X'$ the natural transformation $h_X \to h_{X'}$ that for any scheme Y sends $g \in h_X(Y) = \text{Mor}(Y, X)$ to the composition $f \circ g \in h_{X'}(Y) = \text{Mor}(Y, X')$.

In order for this notion to be of any use at all, a crucial first fact is that the functor of points h_X really does determine the scheme X. This follows from a basic categorical fact.

Lemma VI-1 (Yoneda's Lemma). *Let $\mathscr{C}$ be a category and let X, X' be objects of $\mathscr{C}$.*

(a) *If F is any contravariant functor from $\mathscr{C}$ to the category of sets, the natural transformations from $\text{Mor}(-, X)$ to F are in natural correspondence with the elements of $F(X)$.*

(b) *If the functors $\text{Mor}(-, X)$ and $\text{Mor}(-, X')$ from $\mathscr{C}$ to the category of sets are isomorphic, then $X \simeq X'$. More generally, the maps of functors from $\text{Mor}(-, X)$ to $\text{Mor}(-, X')$ are the same as maps from X to X'; that is, the functor*

$$h : \mathscr{C} \to \text{Fun}(\mathscr{C}^\circ, (\text{sets}))$$

sending X to h_X is an equivalence of $\mathscr{C}$ with a full subcategory of the category of functors.

Proof. For part (a), the correspondence sends $\alpha : \mathrm{Mor}(-,X) \to F$ to the element $\alpha(1_X)$, where $1_X : X \to X$ is the identity map. The inverse takes $p \in F(X)$ to the map α sending an element $f \in \mathrm{Mor}(Y,X)$ to $F(f)(p) \in F(Y)$. As for part (b), we can apply the statement of part (a) to the functor $F = \mathrm{Mor}(-,Y)$. $\qquad\qquad\square$

The following improvement of Lemma VI-1 shows that it is enough to look at the functor of points restricted to the category of affine schemes, or, equivalently, to the category (rings)°, the category of commutative rings with the arrows reversed; and the same thing works in the relative setting.

Proposition VI-2. *If R is a commutative ring, a scheme over R is determined by the restriction of its functor of points to affine schemes over R; in fact*

$$h : (R\text{-schemes}) \to \mathrm{Fun}((R\text{-algebras}),(sets))$$

is an equivalence of the category of R-schemes with a full subcategory of the category of functors.

Of course, a contravariant functor on the category of affine schemes is the same as a covariant functor on the category of rings; so given this result, we will generally think of our contravariant representable functors $h_X : (\text{schemes})° \to (\text{sets})$ as covariant functors on R-algebras. (If we need to make a distinction, we will denote by $h_X^* : (\text{rings}) \to (\text{sets})$ the functor defined by $h_X^*(A) = h_X(\mathrm{Spec}\, A)$ for any R-algebra A.)

Proof. This is really just the statement that schemes are built up out of affine schemes. Let $S = \mathrm{Spec}\, R$. Write h_X for the functor $\mathrm{Mor}_S(-,X)$ restricted to the category of affine schemes over S. It is enough to show that any natural transformation $\varphi : h_X \to h_{X'}$ comes from a unique morphism f over S from X to X'. To construct f from φ, let $\{U_a\}$ be an affine cover of X, and apply φ to the inclusion maps $U_a \subset X$ to get morphisms $U_a \to X'$. These morphisms satisfy the compatibility conditions necessary to define the desired morphism f. Uniqueness comes down to the observation that two morphisms from X to X' that differ are already different when restricted to one of the U_a. $\qquad\square$

Exercise VI-3. Suppose that X is (like virtually all schemes of interest to us) *locally Noetherian* — that is, covered by spectra of Noetherian rings. Prove that X is determined by the restriction of h_X to the category of Noetherian rings.

VI.1.1   Open and Closed Subfunctors

One reason that thinking of schemes as functors is useful is that it is possible to extend some of the basic notions from the geometry of schemes to functors. We will consider some examples.

We first show how to define an open subfunctor of a functor

$$F \in \text{Fun}((\text{rings}), (\text{sets})).$$

We say that a map $\alpha : G \rightarrow F$ of functors from a category $\mathscr{C}$ to the category of sets is *injective* if for every object X the induced map of sets $G(X) \rightarrow F(X)$ is injective (this corresponds to the standard categorical notion, but we will not need this fact). In this case we will say that $\alpha : G \rightarrow F$ is a *subfunctor* of F. For example, if $U \subset X$ is a subscheme, the functor h_U will be a subfunctor of h_X.

We want to define an open subfunctor of a functor F to be a subfunctor that, when restricted to a representable subfunctor $h_X \subset F$, is of the form $h_U \subset h_X$ for an open subscheme $U \subset X$. To carry this out, we need to introduce the notion of a fibered product of functors.

Definition VI-4. If A, B, and C are functors from some category $\mathscr{C}$ to the category of sets and if $f : A \rightarrow C$ and $g : B \rightarrow C$ are morphisms of functors, the *fibered product* $A \times_C B$ is the functor from $\mathscr{C}$ to (sets) defined by setting, for any object Z of $\mathscr{C}$,

$$(A \times_C B)(Z) = \{(a,b) \in A(Z) \times B(Z) \mid f(a) = f(b) \text{ in } C(Z)\},$$

and defined on morphisms of $\mathscr{C}$ in the obvious way.

Definition VI-5. A subfunctor $\alpha : G \rightarrow F$ in $\text{Fun}((\text{rings}), (\text{sets}))$ is an *open subfunctor* if, for each map $\psi : h_{\text{Spec} R} \rightarrow F$ from the functor represented by an affine scheme $\text{Spec} R$ (that is, each $\psi \in F(R)$), the fibered product

$$
\begin{array}{ccc}
G_\psi & \longrightarrow & h_{\text{Spec} R} \\
\downarrow & & \downarrow{\psi} \\
G & \xrightarrow{\ \alpha\ } & F
\end{array}
$$

of functors yields a map $G_\psi \rightarrow h_{\text{Spec} R}$ isomorphic to the injection from the functor represented by some open subscheme of $\text{Spec} R$.

Exercise VI-6. Let $X = \text{Spec} R$ be an affine scheme. Show that the open subfunctors of h_X are exactly the functors of the form

$$F(T) = \{\varphi \in h_X(T) \mid \varphi^*(I)T = T\},$$

for some ideal $I \subset R$.

Exercise VI-7. Let X be a scheme over the field K. Define a functor $F : (\text{schemes}/K)^\circ \rightarrow (\text{sets})$ as follows: for each K-scheme Y, let $F(Y)$ be the set of closed subschemes $Z \subset X \times_K Y$ flat over Y and such that all the fibers of Z over closed points of Y are subschemes of degree 2 of X. Let G be the subfunctor of F obtained by adding the requirement that the fibers of Z over closed points of Y are reduced. Show that $G \subset F$ is open.

To define closed functors, we proceed similarly. A subfunctor $\alpha : G \to F$ in $\mathrm{Fun}((\mathrm{rings}), (\mathrm{sets}))$ is *closed* if for each map $\psi : h_{\mathrm{Spec}\,R} \to F$ the fibered product of ψ and α is a subfunctor of $h_{\mathrm{Spec}\,R}$ isomorphic to the functor represented by a closed subcheme of $\mathrm{Spec}\,R$.

Exercise VI-8. Let $X = \mathrm{Spec}\,R$ be an affine scheme. Show that the open and closed subfunctors of $h_{\mathrm{Spec}\,R}$ are precisely those represented by open and closed subschemes of $\mathrm{Spec}\,R$. (The same is true, and only a little harder, for arbitrary schemes.)

As usual, a little caution is necessary when using these notions. For example:

Exercise VI-9. Suppose that F is a functor from a category $\mathscr{C}$ to the category of sets, and let G be a subfunctor of F. Show by example that the association $C \mapsto F(C) \setminus G(C)$ may not define a functor. Suppose now that $\mathscr{C}$ is the category of rings and F is represented by a scheme X while G is represented by a closed subscheme Y of X, and H is the functor reprsented by the open subscheme $X \setminus Y$. Can you describe H in terms of G and F?

Exercise VI-10. Let X be a scheme over the field K. Define the functor $F : (\mathrm{scheme}/K)^\circ \to (\mathrm{sets})$ as in Exercise VI-7, and define a subfunctor H of F by letting $H(Y)$, for each K-scheme Y, be the set of closed subschemes $Z \subset X \times Y$ flat over Y and such that all the fibers of Z over closed points of Y are subschemes of degree 2 of X supported at a single point of X. Show that H is not in general a closed subfunctor of F.

We will also use the notion of an *open covering* of a functor. This is a collection of open subfunctors that yields an open covering of a scheme whenever we pull back to a representable functor. More precisely: let $F :$ (schemes) $\to$ (sets) be a functor. Consider a collection $\{G_i \to F\}$ of open subfunctors of F. For each map $h_x \to F$ from a representable functor h_X to F, there are open subschemes $U_i \subset X$ such that the fiber product $h_X \times_F G_i$ of h_X and G_i is h_{U_i}. We say that the collection $\{G_i \to F\}$ is an open covering if, for any such map $h_X \to F$, the corresponding open subschemes $U_i \subset X$ cover X.

One warning: if $\{G_i \to F\}$ is an open covering of F, it is not necessarily the case that $F(T) = \bigcup G_i(T)$ for all schemes T. (For example, consider

$$F = h_{\mathrm{Spec}\,\mathbb{Z}}, \quad G_1 = h_{\mathrm{Spec}\,\mathbb{Z}[1/p]}, \quad G_2 = h_{\mathrm{Spec}\,\mathbb{Z}[1/q]},$$

where p and q are distinct primes. Then $\{G_i \to F\}$ is an open covering, but $F(\mathrm{Spec}\,\mathbb{Z})$ consists of one point (the identity map), whereas $G_i(\mathrm{Spec}\,\mathbb{Z}) = \varnothing$ for $i = 1, 2$.) However, it is the case that an open covering $\{G_i \to F\}$

yields a covering $F(T) = \bigcup G_i(T)$ for sufficiently local flat schemes T. For example:

Exercise VI-11. Let $\{G_i \to F\}$ be a collection of open subfunctors of a functor F : (schemes) $\to$ (sets). Show that this is an open covering if and only if $F(\operatorname{Spec} K) = \bigcup G_i(\operatorname{Spec} K)$ for all fields K.

VI.1.2   K-Rational Points

If X is a scheme over a field K, the K-valued points of X over K are maps $\operatorname{Spec} K \to X$ whose composition with the natural map $X \to \operatorname{Spec} K$ is the identity. We claim that such maps correspond exactly to closed points p of X that are *rational over* K (or *K-rational*) in the sense that the residue class field $\kappa(p)$ is K (via the inclusion map of K into the local ring of X at p). Indeed, since $\operatorname{Spec} K$ has no nontrivial open coverings, a map from $\operatorname{Spec} K$ into X is a map into some affine open subscheme $\operatorname{Spec} T$ of X, and such a morphism is determined by a K-algebra map $T \to K$ — that is, by a maximal ideal of T with residue class field K. Conversely, we may reverse the construction and see that any K-rational closed point p gives rise to a unique morphism $\operatorname{Spec} K \to X$ of K-schemes.

We reiterate the warning about working in the category of S-schemes rather than the category of all schemes, where applicable. For example, when working with complex varieties, one would expect $h_{\operatorname{Spec} \mathbb{C}}(\operatorname{Spec} \mathbb{C})$ to be a single point (the identity map) — and this is true in the category of $\mathbb{C}$-schemes. But in the category of schemes, this set is very large!

Exercise VI-12. Let $X = \operatorname{Spec} \mathbb{C}$, considered as an abstract scheme, that is, a scheme over $\mathbb{Z}$. Describe the set $h_X(\operatorname{Spec} \mathbb{C})$ of all $\mathbb{C}$-valued points of $\operatorname{Spec} \mathbb{C}$.

VI.1.3   Tangent Spaces to a Functor

Sometimes it is much easier to compute geometric information about a scheme if one knows its functor of points than if one knows its equations! A typical example occurs with the Zariski tangent space. (We will see this applied in Section VI.2.3).

We will work with schemes over a fixed field K, and all morphisms will be morphisms over K.

Recall that if X is a scheme, then for any K-rational point $p \in X$ the Zariski tangent space to X at p is $\operatorname{Hom}_K(\mathfrak{m}/\mathfrak{m}^2, K)$, where $\mathfrak{m} = \mathfrak{m}_{X,p}$ is the maximal ideal in the local ring of X at p and $K = \kappa(p) = \mathscr{O}_{X,p}/\mathfrak{m}_{X,p}$ is the residue field of X at p. Now let X be a scheme over K. Let X be the affine scheme $\operatorname{Spec} K[\varepsilon]/(\varepsilon)^2$. We claim that *a $K[\varepsilon]/(\varepsilon)^2$-valued point of X is the same as a K-rational closed point p of X together with an element of the Zariski tangent space to X at p.*

To check this assertion, note first that because of the inclusion map $\iota : \operatorname{Spec} K \to \operatorname{Spec} K[\varepsilon]/(\varepsilon^2)$ induced by the algebra map $K[\varepsilon]/(\varepsilon^2) \to K$, a morphism $\operatorname{Spec} K[\varepsilon]/(\varepsilon^2) \to X$ determines a morphism $\operatorname{Spec} K \to X$ and thus a closed, K-rational point p of X. An extension of such a morphism $\operatorname{Spec} K \to X$ consists of a lifting of the K-algebra homomorphism $\mathscr{O}_{X,p} \to \kappa(p) = K$ to a (local) homomorphism $\mathscr{O}_{X,p} \to K[\varepsilon]/(\varepsilon^2)$. Such a homomorphism induces a map from the maximal ideal $\mathfrak{m}_{X,p}$ to the maximal ideal (ε) of $K[\varepsilon]/(\varepsilon^2)$, and since $(\varepsilon^2) = 0$, this map factors through a map $t : \mathfrak{m}_{X,p}/\mathfrak{m}_{X,p}^2 \to (\varepsilon) \cong K$. The map t is an element of $(\mathfrak{m}_{X,p}/\mathfrak{m}_{X,p}^2)^*$, the Zariski tangent space to X at p.

Conversely, given a K-rational point $p \in X$ and $t : \mathfrak{m}_{X,p}/\mathfrak{m}_{X,p}^2 \to (\varepsilon) \cong K$ we may construct a map $\mathscr{O}_{X,p} \to K[\varepsilon]/(\varepsilon^2)$ from the map $\pi : \mathscr{O}_{X,p} \to \kappa(p) = K$ corresponding to p, as follows: π and the K-algebra structure map $K \to \mathscr{O}_{X,p}$ define a K-vector space splitting of $\mathscr{O}_{X,p}/\mathfrak{m}_{X,p}^2$ into $K \oplus \mathfrak{m}_{X,p}/\mathfrak{m}_{X,p}^2$; and we use the identity map on K and the map t on $\mathfrak{m}_{X,p}/\mathfrak{m}_{X,p}^2$ to define a map

$$\mathscr{O}_{X,p}/\mathfrak{m}_{X,p}^2 \to K[\varepsilon]/(\varepsilon^2),$$

which by composition with the projection gives a map $\mathscr{O}_{X,p} \to K[\varepsilon]/(\varepsilon^2)$. This map determines the desired morphism of schemes $\operatorname{Spec} K[\varepsilon]/(\varepsilon^2) \to X$.

In Section VI.2.1 we will see how this characterization of tangent vectors to a scheme may be used to give an intrinsic description of tangent vectors to projective space, completing the discussion in Section III.2.4.

If now F is a functor from the category of K-algebras to (sets) and $p \in F(K)$, we may define the tangent space $T_p F$ to F at p to be the fiber over p in $F(K[\varepsilon]/(\varepsilon^2)) \to F(K)$. One objection that may be raised to this definition is that it gives us the tangent space as a set, rather than as a vector space over K. One can at least define multiplication by elements of K: if $a \in K$, then the map $\varepsilon \mapsto a\varepsilon$ induces an endomorphism of the algebra $K[\varepsilon]/\varepsilon^2$, making the diagram

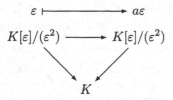

commute; and this induces a map $T_p F \to T_p F$, which is the desired multiplication by a. To make $T_p F$ a vector space, we now need to define an addition map $T_p F \times T_p F \to T_p F$. In general, there seems to be no way to do this; but for those functors F that, like representable functors, preserve fibered products, we can.

To do this, consider the scheme $\operatorname{Spec} K[\varepsilon', \varepsilon'']/(\varepsilon', \varepsilon'')^2$ —that is, the closed subscheme of the plane given by the square of the maximal ideal

of a point. The commutative diagram

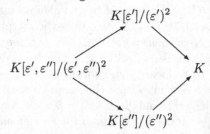

given by the projections expresses $K[\varepsilon', \varepsilon'']/(\varepsilon', \varepsilon'')^2$ as the fibered product over K of $K[\varepsilon']/(\varepsilon')^2$ and $K[\varepsilon'']/(\varepsilon'')^2$. On the other hand, there is a third map, $\sigma : K[\varepsilon', \varepsilon'']/(\varepsilon', \varepsilon'')^2 \to K[\varepsilon]/(\varepsilon^2)$, that takes *both* ε' and ε'' to ε. Thus if the functor F preserves fibered products in the category of algebras, or even just the fibered product given above, we can form the diagram

$$F(K[\varepsilon']/(\varepsilon')^2) \times_{F(K)} F(K[\varepsilon'']/(\varepsilon'')^2) = F(K[\varepsilon', \varepsilon'']/(\varepsilon', \varepsilon'')^2) \xrightarrow{F(\sigma)} F(K[\varepsilon]/(\varepsilon)^2)$$

$$F(K)$$

and taking fibers over $p \in F(K)$, we get the desired addition map.

Exercise VI-13. Verify that these maps make T_pF into a K-vector space, and that this is the old vector space structure in the case where F is a representable functor.

VI.1.4   Group Schemes

It is extremely easy to specify extra structure on a scheme by specifying it on the functor. For example, we may define a *group scheme* as a scheme G and a factorization of the functor $h_G(\text{rings}) \to (\text{sets})$ through the forgetful functor $(\text{groups}) \to (\text{sets})$ — that is, a group scheme is a scheme and a natural way of regarding $\text{Mor}(X, G)$ as a group for each X.

By Yoneda's Lemma (VI-1), this is the same thing as giving maps

$$G \times G \to G, \quad G \to G \quad \text{and} \quad \text{Spec } K \to G$$

representing the multiplication, inverse, and identity element, respectively, and satisfying the usual laws (associativity and so on), but is often much simpler. For example, GL_n can be defined as the affine scheme of invertible integral $n \times n$ matrices,

$$\text{Spec } \mathbb{Z}[x_{ij}][\det(x_{ij})^{-1}],$$

but one usually thinks of it as a functor that associates to every ring T the group $\text{GL}_n(T)$. The interesting point here is just that this family of groups already specifies the structure of a scheme and the additional structure maps!

VI.2 Characterization of a Space by its Functor of Points

It is often much easier, and sometimes more illuminating, to describe the functor of points of an interesting scheme than to give a direct construction of it, whether by gluing or by explicit equations in affine or projective space. Typical examples come from the Hilbert scheme and other moduli problems, where one wants a natural space whose points represent some geometric objects. However, the point of view is also useful in discussing much simpler objects, such as fibered products, or projective space itself.

The basic idea in any case is to first define a functor from the category of schemes to the category of sets, and then to prove an existence theorem asserting that there is a scheme of which it is the functor of points. Of course, to carry this procedure out an essential ingredient is a local (and readily verifiable) criterion for a functor to be representable, and we start by giving such a criterion.

VI.2.1 Characterization of Schemes among Functors

To the extent that we want to define and/or construct schemes first as functors, we run into a fundamental problem, that of determining when a functor comes from a scheme. Here is one characterization.

We say that $F : (\text{rings}) \to (\text{sets})$ is a *sheaf in the Zariski topology* if for each ring R and each open covering of $X = \operatorname{Spec} R$ by distinguished open affines $U_i = \operatorname{Spec} R_{f_i}$ the functor F satisfies the sheaf axiom for the open covering $\bigcup U_i = X$. That is, for every collection of elements $\alpha_i \in F(R_{f_i})$ such that α_i and α_j map to the same element in $F(R_{f_i f_j})$, there is a unique element $\alpha \in F(R)$ mapping to each of the α_i.

This is a reasonably easy property to check in practice. It is in fact enough to guarantee that F comes from a scheme if we know already that F is covered by affine schemes in the following sense. The reader may prove the following theorem.

Theorem VI-14. *A functor $F : (\text{rings}) \to (\text{sets})$ is of the form h_Y for some scheme Y if and only if*

(1) F is a sheaf in the Zariski topology, and

(2) there exist rings R_i and open subfunctors

$$\alpha_i : h_{R_i} \to F$$

such that, for every field K, $F(K)$ is the union of the images of $h_{R_i}(K)$ under the maps α_i.

As an easy application, one can use the theorem to show the existence of fibered products. The construction of fibered products in the category of schemes, which was explained in Chapter I, is of fundamental importance, but it is surprisingly clumsy. Using the functorial point of view, we can at least describe the fibered product of two schemes by giving its functor of points directly. If $X \to S$ and $Y \to S$ are morphisms of schemes, the fibered product $X \times_S Y$ is determined by saying that it is the scheme whose functor of points is the fibered product of functors $h_X \times_{h_S} h_Y$. Its existence is established in the following exercise.

Exercise VI-15. (a) Show that if $f : A \to C$ and $g : B \to C$ are morphisms of functors all of which are sheaves in the Zariski topology, the fibered product $A \times_C B$ is a sheaf in the Zariski topology.

(b) Use the open covering of the fibered product suggested in Chapter I and the above theorem to prove the existence of fibered products in the category of schemes.

The next example gives a different way of looking at maps to projective spaces. Theorem III-37 may be translated immediately into our new language:

Theorem VI-16. *If* $Y = \mathbb{P}^n_{\mathbb{Z}}$, *then*

$$h_Y(X) = \left\{ \begin{matrix} \text{locally free subsheaves } \mathscr{F} \subset \mathscr{O}^{n+1}_X \\ \text{that locally are summands of rank } n \end{matrix} \right\}$$

$$= \frac{\{\text{invertible sheaves } P \text{ on } X \text{ with an epimorphism } \mathscr{O}^{n+1}_X \to P\}}{\{\text{isomorphisms}\}},$$

where isomorphism is defined as in Corollary III-42.

By way of an application, we will combine this description of maps to projective space with the characterization in Section VI.1.3 of tangent vectors to a scheme X as maps of $\operatorname{Spec} K[\varepsilon]/(\varepsilon^2)$ to X to compute the Zariski tangent spaces to $\mathbb{P}^n_K$ at K-valued points. By what we have just said, a K-valued point of $\mathbb{P}^n_K$ is a rank n summand $F \subset K^{n+1}$; let L be the quotient $L = K^{n+1}/F$. The Zariski tangent space T at this point is the set of all summands $F' \subset (K[\varepsilon]/(\varepsilon^2))^{n+1}$ that restrict to F modulo ε. We claim that there is a natural isomorphism

$$T \cong \operatorname{Hom}_K(F, L).$$

To produce it, choose a splitting $K^{n+1} = F \oplus L$ and a basis e_i of F. Any summand F' of $(K[\varepsilon]/(\varepsilon)^2)^{n+1}$ that reduces mod ε to F has a basis of the form $\{e_i + \varepsilon s_i + \varepsilon t_i\}$ with s_i in F and t_i in L, say. We associate to K' the map $\alpha : F \to L$ sending e_i to t_i. Conversely, given any map α, we may define F' to be the module spanned by the elements $e_i + \varepsilon\alpha(e_i)$.

Exercise VI-17. Check that these definitions are independent of all the choices made.

Finally, Theorem VI-14 can also be used to prove the existence of the Grassmannian scheme from its functorial description.

Exercise VI-18. For $0 < k < n$, let

$$g = g(k, n) : (\text{rings}) \to (\text{sets})$$

be the *Grassmannian functor*, that is, the functor given by

$$g(T) = \{\text{rank } k \text{ direct summands of } T^n\}.$$

Prove that this is the functor of points of a closed subscheme $G_{\mathbb{Z}}(k, n)$ of projective space $\mathbb{P}_{\mathbb{Z}}^r$, called the *Grassmannian* of k-planes in n-space, as follows.

First, let $r = \binom{n}{k} - 1$, and let $\mathbb{P}_{\mathbb{Z}}^r = \text{Proj } \mathbb{Z}[\ldots X_I \ldots]$ be the projective space with homogeneous coordinates X_I corresponding to the subsets of cardinality $n - k$ in $\{1, 2, \ldots, n\}$. Define a natural transformation $g \to h_{\mathbb{P}_{\mathbb{Z}}^r}$ of functors by sending a summand $M \subset T^n$ to $\bigwedge^k M \subset \bigwedge^k T^n$. Cover $\mathbb{P}_{\mathbb{Z}}^r$ by the usual open affine subschemes $U_I \cong \mathbb{A}_{\mathbb{Z}}^r$; show that these subschemes are represented by the subfunctors

$$U_I(T) = \left\{ \begin{array}{l} \text{rank } r \text{ summands of } T^{r+1} \text{ such that the } I\text{-th} \\ \text{basis vector of } T^{r+1} \text{ generates the cokernel} \end{array} \right\}$$

and that the intersection (fiber product) of U_I and g is the functor

$$(U_I \cap g)(T) = \left\{ \begin{array}{l} \text{rank } k \text{ summands } M \text{ of } T^n \text{ such that the basis} \\ \text{vectors } e_{i_1}, \ldots, e_{i_{n-k}} \in T^n \text{ generate the cokernel} \end{array} \right\}.$$

Check that the intersection $U_I \cap g$ is represented by an affine scheme. Show that g is a sheaf in the Zariski topology, and conclude that g is represented by a scheme.

Note once more that we do *not* have $g(T) = \bigcup (U_I \cap g)(T)$, except on local rings.

Exercise VI-19. Show that this definition of the Grassmannian coincides with the one given in Section III.2.7.

Exercise VI-20. For a field K, give an analogous definition of the Grassmannian $G_K(k, n)$, and show that it coincides with the product $G_{\mathbb{Z}}(k, n) \times \text{Spec } K$.

We will see other examples of the use of this theorem in the next section. The functorial point of view is developed in far greater depth and detail in Demazure and Gabriel [1970], to which the interested reader is referred for more information.

One of the principal goals in Grothendieck's work on schemes was to find a characterization of scheme-functors by weak general properties that could often be checked in practice and so lead to many existence theorems in algebraic geometry (like Brown's theorem in the homotopy category; see Spanier [1966, Chapter 7.7]). It seemed at first that this program would fail completely and that scheme-functors were really quite special; see Hironaka [1962], for instance. Artin, however, discovered an extraordinary approximation theorem exhibiting a category of functors $\mathscr{F}$ only "a little" larger than the scheme-functors that can indeed be characterized by weak general properties. Geometrically speaking, the functors $\mathscr{F}$ are like spaces obtained by dividing affines by "étale equivalence relations" and then gluing. He calls them *algebraic spaces* (for a typical occasion where they arise, see Section VI.2.4). For details, see Artin [1971] and Knutson [1971].

VI.2.2   Parameter Spaces

The Hilbert Scheme. Probably the one area where the notion of the functor of points has had the most impact is in the construction and description of parameter spaces. We have already mentioned in Section III.3.3, for example, that the subschemes of a projective space $\mathbb{P}_K^n$ over a field K having a given Hilbert polynomial P form a scheme, which we will call $\mathscr{H}_P$. It seems surprising that such a statement has an unambiguous meaning. While it is intuitively plausible that the *set* of such objects might form the points of a variety, couldn't they form a variety in several different ways? And in what sense do they form a scheme?

The answers are obtained by making precise what properties we want the correspondence between the set of subschemes and the set of points of $\mathscr{H}_P$ to have. Specifically, note that if $\mathscr{X} \subset \mathbb{P}_K^n \times B \to B$ is any flat family of subschemes of $\mathbb{P}_K^n$ with Hilbert polynomial P, we get a map from the points of B with residue field K to the points of $\mathscr{H}_P$ with residue field K, sending a point $b \in B$ to the point of $\mathscr{H}_P$ corresponding to the fiber X_b of $\mathscr{H}_P$ over b. It is natural to ask that this map come from a regular map $B \to \mathscr{H}_P$. Carrying this a little further, we want $\mathscr{H}_P$ to have the property that for any scheme B over K, the set of flat families of subscheme of $\mathbb{P}_K^n$ with Hilbert polynomial P parametrized by B is naturally identified with the set of maps from B to $\mathscr{H}_P$. Finally, since the problem of parametrizing subschemes of $\mathbb{P}_K^n$ with a given Hilbert polynomial should be in some sense the same for all K, we would like to do this over $\operatorname{Spec} \mathbb{Z}$ — that is (as in the case of the Grassmannian, which is indeed a special case of a Hilbert scheme) define for each P a single object $\mathscr{H}_P$ over $\operatorname{Spec} \mathbb{Z}$ such that for any K the product $\mathscr{H}_P \times \operatorname{Spec} K$ parametrizes subschemes of $\mathbb{P}_K^n$ with Hilbert polynomial P.

To say this a little differently, we make the following definition:

Definition VI-21. The *Hilbert functor* h_P — called the "functor of flat families of schemes in $\mathbb{P}_{\mathbb{Z}}^n$ with Hilbert polynomial P" — is the functor

$$h_P : (\text{schemes})^\circ \to (\text{sets})$$

that associates to any B the set of subschemes $\mathscr{X} \subset \mathbb{P}_B^n$ flat over B whose fibers over points of B have Hilbert polynomial P.

We then want to take the Hilbert scheme $\mathscr{H}_P$ to be the scheme that represents h_P: in other words, the scheme whose functor of points is h_P.

By Yoneda's Lemma (VI-1), this determines the scheme $\mathcal{H}_P$, if one such exists; the key theorem is thus the following.

Theorem VI-22. *There exists a scheme $\mathcal{H}_P$ whose functor of points is the functor h_P.*

Note that for any scheme S we can make an analogous definition of a functor $h_{P,S} : (S-\text{schemes})^{\circ} \to (\text{sets})$ on the category of S-schemes, and that if $\mathcal{H}_P$ is the scheme representing the functor h_P as above, then $h_{P,S}$ is represented by the S-scheme $\mathcal{H}_P \times S$.

In fact, it turns out that $\mathcal{H}_P \times \operatorname{Spec} K$ is often a scheme that is genuinely *not* a variety. We will describe, in Exercises VI-35 through VI-37, a famous example of Mumford [1962] of a Hilbert scheme that is nonreduced even at points corresponding to nonsingular, irreducible projective varieties.

The statement of Theorem VI-22 has another interpretation that is often useful: saying that the functor h_P is representable is the same thing as saying that there exists a *universal family* — that is, a scheme $\mathcal{H}$ and a subscheme $\mathcal{X} \subset \mathbb{P}_{\mathbb{Z}}^n \times \mathcal{H}$ flat over $\mathcal{H}$ with Hilbert polynomial P — such that any subscheme $Y \subset \mathbb{P}_{\mathbb{Z}}^n \times B$ flat over B with Hilbert polynomial P is equal to the fiber product $Y = \mathcal{X} \times_{\mathcal{H}} B \subset \mathbb{P}_{\mathbb{Z}}^n \times B$ for a unique morphism $B \to \mathcal{H}$. Clearly, if a universal family $\mathcal{X} \subset \mathbb{P}_{\mathbb{Z}}^n \times \mathcal{H}$ exists, then $\mathcal{H}$ represents the functor h_P. Conversely, if a scheme $\mathcal{H}$ represents h_P, then the subscheme $\mathcal{X} \subset \mathbb{P}_{\mathbb{Z}}^n \times \mathcal{H}$ associated to the identity map is universal in the above sense.

We will not give a proof of Theorem VI-22 but will indicate how it may be approached; for more details we refer the reader to Mumford [1966] or Kollár [1996].

The idea is easy to summarize: reducing to the case of a subscheme X of projective space $\mathbb{P}_B^n$ over a base of the form $B = \operatorname{Spec} R$ with R a local ring, such a scheme X is determined by its ideal $I(X) \subset S = R[X_0, \ldots, X_n]$, which in turn is determined for m sufficiently large (in a sense depending only on P) by its degree m piece $I(X)_m \subset S_m$. Setting $M = \binom{m+n}{n}$ and $q = P(m)$, this in turn corresponds to a point in the Grassmannian $G_B(q, M)$ parametrizing summands of codimension $P(m)$ in the free R-module $S_m \cong R^M$. In this way $\mathcal{H}_{P,B}$ becomes a subscheme of the Grassmannian $G_B(q, M)$ of such planes.

The key point is that we can choose a single m that has this property uniformly for every subscheme X with Hilbert polynomial P; that is, that for every P, there is an m_0 such that if $m \geq m_0$ and X is a subscheme of $\mathbb{P}_K^n$ with Hilbert polynomial P, then $I(X)_{l \geq m}$ is generated by $I(X)_m$, and the codimension of $I(X)_m$ in S_m is $P_X(m)$. Examining our proof that the Hilbert polynomial is a polynomial, we see that to prove this it is enough to show that the degrees of the generators of the free modules in the minimal free resolution of $I(X)$ can be bounded in terms of the Hilbert polynomial of X. This is done by using the idea of *Castelnuovo regularity* of X; a more complete description would take us too far afield.

We thus have, for every flat family $\mathscr{X} \subset \mathbb{P}^n_B$ over a base $B = \operatorname{Spec} R$ — that is, an element of $h_P(\operatorname{Spec} R)$ — a summand of corank q in R^M — that is, an element of the set $g(q, M)(R)$, where $g = g(q, M)$ is the Grassmannian functor as defined in Exercise VI-18. This association extends to a natural transformation of functors from h^*_P to $g(q, M)$, where $h^*_P : (\text{rings}) \to (\text{sets})$ is the functor $h^*_P(R) = h_P(\operatorname{Spec} R)$, and hence to a natural transformation from the Hilbert functor h_P to the functor respresented by the Grassmannian $G = G_{\mathbb{Z}}(q, M)$.

To finish the argument, one must show that there exists a subscheme $\mathscr{H}_P \subset G$ such that a morphism $\varphi : B \to G$ comes in this way from a flat family $\mathscr{X} \subset \mathbb{P}^n_B$ with Hilbert polynomial P if and only if φ factors through $\mathscr{H}_P$. We will content ourselves here with describing the equations of $\mathscr{H}_P$ as a closed subscheme of G. Let $\mathscr{Y}$ be the universal subbundle on G. We have multiplication maps

$$\operatorname{mult}_k : \mathscr{Y} \otimes S_k \to S_{k+m},$$

where $S = \mathbb{Z}[X_0, \ldots, X_n]$, and we can take $\mathscr{H}_P$ to be the "determinantal" subscheme of $G_{\mathbb{Z}}(q, M)$ defined by the conditions that

$$\operatorname{rank}(\operatorname{mult}_k) \leq \dim S_{k+m} - P(m + k)$$

for all $k \geq 0$. It is immediate that the desired maps φ all factor through this subscheme. Given any point p in this subscheme, one shows that (because m has been chosen so large) the ideal generated by the corresponding linear subspace of S_m defines a scheme with Hilbert polynomial P. This gives us a "tautological family" on $\mathscr{H}_P$ of schemes with Hilbert polynomial P. Given any map φ from a scheme B into $\mathscr{H}_P$, one can "pull back" this family by using the fibered product to get a family over B, and the map φ will be associated to this family. Once all this is verified, the description of $\mathscr{H}_P$ by its functor of points ensures that $\mathscr{H}_P$ does not depend on the choice of m.

Examples of Hilbert Schemes. We will mention, largely in the form of exercises, some examples of Hilbert schemes. To begin with, the Grassmannian $\mathbb{G}_S(k, n)$ is a Hilbert scheme: it parametrizes subschemes X of degree 1 and dimension k (specifically, with Hilbert polynomial $P(m) = \binom{m+k}{k}$) in the projective space $\mathbb{P}^n_S$. The following exercise deals with the simplest special case of this, but the general statement (and proof) differs only numerically.

Exercise VI-23. Let $P(m) = m + 1$ be the Hilbert polynomial of a line. Show that the Hilbert scheme of subschemes of $\mathbb{P}^3_{\mathbb{Z}}$ with Hilbert polynomial P is the Grassmannian introduced in Exercise VI-20.

Another very straightforward example is hypersurfaces. It is a standard observation that the set of hypersurfaces of degree d in projective space

$\mathbb{P}^n_K$ over a field K may be identified with the points of the projective space of homogeneous polynomials of degree d in $n+1$ variables. In fact, this projective space turns out to be the Hilbert scheme of such hypersurfaces. As in the preceding example, the following exercise deals with one typical example.

Exercise VI-24. Let $P(m) = 2m+1$ be the Hilbert polynomial of a conic curve. Show that the Hilbert scheme $\mathscr{H}_P$ of subschemes of $\mathbb{P}^2_{\mathbb{Z}}$ with Hilbert polynomial P is $\mathbb{P}^5_{\mathbb{Z}}$.

Beyond these examples, the geometry of Hilbert schemes is much less well known. Even the Hilbert schemes parametrizing zero-dimensional subschemes of projective space $\mathbb{P}^n_K$ over a field remain mysterious: Iarrobino [1985], for example, has shown (contrary to naive expectations) that such Hilbert schemes are not in general irreducible. In the case of $\mathbb{P}^2_K$, they are in fact irreducible and nonsingular, but their global geometry presents many problems: see, for example, Collino [1988]. One case where we can actually give a description is the following exercise.

Exercise VI-25. Let P be the constant polynomial 2. Show that the Hilbert scheme $\mathscr{H}_P$ parametrizing subschemes of $\mathbb{P}^2_{\mathbb{Z}}$ with Hilbert polynomial P may be obtained by blowing up the product $\mathbb{P}^2_{\mathbb{Z}} \times \mathbb{P}^2_{\mathbb{Z}}$ along the diagonal and then taking the quotient by the involution exchanging factors.

In the general setting our knowledge of Hilbert schemes is minimal. For example, in the case of curves in projective 3-space $\mathbb{P}^3_K$ over a field K — the simplest example of a Hilbert scheme parametrizing schemes that are pure positive-dimensional but not hypersurfaces — we do not have even a guess as to the number of components of $\mathscr{H}_P$, their dimension, or their smoothness or singularity. For a discussion of this case, see Harris and Eisenbud [1982].

Variations on the Hilbert Scheme Construction. We have defined the Hilbert scheme parametrizing subschemes of projective space with given Hilbert polynomial P. In fact, with very little additional effort we can generalize this substantially.

The first thing to notice is that if $X \subset \mathbb{P}^n_S$ is any closed subscheme, we can define a functor

$$h_{P,X} : (S-\text{schemes})^{\circ} \longrightarrow (\text{sets})$$

by associating to any S-scheme B the set of flat families of subschemes of X with Hilbert polynomial P over B, that is,

$$h_{P,X}(B) = \left\{ \begin{array}{l} \mathscr{X} \subset B \times_S X \subset B \times_S \mathbb{P}^n_S = \mathbb{P}^n_B, \text{ flat} \\ \text{over } B, \text{ with Hilbert polynomial } P \end{array} \right\}.$$

The key fact, which is not hard to establish, is the following.

Exercise VI-26. Show that $h_{P,X}$ is a closed subfunctor of h_P.

It follows that there exists a closed subscheme $\mathcal{H}_{P,X} \subset \mathcal{H}_P$ whose functor of points is $h_{P,X}$; this scheme is what we call the *Hilbert scheme of subschemes* of X with Hilbert polynomial P.

As an important example, if we take $P(m) = \binom{k+m}{k}$ to be the Hilbert polynomial of a k-plane, we will see in Section VI.2.3 below that the scheme $\mathcal{H}_{P,X}$ we arrive at is the Fano scheme $F_k(X)$ of k-planes on X, as defined in Section IV.3. Nor is this just an idle observation: apart from giving a more natural definition, this characterization of the Fano schemes will allow us to determine their tangent spaces.

Now suppose we are given two projective S-schemes $X \subset \mathbb{P}_S^m$ and $Y \subset \mathbb{P}_S^n$. We may embed the product $X \times_S Y$ in projective space via the Segre map

$$X \times_S Y \hookrightarrow \mathbb{P}_S^m \times_S \mathbb{P}_S^n \hookrightarrow \mathbb{P}_S^N,$$

where $N = (m+1)(n+1) - 1$. We thus have Hilbert schemes parametrizing subschemes of a product $X \times_S Y$.

This in turn allows us to parametrize morphisms from X to Y, by considering their graphs as subschemes of the product. The two things we need to check are that

(1) the condition on a subscheme $Z \subset X \times_S Y$ that the projection map $\pi_X : Z \to X$ be an isomorphism is an open condition on the Hilbert scheme of subschemes of $X \times_S Y$; and

(2) the Hilbert polynomials of the graphs Γ_φ of morphisms $\varphi : X \to Y$ of bounded projective degree are bounded. (See Harris [1995] for a definition of "projective degree".)

Given this, we see that there are quasiprojective schemes parametrizing the morphisms of given degree from X to Y, and similarly a quasiprojective scheme $\mathrm{Isom}(X, Y)$ parametrizing isomorphisms from X to Y. Again, by "parametrize" we mean represent the functor

$$\mathrm{isom}_{X,Y} : (S - \mathrm{schemes})^\circ \longrightarrow (\mathrm{sets})$$

given by

$$\mathrm{isom}_{X,Y}(B) = \{\text{isomorphisms } \varphi : B \times_S X \to B \times_S Y \text{ as } B\text{-schemes}\}.$$

We should mention one further generalization of the construction of the Hilbert scheme that is very useful in practice. This is the *relative Hilbert scheme*, which parametrizes subschemes of members of a flat family of schemes.

To set this up, let S be a scheme over a field K, and suppose that $X \subset \mathbb{P}_S^n$ is any scheme flat over S. We can then consider the functor

$$h_{P,X/S} : (K - \mathrm{schemes})^\circ \longrightarrow (\mathrm{sets})$$

that associates to any K-scheme B the set of flat families over B of sub-schemes of fibers of X over S — that is,

$$h_{P,X/S}(B) = \left\{ \begin{array}{c} \text{pairs } (\nu, \Sigma) \text{ such that } \nu : B \to S \text{ and } \Sigma \subset B \times_S X \subset \mathbb{P}^n_B \\ \text{is flat over } B \text{ with Hilbert polynomial } P \end{array} \right\}.$$

Once more, we can adapt our basic construction to show that $h_{P,X/S}(B)$ is represented by a K-scheme $\mathscr{H}_{P,X/S}$.

Finally, we can apply the relative Hilbert scheme construction to a pair of flat families to parametrize morphisms between members of two families! Thus, if $X \to S$ and $Y \to S$ are any flat families, we have a quasiprojective scheme $\mathrm{Mor}_{d,X/S,Y/S}$ parametrizing morphisms of given degree from fibers of X over S to corresponding fibers of $Y \to S$, and similarly a scheme parametrizing isomorphisms.

Note that if $X \to S$ and $Y \to T$ are families with possibly different bases, we can parametrize morphisms from fibers of X/S to fibers of Y/T by pulling both back to families over the product $S \times T$ and performing this construction there.

As an application of this construction, we have:

Exercise VI-27. Fix two integers $g, h \geq 2$. Show that there is a number $N(g,h)$ such that for any nonsingular curves C and C' of genera g and h respectively, the number of maps from C to C' is less than $N(g,h)$.

More generally:

Exercise VI-28. Let $X \subset \mathbb{P}^m_S$ and $Y \subset \mathbb{P}^n_T$ be schemes flat over the K-schemes S and T, and suppose that for any pair of closed points $s \in S$ and $t \in T$ the number $n_d(s,t)$ of morphisms of degree d between the corresponding fibers X_s and Y_t is finite. Show that, for fixed d, $n_d(s,t)$ is bounded as s and t vary.

VI.2.3   Tangent Spaces to Schemes in Terms of Their Functors of Points

Tangent Spaces to Hilbert Schemes. One facet of the Hilbert scheme that is best described in terms of its functor of points is its tangent space at a point. We first introduce the notion of a first-order deformation: if Y is any scheme and $X \subset Y$ a closed subscheme, a *first-order deformation of X in Y* is defined to be a flat family $\mathscr{X} \subset Y \times \mathrm{Spec}\, K[\varepsilon]/(\varepsilon^2)$ such that the fiber of $\mathscr{X}$ over the reduced point $\mathrm{Spec}\, K \subset \mathrm{Spec}\, K[\varepsilon]/(\varepsilon^2)$ is X. It then follows, via the characterization of tangent vectors to schemes given in Section VI.1.3, that *the tangent space to the Hilbert scheme $\mathscr{H}_P$ at a point $[X]$ is the space of first-order deformations in $\mathbb{P}^n_K$ of X*; and more generally, if $Y \subset \mathbb{P}^n_K$ is a projective scheme, the tangent space to the Hilbert scheme $\mathscr{H}^Y_P$ at a point $[X]$ is the space of first-order deformations of X in Y.

This is especially useful since the space of first-order deformations may often be calculated, even in circumstances where we have no hope of writing down the equations of $\mathcal{H}_P$. To do this, we introduce the *normal sheaf* $\mathcal{N}_{X/Y}$ to a closed subscheme X of a scheme Y: this is defined to be the sheaf

$$\mathcal{N}_{X/Y} = \mathrm{Hom}_{\mathcal{O}_X}(\mathcal{I}/\mathcal{I}^2, \mathcal{O}_X) = \mathrm{Hom}_{\mathcal{O}_Y}(\mathcal{I}, \mathcal{O}_X)$$

where $\mathcal{I} = \mathcal{I}_{X/Y}$ is the ideal sheaf of X in Y. We then have the following basic theorem.

Theorem VI-29. *Given a closed subscheme X of a scheme Y, the space of first-order deformations of X in Y is the space of global sections of its normal sheaf $\mathcal{N}_{X/Y}$.*

Proof. To begin with, let $\mathcal{X} \subset Y \times \mathrm{Spec}\, K[\varepsilon]/(\varepsilon^2)$ be any subscheme whose intersection with the fiber $Y \cong Y \times \mathrm{Spec}\, K \subset Y \times \mathrm{Spec}\, K[\varepsilon]/(\varepsilon^2)$ is X (do not assume $\mathcal{X}$ is flat). Let $U \subset Y$ be any affine open subset, $V = X \cap U$ the corresponding affine open subset of X, and $\mathcal{V} = \mathcal{X} \cap (U \times \mathrm{Spec}\, K[\varepsilon]/(\varepsilon^2))$. Let $A = \mathcal{O}_Y(U)$ be the coordinate ring of U and $I = I(V)$ the ideal of V in A, so that the restriction to V of the sheaf $\mathcal{N}_{X/Y}$ is the sheaf associated to the A-module $\mathrm{Hom}(I, A/I)$.

The coordinate ring of $U \times \mathrm{Spec}\, K[\varepsilon]/(\varepsilon^2)$ is $A \otimes K[\varepsilon]/(\varepsilon^2)$; we write an element of this ring as $f + \varepsilon g$, with f and $g \in A(U)$. In particular, we may write the ideal $I(\mathcal{V})$ of $\mathcal{V}$ as

$$I(\mathcal{V}) = (f_1 + \varepsilon g_1, f_2 + \varepsilon g_2, \ldots, f_k + \varepsilon g_k)$$

where by hypothesis the elements $f_i \in A$ generate the ideal I. We claim now that *there exists an A-module homomorphism $\varphi : I \to A/I$ carrying f_i to g_i if and only if $\mathcal{V} \to \mathrm{Spec}\, K[\varepsilon]/(\varepsilon^2)$ is flat* (note that if φ exists, it is unique). The theorem follows immediately from this claim: in one direction, if the family $\mathcal{X} \to \mathrm{Spec}\, K[\varepsilon]/(\varepsilon^2)$ is flat, then by uniqueness the homomorphisms φ patch together to give a section of the sheaf $\mathcal{N}_{X/Y}$; while given a global section of $\mathcal{N}_{X/Y}$ we can simply take $\mathcal{X}$ to be given locally by the ideal

$$\{f + \varepsilon \cdot \varphi(f) : f \in I(V)\}$$

To prove the claim, note first that a $K[\varepsilon]/(\varepsilon^2)$-module M is flat if and only if when we tensor the exact sequence of $K[\varepsilon]/(\varepsilon^2)$-modules

$$0 \to (\varepsilon) \to K[\varepsilon]/(\varepsilon^2) \to K \to 0$$

by M, it remains exact. Applying this to the coordinate ring $B = \mathcal{O}_{\mathcal{X}}(\mathcal{V})$ of $\mathcal{V}$, we see that $\mathcal{V}$ will be flat over $\mathrm{Spec}\, K[\varepsilon]/(\varepsilon^2)$ if and only if the map

$$(\varepsilon) \otimes B \to B$$

is injective—that is, if and only if, for any $f \in A$,

$$\varepsilon \cdot f \in I(\mathcal{V}) \Rightarrow f \in I(V)$$

Suppose now that $f \in A$ and $\varepsilon \cdot f \in I(\mathscr{V})$. We can then write

$$\varepsilon \cdot f = \sum (a_i + \varepsilon b_i) \cdot (f_i + \varepsilon g_i) = \sum a_i f_i + \varepsilon \cdot \sum (a_i g_i + b_i f_i)$$

We know that the first term on the right is zero, since the other two terms in the equality are divisible by ε. Now, if there exists a homomorphism $\varphi : I \to A/I$ of A-modules such that $\varphi(f_i) = g_i$, then we can write

$$\sum a_i \cdot g_i = \sum a_i \cdot \varphi(f_i) = \varphi \left(\sum a_i f_i \right) = 0$$

so the existence of a module map φ carrying f_i to g_i implies that the family $\mathscr{V}$ is flat.

Conversely, suppose that $\mathscr{V} \to \operatorname{Spec} K[\varepsilon]/(\varepsilon^2)$ is flat. Then for any collection of $a_i \in A$ such that $\sum a_i f_i = 0$, we have

$$\varepsilon \cdot \sum a_i \cdot g_i = \sum a_i \cdot (f_i + \varepsilon g_i) \in I(\mathscr{V}) \implies \sum a_i \cdot g_i \in I(V).$$

We can thus define an A-module map $\varphi : I \to A/I$ by sending, for any $a_1, \dots, a_k \in A$, the element $\sum a_i f_i \in I$ to the element $\sum a_i g_i \in A/I$; by the last calculation this will be well defined. $\qquad \square$

One trivial but useful consequence of this theorem is the following.

Corollary VI-30. *The dimension of any irreducible component Σ of the Hilbert scheme is at most the dimension of the space of sections of the normal sheaf of any scheme X with $[X] \in \Sigma$.*

In fact, this a priori estimate for the dimension of a component of the Hilbert scheme gives the right answer more often than not, especially when applied to a general point $[X]$ of a component of $\mathscr{H}_P$. The following exercises give examples of this.

Exercise VI-31. Let Σ be the component of the Hilbert scheme whose general member is a complete intersection $X \subset \mathbb{P}^n_K$ of k hypersurfaces of degree d. Calculate the dimension of the space of global sections of $\mathscr{N}_X$ and show that this is equal to the dimension of Σ.

Exercise VI-32. Generalize the preceding exercise to the case of the component of the Hilbert scheme whose general member is a complete intersection $X \subset \mathbb{P}^n_K$ of hypersurfaces of degrees $d_1, \dots, d_k$. (This can get complicated; you may want to stick to the case $k = 2$, which is enough to see how it goes.)

Exercise VI-33. Let $P(m)$ be the polynomial $3m + 1$ and let Σ be the component of the Hilbert scheme $\mathscr{H}_P$ of subschemes of $\mathbb{P}^3_K$ whose general member is a twisted cubic curve C. Show that the dimension of Σ is 12, and that this is equal to the dimension of the space of sections of $\mathscr{N}_C$.

Exercise VI-34. By way of warning, the component Σ of the preceding exercise is not the only component of the Hilbert scheme: there is another component Σ' whose general member is the (disjoint) union of a plane cubic curve and a point. These two intersect in the locus of schemes $C \subset \mathbb{P}^3_K$ such that C is the union of a plane curve C_0 and the double point (that is, the scheme given by the square of the maximal ideal of a point) supported at a singular point of C_0. At a point of their intersection $\mathcal{H}_P$ is singular, and its tangent space will be strictly larger than the dimension of either Σ or Σ'. Verify this.

It is not always the case, however, that the dimension of a component of the Hilbert scheme is equal to the dimension of the space of sections of the normal sheaf of a general member; there are examples of Hilbert schemes that are nonreduced along whole components, even when the general points of those components correspond to nonsingular, irreducible varieties. The first example of this is due to Mumford [1962]; the following series of exercises describes it.

Mumford's example deals with curves of degree 14 and genus 24 in projective 3-space $\mathbb{P}^3_K$ over a field K. There are (as we shall see) several components of the Hilbert scheme parametrizing such curves; we will be concerned with the component whose general member lies on a nonsingular cubic surface. By way of notation, let $P(m) = 14m - 23$ be the Hilbert polynomial of a curve of degree 14 and genus 24, and let $\mathcal{H}$ be the Hilbert scheme parametrizing subschemes of $\mathbb{P}^3_K$ with this Hilbert polynomial. We will denote by Σ the subset of $\mathcal{H}$ corresponding to nonsingular curves $C \subset \mathbb{P}^3_K$ of degree 14 and genus 24 that are contained in a nonsingular cubic surface S and linearly equivalent on S to $4H + 2L$, where H is the hyperplane divisor and L a line on S.

Exercise VI-35. Show that Σ is a constructible subset of $\mathcal{H}$ and that its closure $\bar{\Sigma}$ in $\mathcal{H}$ has dimension 56.

Not all curves of degree 14 and genus 24 in $\mathbb{P}^3_K$ have to lie on cubic surfaces. Thus, it is not *a priori* clear that the subvariety $\bar{\Sigma} \subset \mathcal{H}$ is an irreducible component of $\mathcal{H}$: the curves C parametrized by $\bar{\Sigma}$ could be specializations of other curves not lying on cubics. To see that this is not in fact the case, we make another dimension count.

Exercise VI-36. Let C be a nonsingular, irreducible curve of degree 14 and genus 24 in $\mathbb{P}^3_K$, and assume that C does not lie on a cubic surface. Show that it must lie on two quartic surfaces T, T' not having a common component, and that the residual intersection of T and T' (that is, the union of the irreducible components of $T \cap T'$ other than C) is a curve of degree 2. By analyzing what this residual intersection may look like, show that the set of such curves C is a constructible subset of $\mathcal{H}$ whose closure has dimension at most 56. Deduce that the subvariety $\bar{\Sigma}$ of Exercise VI-35 is indeed an irreducible component of $\mathcal{H}$.

Exercise VI-37. Now let C be a nonsingular curve of degree 14 and genus 24 lying on a nonsingular cubic surface $S \subset \mathbb{P}_K^3$. Using the exact sequence

$$0 \to \mathcal{N}_{C/S} \to \mathcal{N}_{C/\mathbb{P}_K^3} \to \mathcal{N}_{S/\mathbb{P}_K^3} \otimes \mathcal{O}_C \to 0$$

(where for any pair of schemes $X \subset Y$ we write $\mathcal{N}_{X/Y}$ for the normal sheaf $\mathrm{Hom}(\mathcal{I}_{X/Y}, \mathcal{O}_X)$ of X in Y), show that the dimension of the space of sections of the normal sheaf $\mathcal{N}_{C/\mathbb{P}_K^3}$ is 57. Deduce that $\mathcal{H}$ is nowhere reduced along $\bar{\Sigma}$.

Finally, here is an amusing fact about the Hilbert scheme of rational normal curves, generalizing a calculation we made in Section IV.4.

Exercise VI-38. Let K be a field. For any r, let $P(m)$ be the polynomial $rm + 1$, and let Σ be the open subset of the Hilbert scheme $\mathcal{H}_P$ of subschemes of $\mathbb{P}_K^r$ parametrizing rational normal curves of degree r; check that Σ is irreducible of dimenion $r^2 + 2r - 3$. Let $\mathcal{C} \subset \Sigma \times \mathbb{P}_K^r \to \Sigma$ be the universal curve over Σ. Let L be the function field of Σ, and C_L the fiber of $\mathcal{C}$ over the generic point $\mathrm{Spec}\, L$ of Σ. Show that $C_L \cong \mathbb{P}_L^1$ if and only if r is odd.

Tangent Spaces to Fano Schemes. One particularly nice example of schemes that are in many ways best characterized by their functor of points are the Fano schemes $F_k(X) \subset \mathbb{G}_S(k, n)$ of a scheme $X \subset \mathbb{P}_S^n$ described in Section IV.3. We will see, for example, how the description given in Section VI.1.3 of the tangent spaces to a functor allows us to compute the Zariski tangent spaces to a Fano scheme much more readily that we could from the explicit equations introduced in Section IV.3; this will in turn allow us to say in many cases whether a linear space $\Lambda \subset X$ corresponds to a nonsingular or a singular point of $F_k(X)$. (For the following discussion, we will introduce some notation: for a plane $W \subset K^{n+1}$ and the corresponding linear subspace $\Gamma \subset \mathbb{P}_K^n$, we'll write $\Gamma = [W]$ and $W = \tilde{\Gamma}$.)

The characterization of Fano schemes is straightforward: we may define the Fano scheme $F_k(X)$ of a subscheme $X \subset \mathbb{P}_S^n$ to be simply the Hilbert scheme $\mathcal{H}_{P,X}$ of subschemes of X with Hilbert polynomial $P(m) = \binom{k+m}{k}$ — that is, the functor

$$f_k(X) : (S - \mathrm{schemes})^\circ \longrightarrow (\mathrm{sets})$$

that associates to any S-scheme B the set of families of k-planes contained in $X \times_S B$, that is,

$$f_k(X)(B) = \left\{ \begin{array}{l} \Sigma \subset B \times_S X \subset B \times \mathbb{P}_S^n = \mathbb{P}_B^n,\ \text{flat over } B, \\ \text{such that } \Sigma_b \subset \mathbb{P}_{\kappa(b)}^n \text{ is a } k\text{-plane for all } b \in B \end{array} \right\}.$$

We then have:

Proposition VI-39. *The functor $f_k(X)$ is represented by the Fano scheme $F_k(X)$ introduced in Section IV.3.*

Proof. To establish this, we have to exhibit an isomorphism of functors

$$f_k(X)(B) \longrightarrow \mathrm{Mor}_S(B, F_k(X)),$$

that is, for any S-scheme B a natural bijection between two sets, the set of families $\Sigma \subset B \times_S X \subset B \times \mathbb{P}_S^n = \mathbb{P}_B^n$ of k-planes contained in X and morphisms from B to $F_k(X)$ as S-schemes. In fact, the serious work has already been done: the definition of the map and the proof that it is a bijection are not hard, given that we have already characterized the Grassmannian $\mathbb{G}_S(k,n)$ as the scheme representing the functor of families of k-planes in $\mathbb{P}_S^n$.

The point is, we already have an isomorphism of functors

$$g_S(n+1,\, k+1) \longrightarrow \mathrm{Mor}_S(B, G_S(k+1, n+1))$$

where $g_S(n+1,\, k+1)$ is the Grassmannian functor introduced in Section VI.2.1, here in the category of S-schemes. That is, we associate to any family $\Sigma \subset B \times \mathbb{P}_S^n = \mathbb{P}_B^n$, flat over B with k-plane fibers, a morphism $\varphi_\Sigma : B \to \mathbb{G}_S(k,n)$ over S. Now all we need to check is that the subset $f_k(X)(B) \subset g(n+1,\, k+1)(B)$ is carried into the subset $\mathrm{Mor}_S(B, F_k(X)) \subset \mathrm{Mor}_S(B, G_S(k+1, n+1))$, that is, that

$$\Sigma \subset B \times_S X \subset B \times \mathbb{P}_S^n \iff \varphi_\Sigma(B) \subset F_k(X) \subset G_S(k+1, n+1).$$

This is immediate, given the description in Section IV.3 of the defining equations of $F_k(X)$. $\square$

As promised, the characterization of the Fano scheme by its functor of points allows us to determine its tangent spaces readily, and in particular to give us criteria for the smoothness and/or singularity of Fano schemes.

For the following, then, K will be an algebraically closed field, $X \subset \mathbb{P}_K^n$ will be an arbitrary projective scheme over K, $F_k(X) \subset G_K(k+1, n+1)$ the Fano scheme of k-planes contained in X and $\Lambda \in F_k(X)$ a K-valued point of $F_k(X)$. Write Δ for the scheme $\mathrm{Spec}\, K[\epsilon]/(\epsilon^2)$ and $0 \in \Delta$ for the reduced point $\Delta_{\mathrm{red}} \cong \mathrm{Spec}\, K \subset \Delta$. According to our characterization of $F_k(X)$ as the scheme representing the functor of families of k-planes on X, the tangent space to $F_k(X)$ at the point Λ will be

$$T_\Lambda(F_k(X)) = \left\{ \begin{array}{c} \text{subschemes } \Sigma \subset \Delta \times_K X \text{ flat over } \Delta \\ \text{such that } \Sigma \cap (0 \times_K X) = \Lambda \end{array} \right\}.$$

We will now see how to describe this as a subspace of

$$T_\Lambda(G_K(k+1, n+1)) = \mathrm{Hom}(\tilde{\Lambda}, K^{n+1}/\tilde{\Lambda}).$$

Probably the fastest way to do this is simply to observe that a first-order deformation of a plane $\Lambda \subset \mathbb{P}_K^n$ —that is, a subscheme $\Sigma \subset \Delta \times_K \mathbb{P}_K^n$ flat over Δ and such that $\Sigma \cap (0 \times_K X) = \Lambda$—is the union of its tangent vectors, viewed as subschemes of Σ isomorphic to Δ. In other words, Σ will be contained in a subscheme $\Delta \times_K X \subset \Delta \times_K \mathbb{P}_K^n$ if and only if every tangent vector to Σ is a tangent vector to $\Delta \times_K X$.

Now, suppose that the first-order deformation Σ of a plane Λ corresponds to a homomorphism $\varphi : \tilde{\Lambda} \to K^{n+1}/\tilde{\Lambda}$. The tangent vectors to Σ at a point p (more properly, $0 \times_K \{p\}$) will then be the tangent vectors to $\Delta \times_K \mathbb{P}^n_K$ corresponding to homomorphisms $\nu : \tilde{p} \to K^{n+1}/\tilde{p}$ such that

$$\nu(\tilde{p}) \subset \varphi(\tilde{\Lambda}).$$

Similary, to say that the tangent vector corresponding to such a homomorphism ν is tangent to $\Delta \times_K X$ is to say that

$$\nu(\tilde{p}) \subset \widetilde{T_pX} + \tilde{\Lambda}.$$

Thus, to say that the tangent vector to the Grassmannian associated to the homomorphism φ lies in the tangent space to the Fano scheme $F_k(X)$ at Λ is to say that the image of φ is contained in the tangent space to X at each point, i.e.,

$$T_\Lambda(F_k(X)) = \left\{ \begin{matrix} \varphi \in \operatorname{Hom}(\tilde{p}, K^{n+1}/\tilde{\Lambda}) \text{ such that} \\ \varphi(v) \in \widetilde{T_pX} + \tilde{\Lambda} \text{ for all } v \in \tilde{\Lambda} \end{matrix} \right\}.$$

As an application, consider the simplest possible case, the Fano scheme $F_1(S) \subset \mathbb{G}_K(1,3) = G_K(2,4)$ of lines on a surface $S \subset \mathbb{P}^3_K$. Let $L \subset S$ be a line, which we will assume is not contained in the singular locus of S (if it is, the Fano scheme $F_1(S)$ will have four-dimensional tangent space at L — in other words, it will be very singular!). To a general point $p \in L$, then, we may associate the projective tangent space $\widetilde{T_pS}$, which will be a plane in $\mathbb{P}^3_K$ containing L; thus we get a (rational) map

$$\gamma : L \longrightarrow \mathbb{P}(K^4/\tilde{L}) \cong \mathbb{P}^1_K.$$

This map is given by the partial derivatives of the defining polynomial of S, so that if S is nonsingular along L — in other words, if these partials have no common zeroes — the degree of γ will be $d - 1$. If S has any singular points on L, conversely, it will be less.

In fact, the degree of the map γ is precisely what determines the dimension of the tangent space to $F_1(S)$ at L. This is simple to see: if $\varphi : \tilde{L} \to K^4/\tilde{L}$ is any homomorphism, the induced map

$$\bar{\varphi} : L \longrightarrow \mathbb{P}(K^4/\tilde{L}) \cong \mathbb{P}^1_K$$

$$p \longmapsto [\varphi(\tilde{p}) + \tilde{L}]$$

will have degree 1 if φ has rank 2 and degree 0 if φ has rank 1. Thus, if γ has degree $\deg(\gamma) \geq 2$, the tangent space to $F_1(S)$ at L will be zero-dimensional; if γ has degree 1 it will be one-dimensional; and if γ is constant it will be two-dimensional. We get the following corollaries:

- The Fano scheme of lines on a nonsingular quadric surface in $\mathbb{P}^3_K$ is nonsingular, while the Fano scheme of lines on a quadric cone is everywhere nonreduced.

- More generally, the Fano scheme of the cone in $\mathbb{P}_K^3$ over any plane curve is nonreduced.

- The Fano scheme of lines on a nonsingular surface of degree $d \geq 3$ in $\mathbb{P}_K^3$ consists of isolated reduced points.

There is a further corollary dependent on a further fact about Fano schemes of hypersurfaces: if $X \subset \mathbb{P}_B^n$ is any hypersurface of degree d—that is, a closed subscheme of $\mathbb{P}_B^n$, flat over S, whose fiber over each point $b \in B$ is a hypersurface of degree d in $\mathbb{P}_{\kappa(b)}^n$—then the Fano scheme $F_k(X) \subset G_B(k+1, n+1)$ of k-planes on X is flat over the open subset of B where it has the expected fiber dimension $(k+1)(n-k) - \binom{k+d}{k}$. Since $F_k(X)$ may be described as the zero locus of a section of a locally free sheaf on $G_B(k\,n)$, this is a generalization of the fact that complete intersections are flat, but it is beyond the scope of this book. Given it, however, we have the following consequence:

- All cubic surfaces that contain only finitely many lines contain the same number of lines, properly counted; that is, their Fano schemes all have degree 27.

In general, our description of the tangent spaces to the Fano scheme $F_k(X)$ allows us to determine the dimension of $F_k(X)$ at a point Λ in terms of the normal bundle to the plane $\Lambda \subset X$, but it is rare that this will be determined by the singularities of X along Λ. In fact, as soon as we get to hypersurfaces $X \subset \mathbb{P}_K^n$ with $n \geq 4$, we see examples to the contrary.

Exercise VI-40. Use this characterization of the tangent spaces to Fano schemes to give an example of a nonsingular hypersurface $X \subset \mathbb{P}_K^n$ such that the Fano scheme $F_1(X)$ of lines on X is singular.

VI.2.4 Moduli Spaces

A similar situation arises when we want to construct a moduli space of geometric objects. For example, we would like to identify the set of non-singular, projective curves of genus g over a field K with the set of closed points of a "moduli scheme" $\mathcal{M}_g$. To avoid unnecessary complication, we restrict to the case $\mathrm{char}(K) = 0$. Again, the way to express what we want is to introduce the *functor of nonsingular curves of genus* g: this is the functor

$$\mathcal{M}_g^{\mathrm{fun}} : (K - \mathrm{schemes})^\circ \to (\mathrm{sets})$$

that assigns to any scheme B over K the set of flat morphisms $\pi : \mathcal{X} \to B$ whose fibers are nonsingular curves of genus g, up to isomorphism $\mathcal{X} \cong \mathcal{X}'$ as B-schemes. We define $\mathcal{M}_g$ to be the scheme (if any) that represents the functor $\mathcal{M}_g^{\mathrm{fun}}$.

Since schemes are uniquely determined by their functors of points, the only difficulty with the "definitions" above is whether such schemes exist—

that is, whether the given functors are representable. The answer in the case of $\mathscr{H}_P$ is yes; and indeed, as we have pointed out, the characterization of $\mathscr{H}_P$ as the scheme that represents the functor h_P is crucial to the proof of existence.

However, in the case of $\mathscr{M}_g$, the answer is no! For example, even in the case of nonsingular curves of genus 1 over $\mathbb{C}$ there does not exist a moduli space $\mathscr{M}_1$ in this sense. To see this, it suffices to show that there does not exist a universal family — in other words, a flat morphism $\pi : \mathscr{C} \to \mathscr{M}$ of schemes with fiber nonsingular curves of genus 1 such that for every family $\mathscr{Y} \to B$ of nonsingular curves of genus 1 there are unique maps $\varphi : B \to \mathscr{M}$ and $\Phi : \mathscr{Y} \to \mathscr{C}$ forming a fiber product diagram:

$$
\begin{array}{ccc}
\mathscr{Y} & \xrightarrow{\ \ \Phi\ \ } & \mathscr{C} \\
{\scriptstyle \eta}\downarrow & & \downarrow{\scriptstyle \pi} \\
B & \xrightarrow{\ \ \varphi\ \ } & \mathscr{M}
\end{array}
$$

In fact, there does not even exist a tautological family — that is, a morphism $\mathscr{C} \to \mathscr{M}$ and a bijection between the closed points of $\mathscr{M}$ and the set of isomorphism classes of nonsingular curves of genus 1 such that the fiber over each point $p \in \mathscr{M}$ is in the isomorphism class corresponding to the point p. We will exhibit two kinds of obstructions to the existence of a universal family, one local and one global.

For the local obstruction, recall that curves of genus 1 over $\mathbb{C}$ are classified by their j-*invariant*: we can write any such curve as the plane cubic

$$
y^2 = x(x-1)(x-\lambda)
$$

for some complex $\lambda \neq 0, 1$; and two such curves C_λ and $C_{\lambda'}$ will be isomorphic if and only if their j-invariants $j(\lambda)$ and $j(\lambda')$ are equal, where

$$
j(\lambda) = 256 \cdot \frac{(\lambda^2 - \lambda + 1)^3}{\lambda^2 (\lambda - 1)^2};
$$

see Silverman [1986, Chapter III, Proposition 1.7], for example. It can be shown that given a family $\mathscr{X} \to B$ of nonsingular curves of genus 1 with nonsingular base B, the function j is a regular function on B; and locally around any point $b \in B$, λ can be defined as a regular function, too (though it is not unique). It follows that if there did exist a tautological family, there would have to exist one with base the affine line $\mathbb{A}^1_\mathbb{C}$ with coordinate j. But no such family can exist, because at the point $j = 0$ the function $\lambda^2 - \lambda + 1$ would vanish, and thus j would vanish triply. Less obviously, because $j'(-1) = 0$, it also follows that j can only assume the value 1728 with even multiplicity. (Note that the values $j = 0$ and 1728 correspond to the elliptic curves with "extra automorphisms" — that is, whose automorphism groups contain the automorphism group of a general elliptic curve as subgroups of index 3 and 2, respectively.)

Turning to the global obstruction, even if a tautological family exists over a variety $\mathcal{M}$ whose points correspond to isomorphism classes of curves, such a family may not be universal; that is, it may not induce a bijection between families over a base B and maps of B to $\mathcal{M}$. For example, if we simply exclude the curves of j-invariant 0 and 1728 — that is, consider the functor of families of nonsingular curves of genus 1 not isomorphic to C_0 or C_{1728} — we might hope that the punctured j-line $\mathcal{M} = \mathbb{A}^1_{\mathbb{C}} - \{0, 1728\}$ would be a moduli space; and indeed, a tautological family does exist over this open subset of $\mathbb{A}^1$. It is not universal, however: for example, for any fixed λ, let B' be any variety with fixed-point-free involution τ, and consider the family over $B = B'/\langle\tau\rangle$ formed by taking the quotient of the product $E \times B$ by the involution

$$\iota : ((x,y),p) \mapsto ((x,-y), \tau(p))$$

This is a family all of whose fibers are isomorphic to C_λ, and so it can only come from the constant map $B \to \mathcal{M}$; but it can be shown that the family itself is not trivial.

It is the presence of automorphisms of C_λ that is responsible for this phenomenon. Indeed, an analogous argument shows that we can never have a moduli space for schemes modulo isomorphism when some of the objects to be parametrized admit automorphisms. This explains also the discrepancy between the notions of tautological and universal family: in the case of the Hilbert scheme $\mathcal{H}$ parametrizing subschemes of $\mathbb{P}^n_K$ if two families $\mathcal{X}$, $\mathcal{X}' \subset \mathbb{P}^n_K \times_K B$ over a variety B correspond to the same map $B \to \mathcal{H}$ it follows that they are equal fiber by fiber and hence equal. By contrast, in the case of a moduli space, it would follow only that they are isomorphic fiber by fiber; if the fibers admitted automorphisms, those isomorphisms would not be unique and so might not fit together to give an isomorphism $\mathcal{X} \cong \mathcal{X}'$.

How do we deal with these difficulties? The most naive (and least satisfactory) way is simply to exclude all schemes with automorphisms from consideration when trying to construct a moduli space. This works in some contexts: for example, since the family of curves of genus g with automorphisms has in a suitable sense codimension $g - 2$ among all curves, if we are concerned in particular with the divisor theory of the moduli space of curves of genus $g \geq 4$, we can afford to look just at the moduli space $\mathcal{M}^0_g$ of automorphism-free nonsingular curves of genus g, which does exist.

There are two more serious approaches, both of which are in active use. The first is to take $\mathcal{M}_g$ to be the scheme whose functor of points "most closely approximates" $\mathcal{M}^{\text{fun}}_g$. It turns out that there is such a thing, called a *coarse moduli space*, and that it has nice properties; for example, the value of its functor of points at an algebraically closed field K is really the set of isomorphism classes of nonsingular curves over it. The second way out is to enlarge the category of schemes in a different way, to the category of *algebraic stacks*. A discussion of this would take us too far, so we'll just refer the reader to Vistoli [1989, Appendix] for a short treatment and to Behrend et al. [≥ 2001] for a full treatment; see also Mumford [1965] for an introduction to the functorial point of view on moduli spaces.

References

A. Altman and S. Kleiman, *Introduction to Grothendieck duality theory*, Lecture Notes in Mathematics **146**, Springer, Berlin, 1970.

M. Artin, *Algebraic spaces*, Yale Mathematical Monographs **3**, Yale University Press, New Haven, CT, 1971.

M. F. Atiyah and I. G. Macdonald, *Introduction to commutative algebra*, Addison-Wesley, Reading, MA, 1969.

D. Bayer and D. Eisenbud, "Ribbons and their canonical embeddings", *Trans. Amer. Math. Soc.* **347**:3 (1995), 719–756.

K. Behrend, L. Fantechi, W. Fulton, L. Göttsche, and A. Kresch, "Introduction to stacks". In preparation.

E. Bombieri and D. Mumford, "Enriques' classification of surfaces in char. *p*, III", *Invent. Math.* **35** (1976), 197–232.

N. Bourbaki, *Commutative algebra*, Chapters 1-7, Hermann, Paris, 1972. Reprinted by Springer, New York, 1989.

E. Brieskorn and H. Knörrer, *Plane algebraic curves*, Birkhäuser, Basel, 1986.

J. W. S. Cassels and A. Fröhlich (editors), *Algebraic number theory* (Brighton, 1965), Academic Press, London, 1967. Reprinted 1986.

H. Clemens, J. Kollár, and S. Mori, *Higher-dimensional complex geometry*, Astérisque **166**, Soc. math. France, Paris, 1988.

A. Collino, "Evidence for a conjecture of Ellingsrud and Strømme on the Chow ring of $\text{Hilb}_d\mathbb{P}^2_{\mathbb{C}}$", *Illinois Journal of Mathematics* **32** (1988), 171–210.

J. L. Coolidge, *A treatise on algebraic plane curves*, Oxford University Press, 1931.

D. Cox, J. Little, and D. O'Shea, *Ideals, varieties, and algorithms*, Second ed., Springer, New York, 1997. An introduction to computational algebraic geometry and commutative algebra.

P. Deligne, "La conjecture de Weil, I", *Inst. Hautes Études Sci. Publ. Math.* **43** (1974), 273–307.

M. Demazure and P. Gabriel, *Groupes algébriques, I: Géométrie algébrique, généralités, groupes commutatifs*, Masson, Paris, 1970.

D. Eisenbud, *Commutative algebra with a view toward algebraic geometry*, Graduate Texts in Mathematics **150**, Springer, New York, 1995.

G. Faltings, "Die Vermutungen von Tate und Mordell", *Jahresber. Deutsch. Math.-Verein.* **86**:1 (1984), 1–13.

L.-Y. Fong, "Rational ribbons and deformation of hyperelliptic curves", *J. Algebraic Geom.* **2**:2 (1993), 295–307.

O. Forster, *Lectures on Riemann surfaces*, Graduate Texts in Mathematics **81**, Springer, New York, 1981. Translated from the German by Bruce Gilligan.

W. Fulton, *Intersection theory*, vol. 2, Ergebnisse der Mathematik und ihrer Grenzgebiete (3), Springer, Berlin, 1984.

R. Godement, *Topologie algébrique et théorie des faisceaux*, Actualit'es Sci. Ind. **1252**, Hermann, Paris, 1964.

M. L. Green, "Koszul cohomology and the geometry of projective varieties", *J. Differential Geom.* **19**:1 (1984), 125–171. with an appendix by M. Green and R. Lazarsfeld.

M. Green and R. Lazarsfeld, "On the projective normality of complete linear series on an algebraic curve", *Invent. Math.* **83**:1 (1985), 73–90.

P. Griffiths and J. Harris, *Principles of algebraic geometry*, Wiley-Interscience, New York, 1978. Reprinted 1994.

A. Grothendieck, "Éléments de géométrie algébrique, I: Le langage des schémas", *Inst. Hautes Études Sci. Publ. Math.* **4** (1960), 1–228.

A. Grothendieck, "Éléments de géométrie algébrique, II: Étude globale élémentaire de quelques classes de morphismes", *Inst. Hautes Études Sci. Publ. Math.* **8** (1961), 1–222.

A. Grothendieck, "Éléments de géométrie algébrique, III: Étude cohomologique des faisceaux cohérents (première partie)", *Inst. Hautes Études Sci. Publ. Math.* **11** (1961), 1–167.

A. Grothendieck, "Éléments de géométrie algébrique, III: Étude cohomologique des faisceaux cohérents (seconde partie)", *Inst. Hautes Études Sci. Publ. Math.* **17** (1963), 1–91.

A. Grothendieck, "Éléments de géométrie algébrique, IV: Étude locale des schémas et des morphismes de schémas (première partie)", *Inst. Hautes Études Sci. Publ. Math.* **20** (1964), 1–259.

A. Grothendieck, "Éléments de géométrie algébrique, IV: Étude locale des schémas et des morphismes de schémas (seconde partie)", *Inst. Hautes Études Sci. Publ. Math.* **24** (1965), 1–231.

A. Grothendieck, "Éléments de géométrie algébrique, IV: Étude locale des schémas et de morphismes de schémas (troisième partie)", *Inst. Hautes Études Sci. Publ. Math.* **28** (1966), 1–255.

A. Grothendieck, "Éléments de géométrie algébrique, IV: Étude locale des schémas et des morphismes de schémas (quatrième partie)", *Inst. Hautes Études Sci. Publ. Math.* **32** (1967), 1–361.

R. C. Gunning, *Introduction to holomorphic functions of several variables, III: Homological theory*, Wadsworth and Brooks/Cole, Monterey, CA, 1990.

J. Harris, "Galois groups of enumerative problems", *Duke Math. J.* **46**:4 (1979), 685–724.

J. Harris, *Algebraic geometry: A first course*, Graduate Texts in Math. **133**, Springer, New York, 1995. Corrected reprint of the 1992 original.

J. Harris and D. Eisenbud, *Curves in projective space*, Les Presses de l'Université de Montréal, Montreal, 1982.

J. Harris and I. Morrison, *Moduli of curves*, Graduate Texts in Mathematics **187**, Springer, New York, 1998.

R. Hartshorne, *Algebraic geometry*, Graduate Texts in Mathematics **52**, Springer, New York, 1977.

D. Hilbert, "Über die theorie der algebraischen formen", *Mathematische Annalen* **36** (1890), 473–534.

H. Hironaka, "An example of a non-Kählerian deformation", *Annals of Mathematics* **75** (1962), 190–208.

A. Iarrobino, "Compressed algebras and components of the punctual Hilbert scheme", pp. 146–165 in *Algebraic geometry* (Sitges, Barcelona, 1983), edited by E. Casas-Alvero, Lecture Notes in Math. **1124**, Springer, Berlin, 1985.

G. Kempf, F. F. Knudsen, D. Mumford, and B. Saint-Donat, *Toroidal embeddings, I*, Lecture Notes in Mathematics **339**, Springer, Berlin, 1973.

S. Kleiman, "Misconceptions about K_X", *Enseign. Math.* **25** (1979), 203–206.

D. Knutson, *Algebraic spaces*, Lecture Notes in Math. **203**, Springer, Berlin, 1971.

J. Kollár, "The structure of algebraic threefolds: an introduction to Mori's program", *Bull. Amer. Math. Soc. (N.S.)* **17**:2 (1987), 211–273.

J. Kollár, *Rational curves on algebraic varieties*, Ergebnisse der Mathematik, 3. Folge **32**, Springer, Berlin, 1996.

S. Lang, *Algebraic number theory*, 2nd ed., Graduate Texts in Math. **110**, Springer, New York, 1994.

H. Matsumura, *Commutative ring theory*, Cambridge University Press, Cambridge, 1986. Second edition, 1989.

D. Mumford, "Further pathologies in algebraic geometry", *Amer. J. Math.* **84** (1962), 642–648.

D. Mumford, "Picard groups of moduli problems", pp. 33–81 in *Arithmetical Algebraic Geometry* (Purdue Univ., 1963), edited by O. F. G. Schilling, Harper & Row, New York, 1965.

D. Mumford, *Lectures on curves on an algebraic surface*, Annals of Mathematics Studies **59**, Princeton University Press, Princeton, 1966.

D. Mumford, *Algebraic geometry I : complex projective varieties*, Grundlehren der mathematischen Wissenschaften **221**, Springer, Berlin and New York, 1976.

D. Mumford, *The red book of varieties and schemes*, Lecture Notes in Math. **1358**, Springer, Berlin, 1988.

M. Nagata, *Local rings*, Tracts in Pure and Applied Mathematics **13**, Wiley/Interscience, New York and London, 1962.

D. G. Northcott, *Ideal theory*, Cambridge Tracts in Mathematics and Mathematical Physics **42**, University Press, Cambridge, 1953.

M. Raynaud and L. Gruson, "Critères de platitude et de projectivité. Techniques de "platification" d'un module", *Invent. Math.* **13** (1971), 1–89.

M. Reid, *Undergraduate algebraic geometry*, London Mathematical Society Student Texts **12**, Cambridge University Press, Cambridge and New York, 1988.

B. Segre, *The nonsingular cubic surfaces*, Oxford Univ. Press, 1942.

J.-P. Serre, "Faisceaux algébriques cohérents", *Ann. of Math.* (2) **61** (1955), 197–278.

J.-P. Serre, *Groupes algébriques et corps de classes*, 2nd ed., Hermann, Paris, 1975. Translated as *Algebraic groups and class fields*, Graduate texts in mathematics **117**, Springer, New York, 1988.

J.-P. Serre, *Local fields*, Springer, New York, 1979.

I. R. Shafarevich, *Basic algebraic geometry*, vol. 213, Die Grundlehren der mathematischen Wissenschaften, Springer, New York, 1974. Second edition (in two volumes), 1994.

J. H. Silverman, *The arithmetic of elliptic curves*, Springer, New York, 1986.

E. H. Spanier, *Algebraic topology*, McGraw-Hill, New York, 1966.

R. G. Swan, *Theory of sheaves*, Chicago Lectures in Mathematics, University of Chicago Press, Chicago, 1964.

A. Vistoli, "Intersection theory on algebraic stacks and on their moduli spaces", *Invent. Math.* **97**:3 (1989), 613–670.

W. Vogel, *Lectures on results on Bezout's theorem*, Lectures on mathematics and physics **74**, Tata Institute of Fundamental Research, Bombay, and Springer, Berlin, 1984. Notes by D. P. Patil.

R. J. Walker, *Algebraic Curves*, Princeton Mathematical Series **13**, Princeton University Press, Princeton, NJ, 1950. Reprinted by Dover, 1962, and Springer, 1978.

O. Zariski, "The concept of a simple point of an abstract algebraic variety", *Trans. Amer. Math. Soc.* **62** (1947), 1–52.

Index

Graduate Texts in Mathematics

(continued from page ii)